Sedimentation in Oblique-slip Mobile Zones

Sedimentation in Oblique-slip Mobile Zones

EDITED BY PETER F. BALLANCE
AND HAROLD G. READING

SPECIAL PUBLICATION NUMBER 4 OF THE
INTERNATIONAL ASSOCIATION OF SEDIMENTOLOGISTS
PUBLISHED BY BLACKWELL SCIENTIFIC PUBLICATIONS
OXFORD LONDON EDINBURGH
BOSTON MELBOURNE

Published by Blackwell Scientific Publications
Osney Mead, Oxford, OX2 0ES
8 John Street, London, WC1N 2ES
9 Forrest Road, Edinburgh, EH1 2QH
52 Beacon Street, Boston, Massachusetts 02108, USA
214 Berkeley Street, Carlton, Victoria 3053, Australia

First published 1980

British Library
Cataloguing in Publication Data

Sedimentation in oblique-slip mobile zones.—
(International Association of Sedimentologists.
Special publications; no. 4).
1. Sedimentation and deposition—Congresses
2. Faults (Geology)—Congresses
I. Ballance, Peter F II. Reading, Harold G
III. Series
551.3'04 QE571

ISSN 0141-3600
ISBN 0 632 00607 2

Distributed in the U.S.A. by
Halsted Press, a division of
John Wiley & Sons, Inc., New York

Printed and bound in Great Britain by
Burgess & Son (Abingdon) Ltd
Station Road, Abingdon, Oxfordshire

Contents

1 Introduction
Peter F. Ballance and Harold G. Reading

7 Characteristics and recognition of strike-slip fault systems
Harold G. Reading

27 Analysis of pull-apart basin development produced by *en echelon* strike-slip faults
D. Rodgers

43 Basin development along the late Mesozoic and Cainozoic California margin: a plate tectonic margin of subduction, oblique subduction and transform tectonics
D. G. Howell, J. K. Crouch, H. G. Greene, D. S. McCulloch and J. G. Vedder

63 Evolution of a strike-slip fault-controlled basin, Upper Old Red Sandstone, Scotland
B. J. Bluck

79 Late Caledonian (Devonian) basin formation, western Norway: signs of strike-slip tectonics during infilling
R. J. Steel and T. G. Gloppen

105 Deposits associated with a Hercynian to late Hercynian continental strike-slip system, Cantabrian Mountains, northern Spain
A. P. Heward and Harold G. Reading

127 Strike-slip related sedimentation in the Antalya complex, southwest Turkey
A. H. F. Robertson and N. H. Woodcock

147 New Zealand and oblique-slip margins: tectonic development up to and during the Cainozoic
K. B. Spörli

171 Quaternary sedimentation on the Hikurangi oblique-subduction and transform margin, New Zealand
K. B. Lewis

191 The Makara Basin: a Miocene slope-basin along the New Zealand sector of the Australian-Pacific obliquely convergent plate boundary
G. J. van der Lingen and J. R. Pettinga

217 Late Cainozoic sedimentation and tectonics of the East Coast Deformed Belt, in Marlborough, New Zealand
W. M. Prebble

229 Models of sediment distribution in non-marine and shallow marine environments in oblique-slip fault zones
Peter F. Ballance

237 Offshore sedimentary basins at the southern end of the Alpine Fault, New Zealand
R. J. Norris and R. M. Carter

Spec. Publ. int. Ass. Sediment. (1980) **4**, 1–5

Sedimentation in oblique-slip mobile zones: an introduction

PETER F. BALLANCE *and* HAROLD G. READING

Department of Geology, The University of Auckland, Private Bag, Auckland, New Zealand and Department of Geology and Mineralogy, Parks Road, Oxford OX1 3PR

Following the 1906 San Francisco earthquake, during which up to 7 m of horizontal displacement occurred, the social and economic consequences of such movement have made the 1000 km long San Andreas fault a feature of world wide renown. The Alpine fault of New Zealand is now equally famous to geologists but it was not recognized as a continuous feature, 600 km long, until 1942 (Wellman & Willett, 1942). On the San Andreas fault great lateral displacement was first postulated by Noble (1926) who argued for about 40 km of Tertiary movement. Post-Jurassic displacement of 640 km was postulated by Hill & Dibblee (1953). On the Alpine Fault lateral displacement of about 490 km was described in 1952 (Wellman, 1952; Gage, 1952).

That the Dead Sea rift system might be the result of horizontal movement was first suggested by Dubertret (1932) who postulated sinistral motion of about 160 km between the Sinai–Palestine and Arabian blocks. However it was Quennell (1958, 1959) who documented the geological evidence for about 100 km of motion in two stages, in the Lower Miocene and in the late Pleistocene. In the Caribbean, although the idea of strike-slip motion had been suggested by Hess as early as 1933, and had been further developed by Hess & Maxwell (1953), it was Rod (1956) who showed the extent of strike-slip faults on the southern margin of the Caribbean in northern South America. This was also the year of the classic paper by Moody & Hill (1956) which not only described all known wrench faults and demonstrated their widespread occurrence but also introduced the concept of a world-wide system of wrench faults thought to be the result of global N–S compression. Unfortunately criticism of the application of the Moody & Hill hypothesis to known faults and lineaments tended to distract attention from their accumulation of data on strike-slip systems. Evidence for major horizontal displacements in oceans had to await the matching of magnetic anomalies in the Pacific Ocean (Menard & Vacquier 1958; Vacquier, Raff & Warren, 1961) where displacements of up to several 100 km were postulated.

Small-scale pre-Mesozoic strike-slip faults have been recognized at least since the start of the century, but the first convincing demonstration of the evidence for large-scale movement was by Kennedy (1946). Comparing the Great Glen Fault of Scotland to the San Andreas Fault, he considered the former to have been a NE–SW sinistral fault which moved 104 km as a result of N–S compression in Hercynian times.

Though Crowell (1952, 1954) had recognized progressive transcurrent faulting on the San Gabriel and San Andreas faults during sedimentation in the Ridge Basin, analyses of the structural and sedimentational consequences of bends, branchings, endings and offsets in transcurrent fault systems really began in 1958 (Lensen, 1958; Kingma, 1958a, b in New Zealand and Quennell, 1958 for the Dead Sea Fault). In both these areas further ideas were developed on the sedimentary and geomorphological consequences of strike-slip (Freund, 1965; Clayton, 1966; Freund *et al.*, 1970).

The twin hypotheses of sea-floor spreading and plate tectonics placed before the geological community in the late 1960s the possibility that transform and transcurrent mobile zones were on a much larger scale than had hitherto been envisaged. However, in spite of Webb's (1969) interpretation of the Canadian Appalachians as having been deformed by wrench faults, an initial preoccupation with the simple two-dimensional aspects of movement across convergent and divergent plate margins, diverted the attention of most geologists away from transcurrent zones until the mid-1970s. Then Wilcox, Harding & Seely (1973), Moody (1973), Harding (1974) and Crowell (1974a, b) revived and extended the earlier ideas on wrench faulting by demonstrating the importance of strike-slip motion in the development of petroleum traps and of sedimentary basins. Meanwhile Harland (1971) and Lowell (1972) in the Tertiary of Spitzbergen, and Arthaud & Matte (1975) and Reading (1975) in the Hercynian of western Europe, attempted to explain some orogenic belts by strike-slip movement rather than by subduction. Harland (1971) introduced the concepts of transtension (divergent strike-slip) and transpression (convergent strike-slip). In 1976 Steel interpreted a thick Old Red Sandstone basinal succession in western Norway in terms of Crowell's (1974a) strike-slip model for the Ridge Basin of California and Edwards (1976) did the same for a Tertiary basin in SW England.

Towards the end of the 1970s, papers relating sedimentary basins and orogenic belts to strike-slip tectonics appeared at an increasing rate, Mitchell & Reading (1978) proposing a strike-slip cycle of sedimentation and orogeny.

Strike-slip motion has now been shown to control or to have controlled the location, subsidence, shape, sedimentation and deformation of many basins. In New Zealand, Norris, Carter & Turnbull (1978) interpreted the Cainozoic history of the southern end of the Alpine fault system as the response to transtension in the Oligocene, then pure transcurrent motion and finally transpression which has continued to the present day. In California, Blake *et al.* (1978) related the development of sedimentary basins in the California Borderlands, both onshore and offshore to the dextral motion along the faults of the San Andreas fault system.

In a symposium organized by B. F. Windley at the Geological Society of London on January 18 1978, on 'Oceanic and Continental transform faults', Searle (1979) confirmed the occurrence of fracture zone valleys (e.g. Francheteau *et al.*, 1976) running parallel to the trend of oceanic ridge transform faults. He also reported 20 km long obliquely trending *en echelon* tensional scarps which border smaller basins. Wilson & Williams (1979) showed how the type of sedimentary basin on the Atlantic continental margins is governed by the orientation and location of basement continental fracture-zones, the pattern of wrench tectonics and transform fault movements.

A second symposium, convened by M. G. Audley-Charles and J. V. Hepworth at the same venue in December 1978, 'Magmatism and tectonics of SE Asia', emphasized that an area such as SE Asia cannot be understood without consideration of strike-slip faults. In particular Page *et al.* (1979) demonstrated that in the area in and around northern Sumatra strike-slip motion on the Sumatra fault zone has been responsible not only for structural trends, but for the location of volcanic centres, of sedimentary basins and of back-arc spreading centres in the Andaman Sea and for the occurrence of tectonically sheared ophiolitic serpentinites.

It therefore seemed timely to convene a meeting to bring together a number of those concerned with the topic. New Zealand was an ideal venue since strike-slip faulting is as much a part of every New Zealander's upbringing as it is of every Californian's. P. F. Ballance convened a Symposium held during the 1979 Congress of ANZAAS (Australia–New Zealand Association for the Advancement of Science) at Auckland University on January 22, 1979. Here the papers in this volume dealing with New Zealand were delivered, along with H. G. Reading's general review.

The title of the volume presented problems, various alternatives being discussed from 'Sedimentation in strike-slip orogenic belts' to 'Sedimentation in oblique-slip plate margins', to its final form 'Sedimentation in oblique-slip mobile zones'. The term 'oblique-slip' was preferred to strike-slip because 'pure' strike-slip systems are probably rare; almost all involve some degree of perpendicular motion. 'Mobile zones' was preferred because strike-slip faulting is not confined to plate or continental margins and, especially where sedimentary basins form, is not necessarily part of an orogenic belt.

We are indebted to K. B. Spörli for the following definition of an oblique-slip mobile zone.

> Across an oblique-slip mobile zone the relative motion of the blocks which are in contact along the zone is oblique, that is, in plan view there are components of movement both parallel and perpendicular to the zone. The zone may be a plate boundary, in the form of an active continental margin, an island arc, a collision zone between various plate features (e.g. continent/continent, arc/ridge), or an oceanic or continental transform. One may also wish to include rift zones with oblique directions of spreading, at least during their initial stages of opening, especially in the case of narrow back-arc basins and of continental aulacogens. Oblique-slip mobile zones may lie within a non-rigid plate. The width of the mobile zone is usually measured in tens to hundreds of km.
>
> In oblique-slip mobile zones, transpression and/or transtension may occur. If subduction is taking place, the third dimension in the form of a vertical component of movement will also be important, and the definition of oblique-slip then becomes identical with that for oblique-slip faults (Spencer 1977); i.e. oblique-slip subduction has both strike-slip and dip-slip components of movement. Vertical movement is also important at a smaller scale and as a second order component in the oblique-slip mobile zones without subduction.

We wish to express our thanks to the Organizing Committee of ANZAAS for Harold Reading's expenses for the Symposium, to the Bureau of the International Association of Sedimentologists for their agreement to publish the volume, to numerous colleagues for their help in refereeing the contributions, and not least to the authors who responded, with relatively minor flagellation, to our deadlines.

REFERENCES

ARTHAUD, F. & MATTE, PH. (1975) Les décrochements tardi-hercyniens du sud-ouest de l'Europe. Géométrie et essai de reconstitution des conditions de la déformation. *Tectonophysics*, **25**, 139–171.

BLAKE, M.C. JR, CAMPBELL, R.H., DIBBLEE, T.W. JR, HOWELL, D.G., NILSEN, T.H., NORMARK, W.R., VEDDER, J.C. & SILVER, E.A. (1978) Neogene basin formation in relation to plate-tectonic evolution of San Andreas fault system, California. *Bull. Am. Ass. Petrol. Geol.* **62,** 344–372.

CLAYTON, L. (1966) Tectonic depressions along the Hope Fault, a transcurrent fault in North Canterbury, New Zealand. *N.Z. J. Geol. Geophys.* **9,** 95–104.

CROWELL, J.C. (1952) Probable large lateral displacement on the San Gabriel fault, southern California. *Bull. Am. Ass. Petrol. Geol.* **36,** 2026–2035.

CROWELL, J.C. (1954) Geology of the Ridge Basin area, Los Angeles and Ventura counties, California. *Bull. California Div. Mines.* 170, map sheet 7.

CROWELL, J.C. (1974a) Sedimentation along the San Andreas Fault, California. In: *Modern and Ancient Geosynclinal Sedimentation* (Ed. by R. H. Dott Jr. & R. H. Shaver). *Spec. Publ. Soc. econ. Paleont. Miner. Tulsa,* **19,** 292–303.

CROWELL, J.C. (1974b) Origin of late Cenozoic basins in southern California. In: *Tectonics and Sedimentation* (Ed. by W. R. Dickinson). *Spec. Publ. Soc. econ. Paleont. Miner. Tulsa,* **22,** 190–204.

DUBERTRET, L. (1932) Les formes structurales de la Syrie et de la Palestine; leur origine. *C.R. Acad. Sci. Colon., Paris,* 195.

EDWARDS, R.A. (1976) Tertiary sediments and structure of the Bovey Basin, south Devon. *Proc. Geol. Ass.* **87,** 1–26.

FRANCHETEAU, J., CHOUKROUNE, P., HEKINIAN, R., LE PICHON, X. & NEEDHAM, H.D. (1976) Oceanic fracture zones do not provide deep sections in the crust. *Can. J. Earth Sci.* **13,** 1223–1235.

FREUND, R. (1965) A model of the structural development of Israel and adjacent areas since Upper Cretaceous times. *Geol. Mag.* **102,** 189–205.

FREUND, R., GARFUNKEL, Z., ZAK, I., GOLDBERG, M., WEISSBROD, T. & DERIN, B. (1970) The shear zone along the Dead Sea Rift. *Phil. Trans. R. Soc.* **A267,** 107–130.

GAGE, M. (1952) Transcurrent faulting in New Zealand tectonics (Abstract). *Bull. geol. Soc. Am.* **63,** 1380.

HARDING, T.P. (1974) Petroleum traps associated with wrench faults. *Bull. Am. Assoc. Petrol. Geol.* **58,** 1290–1304.

HARLAND, W.B. (1971) Tectonic transpression in Caledonian Spitzbergen. *Geol. Mag.* **108,** 27–42.

HESS, H.H. (1933) Interpretation of geological and geophysical observations, Navy-Princeton Gravity Expedition to the West Indies, 1932. *U.S. Navy, Hydrographic Office.*

HESS, H.H. & MAXWELL, J.C. (1953) Caribbean Research Project. *Bull. geol. Soc. Am.* **64,** 1–6.

HILL, M.L. & DIBBLEE, T.W. JR (1953) San Andreas, Garlock and Big Pine Faults, California. *Bull. geol. Soc. Am.* **64,** 443–458.

KENNEDY, W.Q. (1946) The Great Glen fault. *Q.J. geol. Soc. Lond.* **102,** 41–76.

KINGMA, J.T. (1958a) The Tongaporutuan sedimentation in Central Hawke's Bay. *N.Z. J. Geol. Geophys.* **1,** 1–30.

KINGMA, J.T. (1958b) Possible origin of piercement structures, local unconformities and secondary basins in the Eastern Geosyncline, New Zealand. *N.Z. J. Geol. Geophys.* **1,** 269–274.

LENSEN, G.J. (1958) A method of graben and horst formation. *J. Geol.* **66,** 579–587.

LOWELL, J.D. (1972) Spitsbergen Tertiary orogenic belt and the Spitsbergen fracture zone. *Bull. geol. Soc. Am.* **83,** 3091–3102.

MENARD, H.W. & VACQUIER, V. (1958) Magnetic survey of part of the deep sea floor off the coast of California. *Office of Naval Research, Research Reviews*, June 1–5.

MITCHELL, A.H.G. & READING, H.G. (1978) Sedimentation and tectonics. In: *Sedimentary Environments and Facies* (Ed. by H. G. Reading), pp. 439–476. Blackwell Scientific Publications, Oxford.

MOODY, J.D. (1973) Petroleum exploration aspects of wrench-fault tectonics. *Bull. Am. Ass. Petrol. Geol.* **57,** 449–476.

MOODY, J.D. & HILL, M.J. (1956) Wrench fault tectonics. *Bull. geol. Soc. Am.* **67,** 1207–1248.

NOBLE, L.F. (1926) The San Andreas rift and some other active faults in the desert region of south-eastern California. *Carnegie Inst. Washington, Year Book.* **25,** 416–422.

NORRIS, R.J., CARTER, R.M. & TURNBULL, I.M. (1978) Cainozoic sedimentation in basins adjacent to a major continental transform boundary in southern New Zealand. *J. geol. Soc. Lond.* **135,** 191–205.

PAGE, B.G.N., BENNETT, J.D., CAMERON, N.R., BRIDGE, D. McC., JEFFERY, D.H., KEATS, W. & THAIB, J. (1979) A review of the main structural and magmatic features of northern Sumatra. *J. geol. Soc. Lond.* **136,** 569–578.

QUENNELL, A.M. (1958) The structural and geomorphic evolution of the Dead Sea Rift. *Q. J. geol. Soc. Lond.* **114,** 1–24.

QUENNELL, A.M. (1959) Tectonics of the Dead Sea Rift. *Int. geol. Congr.* 1956. **20,** 385–405.

READING, H.G. (1975) Strike-slip fault systems: an ancient example from the Cantabrians. *Ninth Intern. Cong. Sedimentology, Nice 1975*, Thème **4,** pt. 2, 287–291.

ROD, E. (1956) Strike-slip faults of northern Venezuela. *Bull. Am. Ass. Petrol. Geol.* **40,** 457–476.

SEARLE, R.C. (1979) Side-scan sonar studies of North Atlantic fracture zones. *J. geol. Soc. Lond.* **136,** 283–291.

SPENCER, E.W. (1977) *Introduction to the Structure of the Earth.* McGraw-Hill, New York. 640 pp.

STEEL, R.J. (1976) Devonian basins of western Norway—sedimentary response to tectonism and varying tectonic context. *Tectonophysics*, **36,** 207–224.

VACQUIER, V., RAFF, A.D. & WARREN, R.E. (1961) Horizontal displacements in the floor of the northeastern Pacific Ocean. *Bull. geol. Soc. Am.* **72,** 1251–1258.

WEBB, G.W. (1969) Paleozoic wrench faults in the Canadian Appalachians. In: *North Atlantic Geology and Continental Drift* (Ed. by M. Kay), *Mem. Am. Ass. Petrol. Geol.* **12,** 754–786.

WILCOX, R.E., HARDING, T.P. & SEELY, D.R. (1973) Basic wrench tectonics. *Bull. Am. Ass. Petrol. Geol.* **57,** 74–96.

WILSON, R.C.L. & WILLIAMS, C.A. (1979) Oceanic transform structures and the development of Atlantic continental margin sedimentary basins—a review. *J. geol. Soc. Lond.* **136,** 311–320.

WELLMAN, H.W. (1952) The Alpine Fault in detail: river terrace displacement at Maruia River. *N.Z. J. Sci. Technol.* **B33,** 409–414.

WELLMAN, H.W. & WILLETT, R.W. (1942) The Geology of the West Coast from Abut Head to Milford Sound—Part 1. *Trans. Roy. Soc. N.Z.* **71,** 282–306.

Spec. Publ. int. Ass. Sediment. (1980) **4,** 7–26

Characteristics and recognition of strike-slip fault systems

HAROLD G. READING

Department of Geology and Mineralogy, Parks Road, Oxford OX1 3PR U.K.

ABSTRACT

Strike-slip fault systems are major tectonic features at the present day. Some form plate boundaries and some are long-lived fundamental faults which may be related to megashears whose history goes back to early Precambrian times. On a regional scale strike-slip systems may have been subjected to periods of transtension (divergent strike-slip) or transpression (convergent strike-slip). On a local scale curvature, braiding and side-stepping of faults result in contemporaneous formation of closely spaced zones of extension and compression. The main effect of strike-slip motion on sedimentation occurs when dip-slip motion causes rapid sinking of sedimentary basins and uplift and erosion of mountains. Sedimentary facies are therefore very varied, both laterally and vertically, but lateral migration of facies is limited. Ancient strike-slip belts may be recognized by (1) lateral matching of displaced palaeogeographies across faults (2) discordance between size and materials of alluvial fans and possible source areas (3) thick, but not laterally extensive, sedimentary piles deposited very rapidly (4) localized uplift and erosion giving rise to unconformities of the same age as thick sedimentary fills nearby (5) extreme lateral facies variations (6) simultaneous development of both extensional and compressional tectonics within the same tectonic belt (7) a wrench fault style of structural deformation, in particular *en echelon* folds (8) little or no metamorphism (9) sparse igneous activity, except locally in zones of transtension. The strike-slip cycle of transtension→basin filling→transpression without large oceans or subduction is an alternative to the Wilson cycle of sea-floor spreading→subduction→continental collision to explain the classic geosynclinal cycle of pre-flysch→flysch→molasse. Strike-slip motion may have been important during the Upper Palaeozoic in western Europe and eastern North America and during the development of much of the Caledonides. It can provide ideal conditions for the occurrences of hydrocarbons, economically significant lacustrine deposits, and mineralization.

INTRODUCTION

Plate tectonics has provided plausible explanations to many previously unresolved geological problems, in particular the origins of ophiolites, calc-alkaline magmatism and paired metamorphic belts. It has helped to explain some orogenic belts, especially those associated with fore-arc regions where evidence for subduction is clear and

0141-3600/80/0904-0007$02.00

distinction can be made between old ocean floor, trenches, fore-arc basins and outer arc troughs. Even though it is not always possible to provide a unique plate tectonic model for ancient rocks, at least the major problems have been identified and possible models clarified.

In many tectonic belts subduction-related models fail to satisfy the data and alternatives need to be explored. One such model is that of the strike-slip orogenic belt along which lithosphere is essentially conserved and neither accreted nor consumed. Seismic activity and deformation are intense along strike-slip zones; differential movements are considerable; metamorphism is feeble; magmatic activity is minor. Sedimentation, on the other hand is both rapid and varied. By examining some modern strike-slip orogenic belts and concentrating on sedimentation, deformation, uplift and erosion within these belts, it is hoped to provide an alternative, and perhaps complementary, model to the subduction-related model. The strike-slip model is then applied in the interpretation of selected ancient orogenic belts. It is also hoped to inspire the reader to consider what is probably the most intriguing question in global tectonics today—whether the primary features of global tectonics are spreading centres and subduction zones, or whether they are the long-lived 'lineaments' expressed as fundamental faults in continents and fracture zones in oceans.

PRESENT-DAY STRIKE-SLIP FAULTS

Strike-slip faults are those whose primary motion is horizontal and parallel to the fault trace. Today they mark regions not only of seismic activity, but also of uplift and erosion of mountains and of sedimentation in rapidly subsiding basins. The better-known strike-slip faults may also be *fundamental* faults (de Sitter, 1964) which both extend deep into the crust and have a long history of movement. Such faults have been described as *transcurrent* if they are considered to rupture the whole lithosphere (Harland, 1978 in discussion of Norris, Carter & Turnbull, 1978), though other writers (e.g. Moody, 1973) have used the term transcurrent synonymously with wrench and strike-slip. Some fundamental faults are boundaries to lithospheric plates; some do not appear to mark plate boundaries and with others it is difficult to determine whether they do or do not. *Transform faults* (Wilson, 1965) terminate either at a spreading ridge or at a subduction zone. In oceans they are of two types (Gilliland & Meyer, 1976). Those which are the result of spreading, *ridge transform faults,* are considered to be secondary. Those which may be the result of earlier fractures in continental crust are *primary transforms* or *boundary transform faults.* They may also be fundamental faults.

Modern boundary transform faults within oceans are generally known as fracture zones, some extending into continental margins (Sheridan, 1974). Examples are the Charlie-Gibbs, Newfoundland, Azores, Vema, Romanche, Falkland–Agulhas Fracture Zones in the Atlantic (Wilson & Williams, 1979); the Owen Fracture Zone and Ninety East Ridge in the Indian Ocean.

Modern transform faults which fracture continental crust include (Fig. 1) (a) the dextral San Andreas fault system which is essentially a continental transform fault connecting the Juan de Fuca spreading centre with that of the Gulf of California; (b) the dextral Alpine fault system of New Zealand linking two oppositely dipping subduction zones, that of the Tonga–Kermadec trench to that of the Puysegur trench to

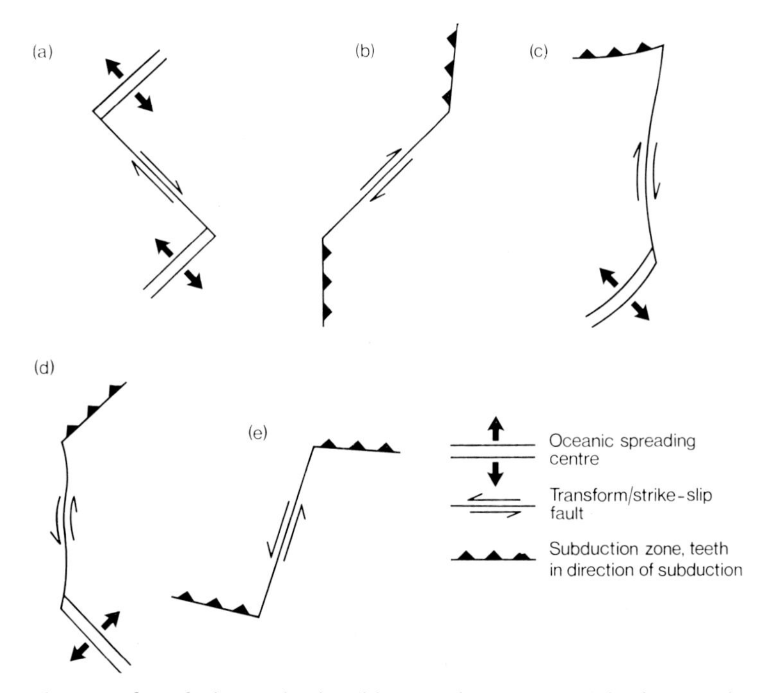

Fig. 1. Boundary transform faults terminating either at a ridge or at a subduction zone (not to scale): (a) San Andreas Fault; Tertiary Fault of Spitzbergen, (b) Alpine Fault of New Zealand, (c) Chugach–Fairweather–Queen Charlotte Island Fault, (d) Dead Sea (Levant) Fault, (e) Kirthar–Sulaiman (Quetta–Chaman) Fault.

the south (Lewis, 1980; Spörli, 1980); (c) the dextral Chugach–Fairweather–Queen Charlotte Islands fault system which connects the Juan de Fuca ridge with the Aleutian subduction zone (Naugler & Wageman (1973); (d) the sinistral Levant or Dead Sea fault system connecting the Red Sea spreading centre to the Taurus zone of plate convergence (Freund, 1965; Garfunkel, 1978); (e) the sinistral Kirthar–Sulaiman (Quetta–Chaman) fault system connecting the Himalayan collision zone with the Makran subduction zone of southern Iran (Abdel-Gawad, 1971); (f) the dextral El Pilar–Oca fault system and the sinistral Greater Antillean–Cayman system bounding the Caribbean plate; (g) the sinistral Sorong fault of New Guinea connecting the Mindanao–Banda trenches with the New Hebrides trench (Froidevaux, 1978); (h) the dextral Guayaquil fault connecting the Galapagos ridge with the Peru Trench.

Long-lived fundamental faults, not apparently related to present plate boundaries, are mainly found within continental crust. They include the Great Glen Fault and possibly also the Highland Boundary Fault and Southern Uplands Fault of Scotland, the North Pyrenean Fault Zone, the Insubric Line of the Alps, the complex strike-slip faults of northwest South America, the Cabot Fault of Newfoundland, and the currently sinistral Rhine graben rift system.

Another group of strike-slip faults, possibly also fundamental faults, are those that occur on the overriding plate, parallel to, and about 200 km from, a subduction zone (Fitch, 1972). These include the dextral Semangko or Sumatra fault system (Fig. 2; Page *et al.*, 1979), the dextral Median Tectonic Fault line of southern Japan (Kaneko, 1966), the dextral Atacama Fault of Chile (Allen, 1965) and the sinistral Philippine

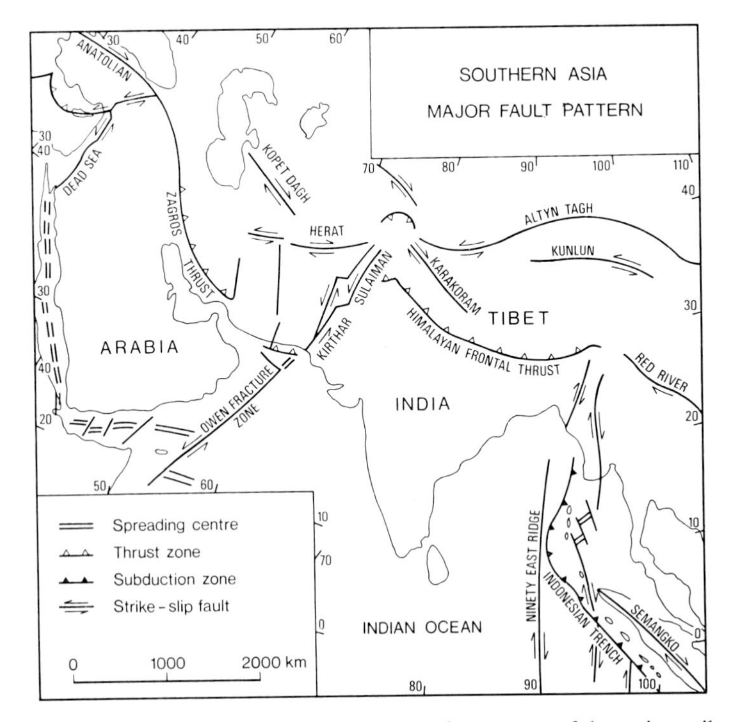

Fig. 2. Map of southern Asia showing position and sense of movement of the major strike-slip faults, subduction zones and spreading centres (after McKenzie, 1972; Nowroozi, 1972; Tapponnier & Molnar, 1975; Mohajer-Ashjai, Behzadi & Berberian, 1975; Page *et al.*, 1979).

Fault (Allen, 1962). Extensions of the Alpine Fault in the North Island of New Zealand lie in a similar position (Lewis, 1980). Attention has been drawn by Hepworth (1979) to ultramafics along some such faults, formed by a process of transduction.

Following continental collision a complex fault pattern is developed (McKenzie, 1972; Molnar & Tapponnier, 1975). As a protuberance on one continent collides with the other continent, large strike-slip faults develop (Fig. 2), their orientation depending on the shape of the protuberance (Tapponnier & Molnar, 1975). Faults of this type include the dextral North Anatolian Fault (Sengör, 1979), the dextral Herat Fault, the sinistral Altyn Tagh Fault, the dextral Karakoram Fault, the sinistral Kangting Fault and the dextral Red River Fault (Fig. 2). While it is possible that these faults were *initiated* at the time of continental collision, it is also likely that many are older lineaments which were *reactivated* by collision and are therefore fundamental faults. Some may also be plate boundaries, for example the Quetta–Chaman fault and the North Anatolian Fault which is thought by McKenzie (1972) to be the boundary between two minor plates, the Turkey and the Black Sea plates. However, identification of minor plates becomes extremely difficult in such areas since many of the plates abound with dislocations and various scales of strike-slip faults and their associated folds and thrusts. Parts of them are even actively volcanic. Thus the concept of rigid plates has to be questioned and there is a continuum from minor plates through to fault blocks (Mohajer-Ashjai, Behzadi & Berberian, 1975). In fact the whole belt of continental collision from Gibraltar to SE Asia contains not only major strike-slip faults bounding minor plates but also innumerable minor strike-slip faults. The width of this belt is at least 1000 km (Fig. 2), and is comparable in size with the Hercynian fold belt (see p. 20).

There is growing evidence from both regional mapping and Landsat imagery that major lineaments, in some areas forming shear patterns extending for many hundreds of kilometres, are an important feature of both continental and oceanic crust. Many of these lineaments yield evidence that they are or were strike-slip faults. Moody (1973) has argued that they represent a world-wide shear pattern which developed early in the Earth's history by lateral compression and has subsequently controlled deformation. One does not have to follow his views completely to be open to the postulate that megashears are as important a feature of global tectonics as are plate boundaries, particularly in continental crust.

Supporting evidence that megashears are long-lived features comes from exposed Precambrian rocks in western Greenland. Here, shear belts which were active as early as 2800 m.y. ago operated in the Jurassic and are probably still active today. As rocks now seen at the surface rose from below 15 km, there has been a progressive change from ductile deformation at lower levels to brittle deformation at higher levels (Bak *et al.*, 1975; Grocott, 1977). Thus studies in the Precambrian support the idea that movement on present day fault systems changes from brittle deformation to ductile deformation at about 15 km depth. Seismic activity (brittle deformation) on the San Andreas fault ceases below about 15 km (Eaton, Lee & Pakiser, 1970). However, only some of these major lineaments are today acting as strike-slip faults, and even they may not always have done so. Contemporary movement along mega-shear zones depends on the current regional stress pattern. Therefore at different times such zones may have acted as normal, reverse or strike-slip faults, even to the extent of having suffered separate phases of sinistral or dextral movement.

Simple strike-slip systems

Along a major strike-slip fault system, there may be quite small-scale alternate zones of extension and compression. These are due to curvature along a strike-slip fault, the braiding of faults within a strike-slip fault system or side-stepping of *en echelon* faults (Fig. 3). The early ideas of Kingma (1958a, b) and Quennell (1958, 1959) have been pursued by others, in particular Crowell (1974a, b) for the San Andreas Fault system and Freund (1965) and Freund *et al.* (1970) for the Dead Sea (Levant) Fault system.

(1) The curvature of a strike-slip fault gives alternate zones of divergence and convergence (Fig. 3a). The former are areas of extension, where basins develop; the latter are areas of compression, uplift and erosion.

(2) Faults split and rejoin to form an anastomosing pattern. Where two faults converge, the block between them is compressed and uplifted; where they diverge, extension results in sinking (Fig. 3b and c).

(3) Faults die out and motion may be taken up by an adjacent, side-stepping, parallel fault. At the end of a fault either extension or compression may occur (Fig. 3d). Where a parallel fault takes up the motion again, either a pull-apart basin forms like the Dead Sea or sharp restraining folds and thrust faults develop (Fig. 3e).

In addition, along a strike-slip fault system there may be zones where the trend of an existing major, perhaps fundamental, fault differs from that of the regional slip

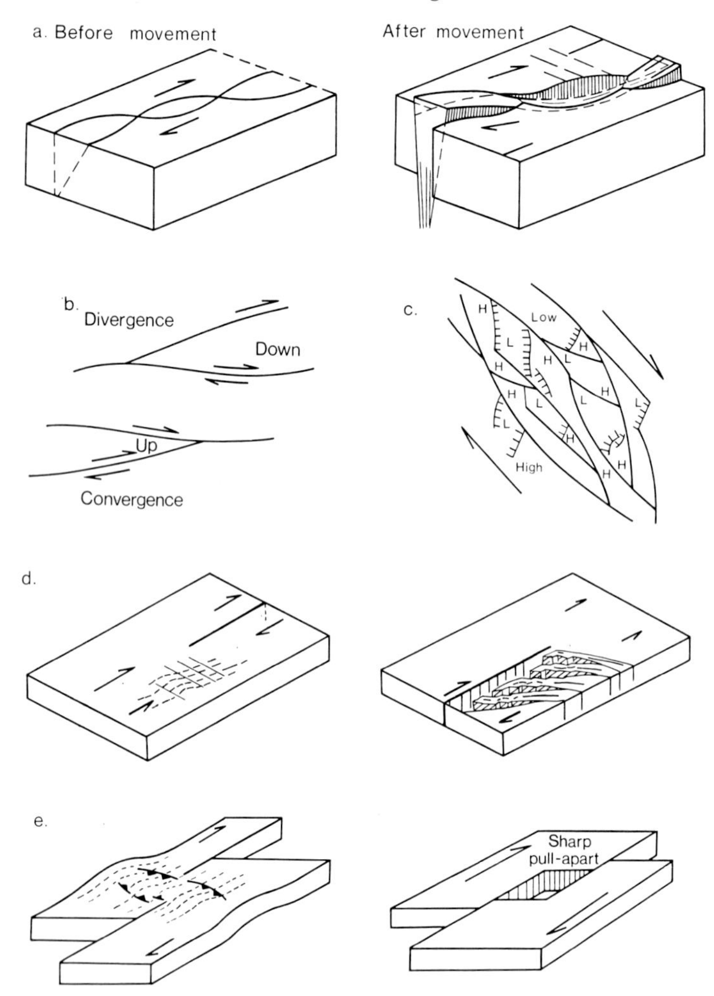

Fig. 3. Types of strike-slip fault pattern that produce extensional subsiding basins and compressional, uplifted blocks: (a) curved fault trace, (b) divergent and convergent fault patterns, (c) anastomosing faults, (d) fault terminations, (e) side-stepping faults (after Kingma, 1958b; Quennell, 1958; Crowell, 1974b).

vector. This happens in the Transverse Ranges of California and along the main Alpine Fault where, in both cases, the zones of uplift are thought to result from obliquity of the fault trends to the regional slip vector (Scholz, 1977).

Modification of simple strike-slip

As pointed out by Harland (1971), it is seldom possible to describe the horizontal movements of plates, even on a plane surface, in terms simply of the three tectonic regimes of extension (divergence), transcurrence (conservation) and compression (convergence). Movement between blocks is normally intermediate or oblique, and strike-slip motion is normally either divergent or convergent (Wilcox, Harding & Seely, 1973).

Harland terms a combination of transcurrence and extension *transtension,* and a combination of transcurrence and compression *transpression.* Transtensile regimes are marked by normal faulting (possibly with drape folding), basin formation and volcanicity. Transpressive regimes are marked by thrust faulting, reverse faulting, folding and uplift.

The nature of movements on the fundamental faults defining plate boundaries are affected by plate interaction. One regime may change to another over periods of several million years. For example the San Andreas Fault system may have changed from a transtensile regime in the Miocene to a transpressive one in the late Pliocene and Pleistocene (e.g. Nardin & Henyen, 1978). The Alpine Fault system may also have changed to its present dominantly transpressive regime in the late Miocene (Carter & Norris, 1976; Norris *et al.*, 1978).

Effects of strike-slip

Although the displacement along strike-slip faults is dominantly horizontal, the most obvious motion at any one place may be dip-slip. It is this vertical movement of blocks relative to one another that produces the main sedimentary effects of a strike-slip fault system. However, there are major differences in the sedimentary effects of transpression and transtension. Transpression leads to folding, thrusting and vertical uplift of mountainous zones on land and of ridges and perhaps islands offshore. Transtension leads to the sinking of basins bordered by closely spaced normal faults.

During periods and within zones of transpression there will be little chance for basins to develop within the strike-slip belt and the eroded sediments are mainly carried outside the mountain belt to provide material for clastic sedimentation elsewhere. Thus the effects are similar to those of partly compressive orogenic belts from which material is removed and carried by superficial processes to distant cratonic or even oceanic areas. For example, sediment from Himalayan continental crust is not only being deposited in the 'molasse' foredeeps to the south but is also being transferred thousands of kilometres onto oceanic crust as the Indus and Bengal submarine fans. Similarly, Tertiary transpressive belts, as in Spitzbergen (Lowell, 1972) or in the Pyrenees, lack internal sedimentary basins. In addition, straight segments of modern transpressive strike-slip faults, such as along the main southern Alps of New Zealand, are not major areas of sedimentation.

During periods and within zones of transtension, basins form within strike-slip belts which thus become important loci for sedimentation. In these circumstances relatively little material may be transported outside the belt. However, since sedimentation requires a source as well as a basin, transtensile belts are areas of substantial sedimentation only on land or at continental margins. In oceanic systems thick accumulations are probably confined to narrow zones immediately adjacent to faults. In addition, basins and uplifted blocks resulting from the splaying, offsetting and anastomosing of faults are essential features for abundant intra-belt sedimentation.

Sedimentary environments and facies

Every type of sedimentary facies may be found in strike-slip basins. Individual facies have a limited lateral extent and there is usually evidence for a nearby active source area, frequently indicated by restricted deposition of locally derived conglomerates and breccias. However, these features are not by themselves diagnostic of strike-slip basins, only of sedimentation in fault-bounded, rapidly subsiding grabens.

On land the most important initial depositional environment within a strike-slip fault zone is lacustrine. Such lakes are long and narrow, subside rapidly and contain thick columns of sediment. Typically they are bordered by alluvial fans and fed from one or perhaps both ends by a river and its delta. As they fill, they pass from deep water to shallow water and may become emergent, but, since they are continually subsiding, particular facies tend to remain in the same zones for long periods of time and lateral migration of facies is limited. The particular facies which form depend a great deal on the climate and chemistry of the lake waters (Hardie, Smoot & Eugster, 1978). As lakes are filled, they become the sites of rivers so that the basin-fill sequence may be ended by fluvial sediments (e.g. the Pliocene Ridge Basin, California, Fig. 5b; Crowell, 1975; Link & Osborne, 1978).

Bordering continental strike-slip fault zones, especially where transpression is dominant, there may be wide alluvial plains where climatically controlled fluvial and interfluvial sediments can form, derived from the uplifted mountains (e.g. the Pleistocene and modern gravels of the Canterbury Plains to the east of the Alpine Fault; Suggate, 1965).

The characteristic marine depositional environment of strike-slip zones is a basin ranging in size from about 10×20 km to about 30×100 km and reaching depths of over 1000 m. These are best known in the California continental borderland and around New Zealand (Moore, 1969; Summerhayes, 1969). Sedimentary facies vary from pelagic or hemipelagic to terrigenous or hemiterrigenous where continental sources are important. Subaqueous mass–gravity processes transport coarser-grained terrigenous clastics into the basin, to be deposited on basin plains, submarine fans and slope-aprons. The basins may be partially bounded by narrow shelves where most terrigenous sediment is collecting at present, owing to the recent rise in sea-level. Occasional interception by submarine canyons allows some sediment to be transported to the submarine fans and basin plains.

Where carbonate banks are developed at basin margins, they may be important sources of detrital carbonates to the basins (e.g. Yallahs Basin off Kingston, Jamaica; Burke, 1967).

ANCIENT STRIKE-SLIP MOBILE BELTS

There are considerable difficulties in recognizing ancient strike-slip belts because unambiguous evidence is seldom preserved and because, since all such belts have elements of transtension and transpression, both extensional and compressional features may be present. Nevertheless some mobile belts are more satisfactorily explained as zones of strike-slip movement than of major regional compression.

Lateral motion

As in present-day strike-slip belts, strike-slip motion is best recognized directly by the matching of particular rock types or sedimentary facies which have been displaced along a fault. Reconstructions of palaeogeography also may indicate that a lateral shift has occurred (Nelson & Nilsen, 1974). This method is preferable to the matching of individual facies since variability and repetition of facies are marked along all strike-slip fault zones.

The petrography (e.g. large clast, sandstone, or heavy mineral composition) of alluvial fans, may indicate a source area of a particular kind. This may no longer be present on the neighbouring upthrown side of the fault (Heward & Reading, 1980) but perhaps can be located some distance away. Alternatively, palaeogeography may suggest a lateral shift is required to provide a source area of a size commensurate with the size and shape of alluvial or submarine fans (Steel & Gloppen, 1980; Norris & Carter, 1980).

Vertical motion

Large and rapid vertical movement is common along oblique-slip faults. The larger syn-sedimentary faults may be difficult to detect except by finding sudden lateral thickness or facies changes; and, where sediment supply is abundant, facies changes may not coincide with fault lines and only thickness changes occur. Evidence that sediment was derived from the upthrown side is rare, except where alluvial or submarine fans exist. Palaeocurrents and palaeogeographical evidence from fluvial, deltaic or marine environments are more likely to show sediment transport parallel to the strike-slip fault than perpendicular to it. However, small-scale structures showing down-basin extension and the palaeo-sedimentary slope may be present. Because of the differential movement, sedimentary piles are usually several kilometres thick and have been deposited very rapidly.

Lateral facies changes

Lateral environmental change is a characteristic of all small, quickly subsiding basins. Unless a major delta, fed from an extensive area of erosion, enters one end of the basin, most sediment is supplied from local sources and the limited points of supply restrict the extent of any particular facies (Link & Osborne, 1978; Steel & Gloppen, 1980). Where sedimentation is controlled by active faults, the migration of facies is limited and facies units are thick and of restricted lateral extent. This contrasts with tectonically stable regions where facies units tend to be thin and extensive because facies belts migrate continuously on account of the fine interplay of sediment supply, base level and climatic changes, the over-extension of deltas and aggradation of rivers.

Compressional factors

The features described above result from rapid vertical movement and, by themselves, are indicative only of extension and normal faulting. They may be found in any down-faulted graben. However, all strike-slip belts, except very transtensile ones, have zones of compression. Somewhere along the belt there should be evidence of localized compression resulting in thrusting, uplift and perhaps the formation of nappes. Thus the view of the belt needs to be regional rather than local. Uplift also leads to erosion and the development of unconformities, commonly with strong angular discordance. Such deformation should be contemporaneous with sedimentation nearby. Unfortunately, in most ancient orogenic belts, especially in continental deposits where stratigraphical indices are sparse, correlations are frequently based upon the assumption that unconformities or 'tectonic events' are synchronous. The same is true of subsurface studies in petroleum provinces where 'correlations' are made by linking discordant surfaces. However, if the evidence is looked at critically many of these 'correlations' are found to be of doubtful value, and frequently there is a discrepancy between correlations based on orogenic events and on biostratigraphy. Consequently many classical correlations in orogenic belts are suspect. Unconformities and fold phases are, by the very nature of strike-slip orogenic belts, bound to be limited in extent and contemporaneous with a continuous sedimentary column not very far away.

While sedimentation and deformation are simultaneous on a local scale due to curvature, side-stepping and braiding of faults, modern strike-slip fault zones may change, over periods of a few million years, between transpression and transtension. Consequently there may also be longer periods when either sedimentation or deformation is dominant throughout an ancient belt. Therefore not all classical conclusions on the separation of phases of sedimentation and folding are mistaken. The problem is to distinguish between those features resulting from long-term and regional phases of transtension and transpression and those due to local complexities in the fault pattern.

Structural pattern

Clay model experiments (e.g. Cloos, 1955; Tchalenko, 1970; Wilcox *et al.*, 1973) show a characteristic pattern of drag folds and secondary faults along strike-slip fault zones (Fig. 4). Folds form at an oblique angle to the strike-slip fault in an *en echelon* pattern, though if transpression is dominant the folds are rotated so that they nearly parallel the strike-slip fault (Harland, 1971). *En echelon* folds are therefore important indicators in plastic cover rocks of strike-slip in more rigid basement rocks at depth. They are also important petroleum reservoirs (Harding, 1974).

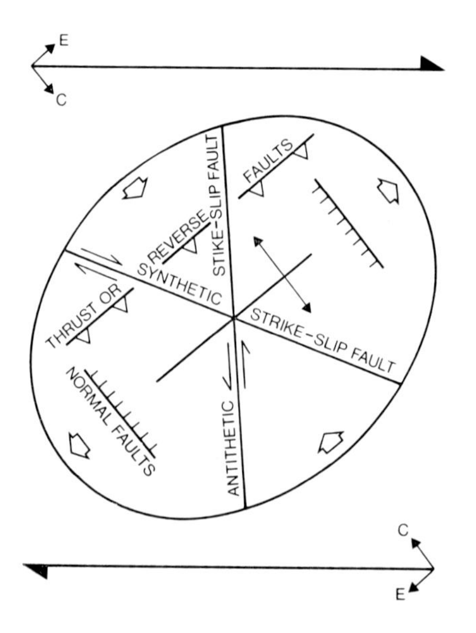

Fig. 4. Structural pattern resulting from simple shear produced by an E–W dextral shear couple (after Harding, 1974).

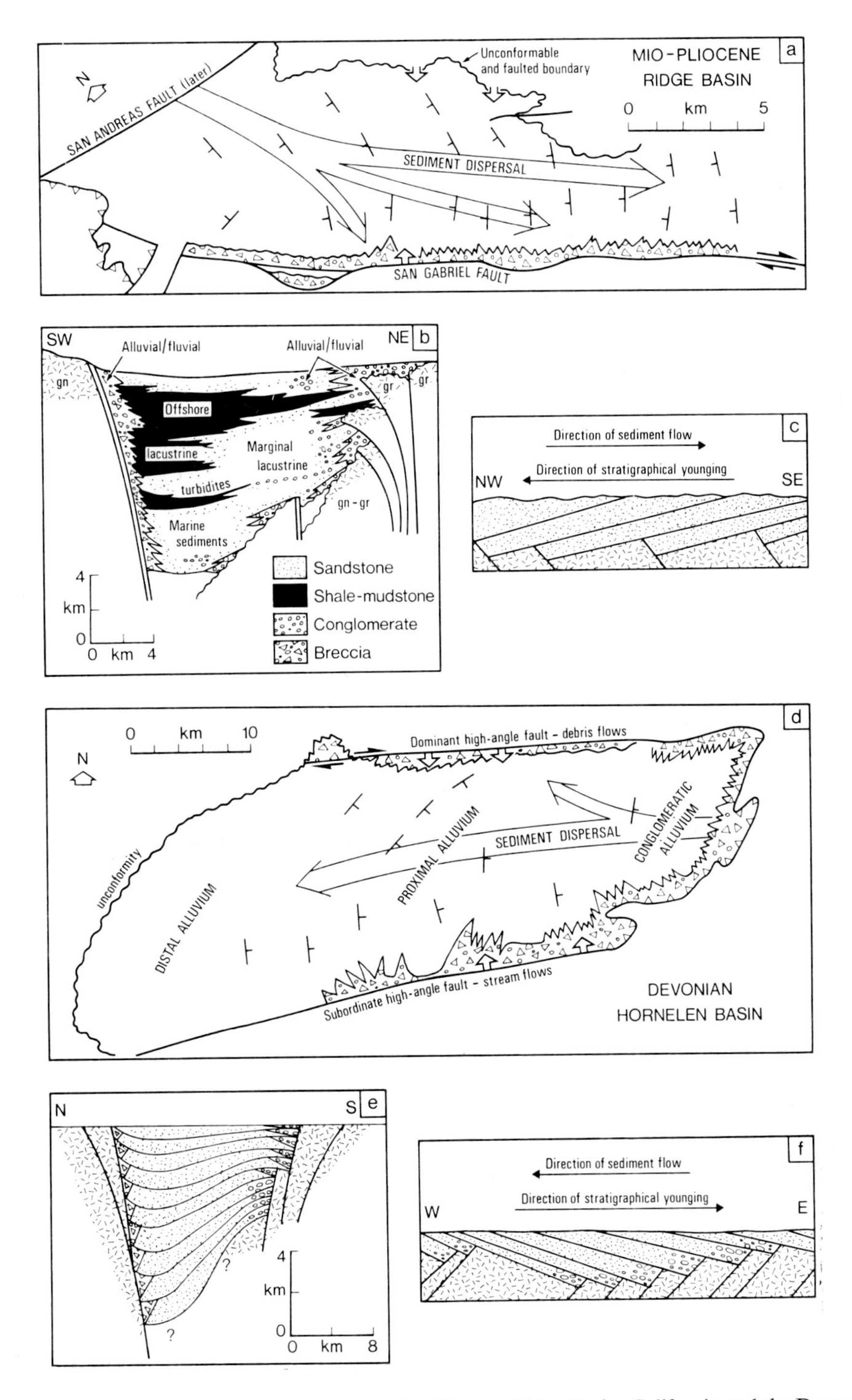

Fig. 5. Structural and sedimentary patterns in the Pliocene Ridge Basin, California and the Devonian Hornelen Basin, Norway (after Crowell, 1975; Link & Osborne, 1978; Steel & Aasheim, 1978; Steel & Gloppen, 1980).

Metamorphism

Regional metamorphism is normally absent or low grade, even in upthrust basement blocks of underlying continental crust (e.g. Lowell, 1972; Heward & Reading, 1980). If it is present, it is probably the result of an earlier phase of burial or deformation.

Igneous activity

Igneous rocks are generally sparse, particularly in transpressive belts. However, they may be more voluminous where extension is important and hence mark periods or zones of transtension. When the extensional component is very large, ophiolites may be formed at spreading centres. Serpentinites too are frequently found along strike-slip faults (Page *et al.*, 1979; Robertson & Woodcock, 1980) and they are common in oceanic fracture zones (DeLong, Dewey & Fox, 1979). Calc-alkaline rocks are absent except where there is an associated subduction zone as in Sumatra and at the northern end of the Alpine Fault system. Per-alkaline granites occur in strike-slip belts but their significance is still not understood.

Sedimentary fill of strike-slip basins

Strike-slip basins are elongate parallel to the strike-slip system and are commonly very deep in relation to their width (Kingma, 1958a; Clayton, 1966; Crowell, 1974a; Link & Osborne, 1978). Cross-sections are usually asymmetrical since vertical movements on both sides are unlikely to be the same. For example the Ridge Basin (Fig. 5) has one simple but dominant downfaulted margin; the other is more complex with a combination of unconformities, upthrusts and dip-slip faults. The Hornelen Basin of Norway (Steel & Aasheim, 1978; Steel & Gloppen, 1980) has a dominant strike-slip margin at the north with a narrow zone of debris flow fans. This margin contrasts with the subordinate or normal southern margin where deposition on the alluvial fans is by gentler stream flow across a wider zone (Fig. 5). In both these basins the locus of sedimentation migrated away from the direction of strike-slip motion of the basin and sediments therefore young in this direction, as shown by the stratal dip. This is opposite to that of palaeoflow and sediment fining showing that the sediment source lay in the direction of and possibly migrated backwards in the same direction (Fig. 5).

Although the Hornelen Basin is entirely alluvial whereas the Ridge Basin was marine and lacustrine before it became alluvial, their structural and palaeocurrent patterns are similar. However, this model may not be a general one for all strike-slip basins (cf. Ballance, 1980; Bluck, 1980).

Geosynclinal cycle

The classical geosynclinal cycle of (1) *pre-flysch* (starved basin phase, with or without ophiolites), (2) *flysch* (deep marine clastic phase), (3) *molasse* (continental clastic phase) is now generally interpreted as the result of the Wilson (1966, 1968) cycle of (1) oceanic opening by sea-floor spreading, (2) oceanic closure by subduction, (3) continental collision. Variations are conceded and debate centres on such problems as whether the cycle is the result of opening and closing of a 'true' ocean or a small back-arc basin, or whether the flysch was deposited on ocean floor, in a trench or in a fore-arc or back-arc basin.

Nevertheless the emphasis has been on subduction and many geologists appear to think that all orogenic belts must have been the result of subduction, the only questions remaining being precisely where to draw the palaeo-subduction zones and in which direction they dipped.

The geosynclinal sequence can also be explained by the strike-slip cycle (Mitchell & Reading, 1978) which has three major phases.

(1) *Phase of transtension* leading to the formation of small basins. If these are formed within continental crust and extension is not too great, there is, initially, slumping and landslipping into the basin and deposition of conglomerates and breccias on the margins passing into fine-grained lacustrine sediments in the centre of the basin. As extension proceeds, crustal thinning may allow magma to rise and igneous rocks, probably basalt, to be emplaced. The basin may sink below sea level (e.g. Dead Sea and Salton Trough), and evaporites may form. As extension proceeds further the sea may penetrate the basin. Alternatively the basin may start at or even below sea level and be marine from the outset. Further extension allows true oceanic crust to form, giving rise to ophiolites. The Gulf of California is the best example today of this situation. Its spreading ridges are offset by transform faults and are oriented in a N–S direction which is that of extension in a NW–SE trending dextral fault system. Another example may be the Andaman Sea, behind the northward extension of the Java Trench (Fig. 2), where spreading centres are aligned in a NE–SW direction (Curray *et al.*, 1979) compatible with the northward dextral motion of the Indian plate relative to the SE Asian plate (Page *et al.*, 1979).

(2) *Phase of basin filling* by various processes of mass gravity transport. This phase overlaps with the extending phase. Conglomerates and breccias may continue to form at faulted margins, but away from the margins deposition will be mainly by turbidity currents. If the basin continues to deepen, sediments slide and are deformed towards the basin centre. As sediment cover increases, extensional growth faulting, gravity folds and faults also displace sediment into the basin centre. At faulted margins there may be small thrusts and overfolds. When extension ceases, filling continues and the sediments may extend beyond the original fault-bounded basin.

(3) *Phase of transpression* overlapping with the basin fill phase. Initially sedimentation continues, but as the basin fills up and is perhaps uplifted above sea level, there is a change from marine turbidites to lacustrine and/or fluvial sediments. Compressive structures increase near the margins and the basin sediments become folded. Continuing uplift leads to erosion of the deformed sediments.

The three phases give a sequence for each basin which passes from pre-flysch, possibly ophiolitic, and sometimes initially non-marine, through flysch, with local wildflysch, to molasse. Deformation is initially extensional, becoming increasingly compressional as the cycle is completed.

Over a large area, several of these basins may form, partly simultaneously and partly diachronously. They can then give the impression of great stratigraphical thicknesses of sediment. If the belt passes from an early period of transtension to one of transpression then there will be an overall stratigraphic change through the classic geosynclinal cycle. However, neither wide oceans nor subduction are necessary. All that is needed is the fragmentation of essentially continental areas such as the present Mediterranean, the seas and islands of SE Asia, and the Continental Borderlands and Gulf of California of southwestern North America.

Every gradation is possible from belts where a relatively small amount of transtension is followed by transpression, with no formation of ocean-floor and entirely within continental crust, through those where there is some ocean-floor spreading and some loss of ocean by subduction, to those where there is major opening and closing, with very minor strike-slip. Thus the Wilson cycle and the strike-slip cycle are end-members of a continuum. Most ancient orogenic belts have elements of both.

Some ancient examples

(1) The Hercynian fold belt of western Europe has been the subject of much discussion, particularly on the positioning of subduction zones. They have been placed across southwest England, in the English Channel, in France etc. Yet there is little positive evidence for Hercynian subduction. Calc-alkaline volcanics, paired metamorphic belts and ophiolites are lacking. Arthaud & Matte (1975, 1977) have shown that the late-Variscan structural pattern of western Europe is one of essentially NW or WNW trending dextral wrench faults and NE trending sinistral faults with associated second order faults. They interpret this pattern as part of a 2000 km wide right-lateral shear zone induced by the relative motion of a northern plate which includes the Canadian Shield, Greenland and northern Europe and a southern plate which includes the African Shield.

Within this belt late Hercynian basins formed, such as the Alès coal basin in southern France (Gras, 1972; Arthaud & Matte, 1977, Fig. 8). Here 13 km of sinistral movement along a primary fault led to sedimentation and the sliding of sedimentary nappes. A later compressional phase caused folding, thrusting, and plastic deformation along the wrench fault. Similar small late Hercynian basins abound in western Europe and the possibility that they too are strike-slip basins should be considered. Arthaud & Matte (1977) argue that strike-slip becomes important in Europe only during late Hercynian (Stephanian) times. However, Heward & Reading (1980) suggest that earlier (Westphalian) sedimentary basins and deformation in the Cantabrian Mountains of northern Spain, can be interpreted as the result of strike-slip motion. This concept may well be extended to the Carboniferous in other parts of Europe and also to North America where, in the Canadian Appalachians, strike-slip faulting with offsets as much as 200 km occurred during the Carboniferous (Webb, 1969). A characteristic feature throughout the Upper Carboniferous and in many places in the Lower Carboniferous, especially of Scotland, the Maritime Provinces of Canada and Newfoundland (Belt, 1968, 1969) is the abundance of small rapidly subsiding basins, with complex facies patterns between local source areas. Belt (1969) showed how strike-slip motion was contemporaneous with sedimentation in the Canadian regions, and, as pointed out by Davis & Ehrlich (1975), the Appalachian geosyncline is a mosaic of loosely connected depositional basins, each possessing its own source terrain.

Thus the sedimentary and structural patterns within the Variscan belt might be explained by considering it as a major right-lateral strike-slip belt within which local extension and compression gave rise to sedimentary basins, source areas, unconformities and deformation. Oceanic crust may occasionally have been developed but normally basin extension was insufficient for it to form and the results of extension were limited to a lull in sedimentation and some volcanic activity.

(2) In the Caledonides the evidence is strong for the existence of a major ocean, the Iapetus Ocean during the Lower Palaeozoic (Harland & Gayer, 1972). During its

closing stages, as it was progressively destroyed by subduction, lateral motion was also important (Phillips, Stillman & Murphy, 1976). In addition, earlier, Ordovician, basins such as the South Mayo Trough of western Ireland (Dewey, 1963) may have been transtensional basins. On the southern margin of the ocean, in England and Wales, southeastward subduction explains the broader features, especially the volcanism (Fitton & Hughes, 1970). However, the detailed pattern of volcanic activity and of sedimentary basins such as the Welsh basin is controlled very largely by contemporaneous NE–SW trending faults (Rast, 1969). By analogy with those which occur on the overriding plate behind present day subduction zones (see pp. 9–10 and cf. Walcott, 1978), these faults may have been strike-slip faults taking up lateral motion between the two obliquely converging plates.

Whilst these suggestions are speculative at present, there is well established evidence that strike-slip is an important aspect of Caledonian continental collision. In Newfoundland, Ireland and Scotland, Belt (1969) showed how Lower Carboniferous strike-slip faults were initiated in Late Devonian times. Steel & Aasheim (1968) and Steel & Gloppen (1980) regard the early to middle Devonian Old Red Sandstone basins of Norway as the result of either extensional or transtensile tectonics. Bluck (1978, 1980) has also argued that both the Lower and Upper Old Red Sandstone of the Midland Valley of Scotland accumulated in strike-slip basins.

(3) A well described Tertiary fluvio-lacustrine rift valley basin is the Bovey Basin of south Devon, England (Edwards, 1976). It is the largest of three basins that lie along the Sticklepath–Lustleigh NW–SE trending dextral fault, one of a number of similar trending faults which rotated southwestern England dextrally in Tertiary times. About 2 km of movement along side-stepping faults produced a basin 4 km across and 10 km long and with an estimated sediment thickness of 1200 m. Only the upper 300 m of sediment is exposed and this consists of kaolinitic clays, lignites and sands deposited under fluvial and lacustrine conditions in the centre of the basin which was bordered by coarse-grained alluvial fans. At a late stage of the fill, sediments spread outwards over the margins of the basin. At the southern margin of the basin Devonian basement is thrust over Tertiary basinal sediments, indicating more or less contemporaneous deformation and sedimentation.

Economic importance of strike-slip

Strike-slip systems are both actually and potentially important areas for petroleum (Moody, 1973; Harding, 1974; Holmgren, Moody & Emmerich, 1975). California, for example, provides some 800,000 bbls of oil a day and west of the San Andreas Fault more than 90% of discovered petroleum has accumulated in late Tertiary basins that formed during the strike-slip regime of the last 15 m.y. (Blake *et al.*, 1978). The reasons are that strike-slip fault systems provide ideal conditions for the genesis of source rocks, reservoir rocks, seals and traps.

The early stage of small extensional, starved basins within continental crust allows organic-rich pelagic and hemipelagic sediments, including organic-rich turbidites, to be deposited under partially reducing conditions. Source rock maturation is encouraged by rapid burial and subsidence and perhaps higher than normal heat flow (Wilde, Normark & Chase, 1978). Sediment supply is rapid and various forms of sand body are deposited to form reservoir rocks. The most important sand body is probably the submarine fan (MacPherson, 1978), but locally shelf sands and small deltas may occur. As the basins become continental, non-marine reservoir sediments, especially fluvial sands, may become important and lacustrine sediments themselves may develop as further source rocks.

Traps tend to be small but very abundant, partly because of the irregular facies distribution and the many local unconformities. However, the most important traps are structural, in the form of *en echelon* anticlinal folds above and on the flanks of strike-slip faults and in folds associated with reverse faults in zones of local compression (Harding, 1974; Blake *et al.*, 1978; Pilaar & Wakefield, 1978). In detail the structural pattern varies according to the stage of strike-slip deformation and the amount of displacement (Harding, 1974). Growth faults and related compaction anticlines may also be important (MacPherson, 1978).

The formation of thick lacustrine sediments is also significant economically. Important successions of evaporites (Neev & Emery, 1967), ball clays (Edwards, 1976), oil shales, lignites and coals (Heward & Reading, 1980) are all found in strike slip fault zones.

Although hydrothermal gradients are lower than those at spreading centres, thermal springs are commonly associated with strike-slip zones. Mineralization occurs along oceanic fracture zones and deposits of pyrite, gypsum and barite have been reported along strike-slip faults from on-land and off-shore Californian Borderlands (Lonsdale, 1979). In addition many workers believe that ore bodies are located along the line of major lineaments or megashears (e.g. O'Driscoll, 1968; Scott, 1978).

CONCLUSIONS

Strike-slip motion is an important feature of global geology today. Some modern strike-slip fault systems are related to fundamental crustal weaknesses with a history going back as far as the geological data allow and have similarities with the major shear-zones of Precambrian terrains. These fundamental faults may not always have operated as strike-slip faults since their inception and trend are the result of earlier events. Essentially they accommodate the global movements, both inter-plate and intra-plate, of the time. They may therefore have moved in different directions at different times and had periods when they were subject to extensional or compressional stress.

Strike-slip is part of a continuum with extension at one end and compression at the other and normally it will be associated with one or the other stress regime on a regional scale. During transtension, basins will tend to form and sedimentation will be largely within the belt. During transpression, uplift will dominate and sediments will be transported outside the belt. On a local scale the complexity of a strike-slip fault system will give rise to a complex of upthrust eroding blocks and down-faulted basins.

The strike-slip model is proposed as an alternative to the Wilson cycle of geosynclinal/orogenic development. Both models are equally valuable as aids to understanding global processes, the Wilson model perhaps more applicable to oceanic plates and their subducting margins, the strike-slip model more applicable to continental plates and smaller basinal features. The two models are thus complementary.

ACKNOWLEDGMENTS

This paper has been greatly improved by constructive criticism of early drafts from P. F. Ballance, R. M. Carter, J. V. Hepworth, R. J. Norris, A. P. Heward, C. J. Pudsey, D. A. Rodgers, K. B. Spörli, and R. Steel. Its inadequacies, however, are the responsibility of the author.

REFERENCES

Abdel-Gawad, M. (1971) Wrench movements in the Baluchistan arc and relation to Himalayan–Indian Ocean tectonics. *Bull. geol. Soc. Am.* **82,** 1235–1250.

Allen, C.R. (1962) Circum-Pacific faulting in the Philippines–Taiwan region. *J. geophys. Res.* **67,** 4795–4812.

Allen, C.R. (1965) Transcurrent faults in continental areas. Continental Drift Symposium. *Phil. Trans. R. Soc.* **A258,** 82–89.

Arthaud, F. & Matte, Ph. (1975) Les décrochements tardi-hercyniens du sud-ouest de l'Europe. Géométrie et essai de reconstitution des conditions de la déformation. *Tectonophysics,* **25,** 139–171.

Arthaud, F. & Matte, Ph. (1977) Late Paleozoic strike-slip faulting in southern Europe and northern Africa: Result of a right-lateral shear zone between the Appalachians and the Urals. *Bull. geol. Soc. Am.* **88,** 1305–1320.

Bak, J., Sørensen, K., Grocott, J., Korstgård, J., Nash, D. & Watterson, J. (1975) Tectonic implications of Precambrian shear belts in western Greenland. *Nature (phys. Sci.),* **254,** 566–569.

Ballance, P.F. (1980) Models of sediment distribution in non-marine and shallow marine environments in oblique-slip fault zones. In: *Sedimentation in Oblique-Slip Mobile Zones* (Ed. by P. F. Ballance and H. G. Reading). *Spec. Publ. int. Ass. Sediment.* **4,** 229–236.

Belt, E.S. (1968) Carboniferous continental sedimentation, Atlantic Provinces, Canada. In: *Symposium on Continental Sedimentation, Northeastern North America* (Ed. by G. de V. Klein). *Spec. Pap. geol. Soc. Am.* **106,** 127–176.

Belt, E.S. (1969) Newfoundland Carboniferous stratigraphy and its relation to the Maritimes and Ireland. In: *North Atlantic—Geology and Continental Drift* (Ed. by M. Kay). *Mem. Am. Ass. Petrol. Geol.* **12,** 734–753.

Blake, M.C. Jr, Campbell, R.H., Dibblee, T.W. Jr, Howell, D.G., Nilsen, T.H., Normark, W.R., Vedder, J.C. & Silver, E.A. (1978) Neogene basin formation in relation to plate-tectonic evolution of San Andreas fault system, California. *Bull. Am. Ass. Petrol. Geol.* **62,** 344–372.

Bluck, B.J. (1978) Sedimentation in a late orogenic basin: the Old Red Sandstone of the Midland Valley of Scotland. In: *Crustal evolution in northwestern Britain and adjacent regions* (Ed. by D. R. Bowes and B. E. Leake). *Spec. Issue geol. J.* **10,** 249–278.

Bluck, B.J. (1980) Evolution of a strike-slip fault-controlled basin, Upper Old Red Sandstone, Scotland. In: *Sedimentation in Oblique-Slip Mobile Zones* (Ed. by P. F. Ballance and H. G. Reading). *Spec. Publ. int. Ass. Sediment.* **4,** 63–78.

Burke, K. (1967) The Yallahs Basin: a sedimentary basin southeast of Kingston, Jamaica. *Mar. Geol.* **5,** 45–60.

Carter, R.M. & Norris, R.J. (1976) Cainozoic history of southern New Zealand: an accord between geological observations and plate-tectonic predictions. *Earth Planet. Sci. Letts,* **31,** 85–94.

Clayton, L. (1966) Tectonic depressions along the Hope Fault, a transcurrent fault in North Canterbury, New Zealand. *N.Z.J. Geol. Geophys.* **9,** 95–104.

Cloos, E. (1955) Experimental analysis of fracture patterns. *Bull. geol. Soc. Am.* **66,** 241–256.

Crowell, J.C. (1974a) Sedimentation along the San Andreas Fault, California. In: *Modern and Ancient Geosynclinal Sedimentation* (Ed. by R. H. Dott Jr and R. H. Shaver). *Spec. Publ. Soc. econ. Paleont. Miner., Tulsa,* **19,** 292–303.

Crowell, J.C. (1974b) Origin of late Cenozoic basins in southern California. In: *Tectonics and Sedimentation* (Ed. by W. R. Dickinson). *Spec. Publ. Soc. econ. Paleont. Miner., Tulsa,* **22,** 190–204.

CROWELL, J.C. (1975) The San Gabriel fault and Ridge Basin, southern California. In: *San Andreas Fault in Southern California* (Ed. by J. C. Crowell). *Spec. Rep. California Div. Mines Geol.* **118,** 208–233.

CURRAY, J.R., MOORE, D.G., LAWVER, L.A., EMMEL, F.J., RAITT, R.W., HENRY, M. & KIECKHEFER, R. (1979) Tectonics of the Andaman Sea and Burma. In: *Geological and Geophysical Investigations of Continental Margins* (Ed. by J. S. Watkins, L. Montadert and P.W. Dickerson). *Mem. Am. Ass. Petrol. Geol.* **29,** 189–198.

DAVIS, M.W. & EHRLICH, R. (1975) Late Paleozoic crustal composition and dynamics in the southeastern United States. In: *Carboniferous of the Southeastern United States* (Ed. by G. Briggs). *Spec. Pap. geol. Soc. Am.* **148,** 171–185.

DELONG, S.E., DEWEY, J.F. & FOX, P.J. (1979) Topographic and geologic evolution of fracture zones. *J. geol. Soc. Lond.* **136,** 303–310.

DEWEY, J.F. (1963) The Lower Palaeozoic stratigraphy of central Murrisk, County Mayo, Ireland and the evolution of the South Mayo trough. *J. geol. Soc. Lond.* **119,** 313–344.

EATON, J.P., LEE, W.H.K. & PAKISER, L.C. (1970) The use of microearthquakes in the study of the mechanism of earthquake generation along the San Andreas fault in California. *Tectonophysics,* **9,** 259–282.

EDWARDS, R.A. (1976) Tertiary sediments and structure of the Bovey Basin, south Devon. *Proc. Geol. Ass.* **87,** 1–26.

FITCH, T.J. (1972) Plate convergence, transcurrent faults, and internal deformation adjacent to southeast Asia and the western Pacific. *J. geophys. Res.* **77,** 4432–4460.

FITTON, J.G. & HUGHES, D.J. (1970) Volcanism and plate tectonics in the British Ordovician. *Earth Planet. Sci. Letts,* **8,** 223–228.

FREUND, R. (1965) A model of the structural development of Israel and adjacent areas since Upper Cretaceous times. *Geol. Mag.* **102,** 189–205.

FREUND, R., GARFUNKEL, Z., ZAK, I., GOLDBERG, M., WEISSBROD, T. & DERIN, B. (1970) The shear zone along the Dead Sea rift. *Phil. Trans. R. Soc.* **A267,** 107–130.

FROIDEVAUX, C.M. (1978) Tertiary tectonic history of Salawati area, Irian Jaya, Indonesia. *Bull. Am. Ass. Petrol. Geol.* **62,** 1127–1150.

GARFUNKEL, Z. (1978) The Negev: regional synthesis of sedimentary basins. *10th Int. Congr. Sediment. Guidebook Pt. I: Pre-congress, Israel,* pp. 35–110.

GILLILAND, W.N. & MEYER, G.P. (1976) Two classes of transform faults. *Bull. geol. Soc. Am.* **87,** 1127–1130.

GRAS, H. (1972) *Étude géologique détaillée du Bassin houiller des Cévennes* (*Massif Central Français*). Unpublished Ph.D. Thesis, University of Clermont-Ferrand. 300 pp.

GROCOTT, J. (1977) The relationship between Precambrian shear belts and modern fault systems. *J. geol. Soc. Lond.* **133,** 257–262.

HARDIE, L.A., SMOOT, J.P. & EUGSTER, H.P. (1978) Saline lakes and their deposits: a sedimentological approach. In: *Modern and Ancient Lake Sediments* (Ed. by A. Matter and M. E. Tucker). *Spec. Publ. int. Ass. Sediment.* **2,** 7–41.

HARDING, T.P. (1974) Petroleum traps associated with wrench faults. *Bull. Am. Ass. Petrol. Geol.* **58,** 1290–1304.

HARLAND, W.B. (1971) Tectonic transpression in Caledonian Spitzbergen. *Geol. Mag.* **108,** 27–42.

HARLAND, W.D. & GAYER, R.A. (1972) The Arctic Caledonides and earlier Oceans. *Geol. Mag.* **209,** 289–314.

HEPWORTH, J.V. (1979) In discussion to Page, B.N.G., Bennett, J.D., Cameron, N.R., Bridge, D. McC., Jeffrey, D.H., Keats, W. and Thaib, J. A review of the main structural and magmatic features of northern Sumatra. *J. geol. Soc. Lond.* **136,** 578–579.

HEWARD, A.P. & READING, H.G. (1980) Deposits associated with a Hercynian to late Hercynian continental strike-slip system, Cantabrian Mountains, northern Spain. In: *Sedimentation in Oblique-Slip Mobile Zones* (Ed. by P. F. Ballance and H. G. Reading). *Spec. Publ. int. Ass. Sediment.* **4,** 105–125.

HOLMGREN, D.A., MOODY, J.D. & EMMERICH, H.H. (1975) The structural settings for giant oil and gas fields. *Proc. 9th World Petrol. Congr.* **2,** 45–54.

KANEKO, S. (1966) Transcurrent displacement along the Median Line, southwestern Japan. *N.Z.J. Geol. Geophys.* **9,** 45–59.

KINGMA, J.T. (1958a) The Tongaporutuan sedimentation in Central Hawke's Bay. *N.Z.J. Geol. Geophys.* **1,** 1–30.

KINGMA, J.T. (1958b) Possible origin of piercement structures, local unconformities and secondary basins in the Eastern Geosyncline, New Zealand. *N.Z.J. Geol. Geophys.* **1,** 269–274.

LEWIS, K.B. (1980) Quarternary sedimentation on the Hikurangi oblique-subduction and transform margin, New Zealand. In: *Sedimentation in Oblique-Slip Mobile Zones* (Ed. by P. F. Ballance and H. G. Reading). *Spec. Publ. int. Ass. Sediment.* **4,** 171–189.

LINK, M.H. & OSBORNE, R.H. (1978) Lacustrine facies in the Pliocene Ridge Basin Group: Ridge Basin, California. In: *Modern and Ancient Lake Sediments* (Ed. by A. Matter and M. E. Tucker). *Spec. Publ. int. Ass. Sediment.* **2,** 169–187.

LONSDALE, P. (1979) A deep-sea hydrothermal site on a strike-slip fault. *Nature*, **281,** 531–534.

LOWELL, J.D. (1972) Spitsbergen Tertiary orogenic belt and the Spitsbergen fracture zone. *Bull. geol. Soc. Am.* **83,** 3091–3102.

MCKENZIE, D.P. (1972) Active tectonics of the Mediterranean region. *Geophys. J.R. astr. Soc.* **30,** 109–185.

MACPHERSON, B.A. (1978) Sedimentation and trapping mechanism in Upper Miocene Stevens and older turbidite fans of southeastern San Joaquin Valley, California. *Bull. Am. Ass. Petrol. Geol.* **62,** 2243–2274.

MITCHELL, A.H.G. & READING, H.G. (1978) Sedimentation and tectonics. In: *Sedimentary Environments and Facies* (Ed. by H. G. Reading), pp. 439–476. Blackwell Scientific Publications, Oxford.

MOHAJER-ASHJAI, A., BEHZADI, H. & BERBERIAN, M. (1975) Reflections on the rigidity of the Lut block and recent crustal deformation in eastern Iran. *Tectonophysics*, **25,** 281–301.

MOLNAR, P. & TAPPONNIER, P. (1975) Cenozoic tectonics of Asia: effects of a continental collision. *Science*, **189,** 419–426.

MOODY, J.D. (1973) Petroleum exploration aspects of wrench-fault tectonics. *Bull. Am. Ass. Petrol. Geol.* **57,** 449–476.

MOORE, D.G. (1969) Reflection profiling studies of the California continental borderland: structure and Quaternary turbidite basins. *Spec. Pap. geol. Soc. Am.* **107.** 142 pp.

NARDIN, T.R. & HENYEY, T.L. (1978) Pliocene–Pleistocene diastrophism of Santa Monica and San Pedro shelves, California Continental Borderland. *Bull. Am. Ass. Petrol. Geol.* **62,** 247–272.

NAUGLER, F.P. & WAGEMAN, J.M. (1973) Gulf of Alaska: magnetic anomalies, fracture zones and plate interaction. *Bull. geol. Soc. Am.* **84,** 1575–1584.

NEEV, D. & EMERY, K.O. (1967) The Dead Sea: depositional processes and environments of evaporites. *Bull. Israel geol. Surv.* **41.** 147 pp.

NELSON, C.H. & NILSEN, T.H. (1974) Depositional trends of modern and ancient deep-sea fans. In: *Modern and Ancient Geosynclinal Sedimentation* (Ed. by R. H. Dott Jr and R. H. Shaver). *Spec. Publ. Soc. econ. Paleont. Miner, Tulsa*, **19,** 69–91.

NORRIS, R.J. & CARTER, R.M. (1980) Offshore sedimentary basins at the southern end of the Alpine Fault, New Zealand. In: *Sedimentation in Oblique-Slip Mobile Zones* (Ed. by P. F. Ballance and H. G. Reading). *Spec. Publ. int. Ass. Sediment.* **4,** 237–265.

NORRIS, R.J., CARTER, R.M. & TURNBULL, I.M. (1978) Cainozoic sedimentation in basins adjacent to a major continental transform boundary in southern New Zealand. *J. geol. Soc. Lond.* **135,** 191–205.

NOWROOZI, A.A. (1972) Focal mechanism of earthquakes in Persia, Turkey, West Pakistan and Afghanistan and plate tectonics of the Middle East. *Bull. seismol. Soc. Am.* **62,** 823–850.

O'DRISCOLL, E.S.T. (1968) Notes on the structure of the Broken Hill lode and its tectonic setting—Broken Hill Mines. *Proc. Australas. Inst. Min. Metall. Ann Conf.* 1968, 87–102.

PAGE, B.G.N., BENNETT, J.D., CAMERON, N.R., BRIDGE, D. MCC., JEFFERY, D.H., KEATS, W. & THAIB, J. (1979) A review of the main structural and magmatic features of northern Sumatra. *J. geol. Soc. Lond.* **136,** 569–578.

PHILLIPS, W.E.A., STILLMAN, C.J. & MURPHY, T. (1976) A Caledonian plate tectonic model. *J. geol. Soc. Lond.* **132,** 579–609.

PILAAR, W.F.H. & WAKEFIELD, L.L. (1978) Structural and stratigraphic evolution of the Taranaki Basin, offshore North Island, New Zealand. *J. Ass. Petrol. Engin. Aust.* 1978, 93–101.

QUENNELL, A.M. (1958) The structural and geomorphic evolution of the Dead Sea Rift. *Q.J. geol. Soc. Lond.* **114,** 1–24.

QUENNELL, A.M. (1959) Tectonics of the Dead Sea Rift. *Int. geol. Congr.* 1956, **20,** 385–405.

RAST, N. (1969) The relationship between Ordovician structure and volcanicity in Wales. In: *The Pre-Cambrian and Lower Palaeozoic Rocks of Wales* (Ed. by A. Wood), pp. 305–335. University of Wales Press, Cardiff.

ROBERTSON, A.H.F. & WOODCOCK, N.H. (1980) Strike-slip related sedimentation in the Antalya Complex, southwestern Turkey. In: *Sedimentation in Oblique-Slip Mobile Zones* (Ed. by P. F. Ballance and H. G. Reading). *Spec. Publ. int. Ass. Sediment.* **4,** 127–145.

SCHOLZ, C.H. (1977) Transform fault systems of California and New Zealand: similarities in their tectonic and seismic styles. *J. geol. Soc. Lond.* **133,** 215–229.

SCOTT, S.D. (1978) Structural control of the Kuroko deposits of the Hokuroku district, Japan. *Min. Geol.* **28,** 301–311.

SENGÖR, A.M.C. (1979) The North Anatolian transform fault: its age, offset and tectonic significance. *J. geol. Soc. Lond.* **136,** 269–282.

SHERIDAN, R.E. (1974) Atlantic continental margin of North America. In: *The Geology of Continental Margins* (Ed. by C. A. Burk and C. L. Drake), pp. 391–407. Springer Verlag, New York.

SITTER, L.U. DE (1964) *Structural Geology*, 2nd edn. McGraw-Hill, New York. 551 pp.

SPÖRLI, K.B. (1980) New Zealand and oblique-slip margins: tectonic development up to and during the Cainozoic. In: *Sedimentation in Oblique-Slip Mobile Zones* (Ed. by P. F. Ballance and H. G. Reading). *Spec. Publ. int. Ass. Sediment.* **4,** 147–170.

STEEL, R. & AASHEIM, S.M. (1978) Alluvial sand deposition in a rapidly subsiding basin (Devonian, Norway). In: *Fluvial Sedimentology* (Ed. by A. D. Miall). *Mem. Can. Soc. Petrol. Geol.* **5,** 385–412.

STEEL, R. & GLOPPEN, T.G. (1980) Late Caledonian (Devonian) basin formation, western Norway: signs of strike-slip tectonics during infilling. In: *Sedimentation in Oblique-Slip Mobile Zones* (Ed. by P. F. Ballance and H. G. Reading). *Spec. Publ. int. Ass. Sediment.* **4,** 79–103.

SUGGATE, R.P. (1965) Late Pleistocene Geology of the northern part of the South Island, New Zealand. *Bull. geol. Surv. N.Z.* **77,** 91 pp.

SUMMERHAYES, C.P. (1969) Marine Geology of the New Zealand Subantarctic Sea Floor. *Mem. N.Z. Ocean. Inst.* **50.** 92 pp.

TAPPONNIER, P. & MOLNAR, P. (1975) Slip-line field theory and larger-scale continental tectonics. *Nature*, **264,** 319–324.

TCHALENKO, J.S. (1970) Similarities between shear zones of different magnitudes. *Bull. geol. Soc. Am.* **81,** 1625–1640.

WALCOTT, R.I. (1978) Present tectonics and late Cenozoic evolution of New Zealand. *Geophys. J.R. astr. Soc.* **52,** 137–164.

WEBB, G.W. (1969) Paleozoic wrench faults in the Canadian Appalachians. In: *North Atlantic: Geology and Continental Drift* (Ed. by M. Kay). *Mem. Am. Ass. Petrol. Geol.* **12.** 754–786.

WILCOX, R.E., HARDING, T.P. & SEELY, D.R. (1973) Basic wrench tectonics. *Bull. Am. Ass. Petrol. Geol.* **57,** 74–96.

WILDE, P., NORMARK, W.R. & CHASE, T.E. (1978) Channel sands and petroleum potential of Monterey deep-sea fan, California. *Bull. Am. Ass. Petrol. Geol.* **62,** 967–983.

WILSON, J.T. (1956) A new class of faults and their bearing on continental drift. *Nature,* **207,** 343–347.

WILSON, J.T. (1966) Did the Atlantic close and then re-open? *Nature,* **211,** 676–681.

WILSON, J.T. (1968) Static or mobile earth: the current scientific revolution. *Proc. Am. phil. Soc.* **112,** 309–320.

WILSON, R.C.L. & WILLIAMS, C.A. (1979). Oceanic transform structures and the development of Atlantic continental margin sedimentary basins: a review. *J. geol. Soc. Lond.* **136,** 311–320.

Spec. Publ. int. Ass. Sediment. (1980) **4,** 27–41

Analysis of pull-apart basin development produced by *en echelon* strike-slip faults

DONALD A. RODGERS

Cities Service Company, Energy Resources Group, Exploration and Production Research Laboratory, Box 3908, *Tulsa, Oklahoma* 74102, *U.S.A.*

ABSTRACT

Mathematical models are used to study the fault patterns and shapes of pull-apart basins produced by right-stepping *en echelon* right-lateral strike-slip faults. The basin shapes and fault patterns are controlled by: (1) the amount of overlap between the faults; (2) the amount of separation between the faults; and (3) whether or not the tops of the faults intersect the ground surface. Secondary strike-slip faults will develop in some part of any pull-apart basin, but secondary normal faults develop only when the tops of the master faults are near the ground surface. The models suggest that, as pull-apart basins evolve, faults in the basement may change their sense of movement. Fault patterns in the basement may be quite different from those in the sedimentary fill of the basin, but as the basin evolves the faults in the centre of the basin will tend to become strike-slip both in the basement and in the sedimentary fill.

INTRODUCTION

Basins associated with major strike-slip fault zones such as the San Andreas fault zone in California are generally elongate parallel to the strike of the fault zone and can be filled with substantial accumulations of sediment. The size of these basins ranges from that of a sag pond a few hundred metres long and filled with a few metres of sediments to the Salton Trough in southern California which is about 200 km long, 80 km wide, and filled with up to 10 km of sediments.

Wilcox, Harding & Seely (1973), Harding (1976), Norris, Carter & Turnbull (1978) and Blake *et al.* (1978) discuss the structural development associated with a single strike-slip fault but do not consider in detail the problem of *en echelon* faults. Burchfiel & Stewart (1966) and Crowell (1974) discuss qualitatively the effects of *en echelon* faults. As Fig. 1 suggests, left-stepping right-lateral strike-slip faults produce a zone of compression and uplift between the two faults, and right-stepping right-lateral strike-slip faults produce a zone of tension and depression between the two

0141-3600/80/0904-0027$02.00

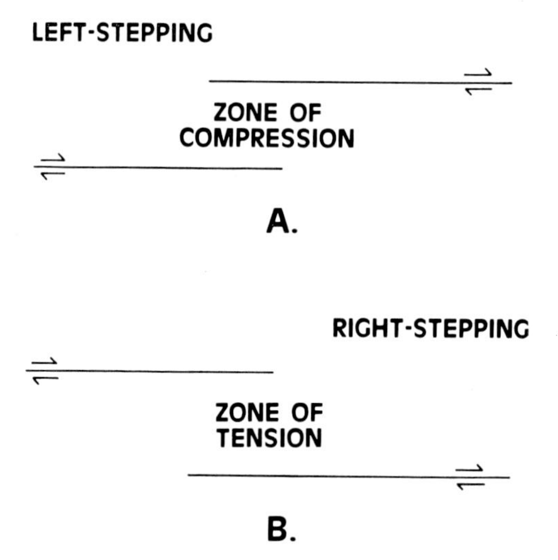

Fig. 1. Terms for *en echelon* right-lateral strike-slip faults.
A. Left-stepping (looking along one *en echelon* fault towards the adjacent fault, the offset is to the left); produces compression between faults.
B. Right-stepping; produces tension between faults.

faults. The zone of tension and depression has been called a pull-apart basin by Burchfiel & Stewart (1966) and Crowell (1974).

Pull-apart basins have been postulated along the Hope fault in New Zealand (Clayton, 1966; Freund, 1971), along the Levant fault zone (Dead Sea Rift) in Israel (Quennell, 1959; Freund, 1965), in Death Valley in California (Hill & Troxel, 1966), and in Norton Sound offshore of western Alaska (Fisher *et al.*, 1979). The crustal thinning, high heat flow, and potentially thick sedimentary accumulations within pull-apart basins make the basins prospective targets for geothermal and hydrocarbon exploration. Thus, it would be useful to study in some detail the structural and sedimentological development of pull-apart basins.

Crowell (1974) presents a qualitative model of the structures within a pull-apart basin (Fig. 2). The model is generalized from field studies of pull-apart basins in California.

In this paper mathematical models of pull-apart basins of various configurations are used to suggest the structures which might develop within the basins. The suggested models will be compared to Crowell's model and field examples, and the effects of the predictions on sedimentation will be discussed.

MODELS

The mathematical models are based on elastic dislocation theory (Chinnery, 1961, 1963). It has been shown that fault models based on this theory produce surface deformations which are similar to those deformations observed after earthquakes (Savage & Hastie, 1969; Alewine, 1974). The fault model is produced by making a cut within a body and moving one side of the cut relative to the other, then welding the sides of the cut together. This results in a discontinuity in the displacement field

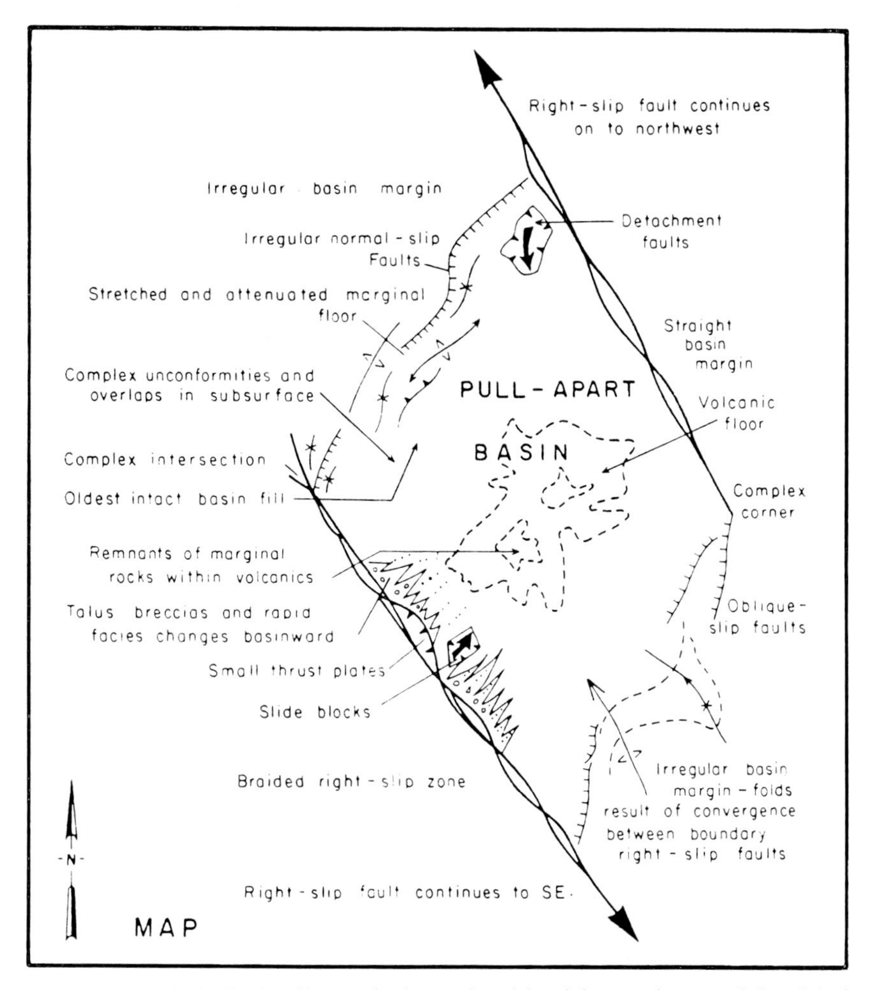

Fig. 2. Sketch map of idealized pull-apart basin produced by right-stepping *en echelon* right-lateral strike-slip faults (from Crowell, 1974, fig. 3).

across the cut. The body containing the deformed cut will be strained, and it is relatively easy mathematically to obtain the displacements, strains and stresses produced by the cut.

The equations of Chinnery (1961, 1963) can be used to calculate the elastic displacement, strain, and stress fields in an isotropic, homogeneous, linearly elastic half-space which are produced by a rectangular displacement discontinuity (or fault) in the half-space. The earth, of course, is not an isotropic, homogeneous, linearly elastic half-space. However, the equations of Chinnery (1961, 1963) are algebraically fairly simple and are fully integrated. Thus, many models can be generated in a small amount of time on any computer. By using a model that is physically simple, we are able to investigate a large number of geologically interesting situations. However, we must always keep in mind the limitations of the model and not attempt to make too detailed predictions from a first-order theory.

In Chinnery's model the offset across the fault is constant over the whole fault and falls instantaneously to zero at the edges of the fault. This, of course, produces a singularity around the edges of the fault. Chinnery & Petrak (1968) have shown that eliminating the singularity with an offset that goes to zero in a finite distance will smooth out the elastic fields produced by the constant offset model but will not affect in any important way the conclusions drawn from the constant offset model. Using a non-constant fault offset introduces complexities into the solutions which require numerical integration of the solutions. Thus, the constant offset model will be used in this paper in order to run a large number of models.

Finally, the equations of Chinnery (1961, 1963) are based on infinitesimal strain theory and thus are strictly valid only when the offsets on the faults are small compared to the fault dimensions. Infinitesimal strain models can describe reasonably well finite strain field problems (Johnson, 1970; Couples & Stearns, 1978; Reches & Johnson, 1978) and laboratory models (Sanford, 1959; Reches & Johnson, 1978). Rodgers (1979) has shown that a dislocation model of the San Andreas fault in southern California can explain the Holocene and Quaternary fault pattern in the Transverse Ranges. Thus it seems reasonable to suggest that the infinitesimal strain theory in this paper can be used to draw valid conclusions about large-scale deformations.

CALCULATED RESULTS

The postulated examples of pull-apart basins are not well documented in terms of the structures within the basin. This is not unexpected, for the thick sedimentary fills of the basins obscure the structural pattern. However, the shapes of the basins are known, and it is possible to compare the calculated and observed basin shapes. In this section the vertical displacement field of the ground surface will first be presented, as calculated from several models, and then the predicted fault patterns determined from the stress fields calculated from the same models. The models are chosen to show the effects of varying the amount of overlap and separation (Fig. 3).

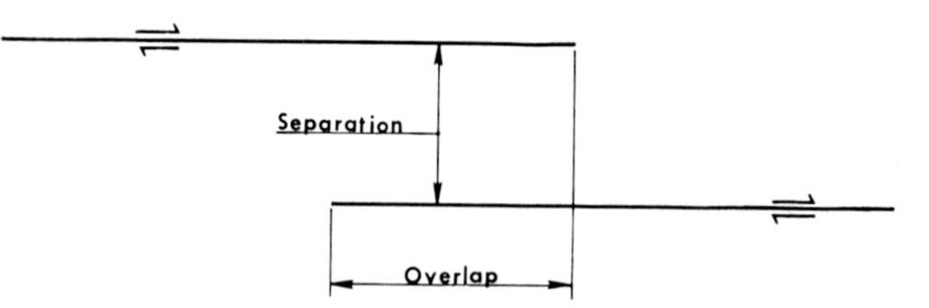

Fig. 3. Definition of separation and overlap for *en echelon* faults.

In order to look at the simplest case, the faults extend to infinity at depth and along strike away from the zone of overlap. This is a theoretical convenience which will not appreciably affect the results in the upper 10–15 km of the crust. In all the models, the master *en echelon* strike-slip faults have equal right-lateral offsets of 1 m. Changing the value of the offsets on the faults affects the magnitude of the calculated displacement and stress fields but does not affect the shapes of the calculated fields. It is possible to change the shapes of the calculated fields by making the offsets on the master faults unequal or by allowing the offsets across the master faults to decay slowly to zero over a finite length of the fault. Unless stated explicitly, the tops of the master faults intersect the ground surface.

Vertical displacements

When the separation of the master faults is twice the overlap, there is a basin between the faults (Fig. 4a). The deepest part of the basin is along a line joining the ends of the two faults with small uplifts at the ends of the master faults. Since the offset

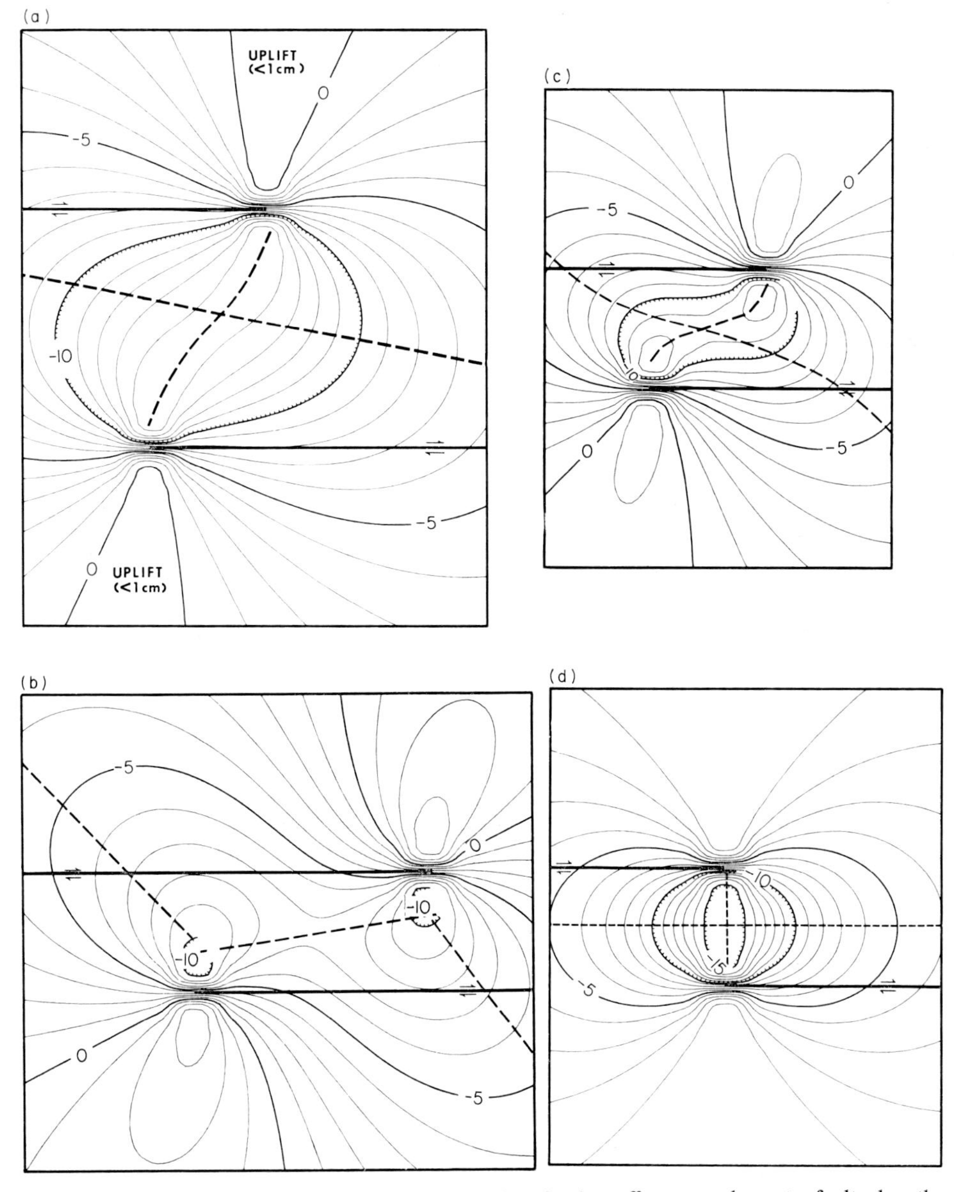

Fig. 4. Vertical displacement in cm of the ground surface for 1 m offset on each master fault when the tops of the master faults intersect the ground surface. Heavy lines show master faults. Negative values are down, and positive values are up. Contour interval 1 cm. Dashed lines show the approximate locations of the axes of the basin.

(a) Separation (20 km) equals twice the overlap; (b) separation (10 km) equals 1/2 the overlap; (c) separation (10 km) equals the overlap; (d) overlap is zero, separation equals 10 km.

on each of the master faults is 1 m, the plotted displacements are percentages of the offset on the master faults. Thus for this model, the depth of the basin is about 15% of the offset on the master faults.

When the overlap is twice the separation of the *en echelon* faults (Fig. 4b), two basins develop between the ends of the two faults, and the uplifts at the ends of the faults are larger than in Fig. 4a. Basin depths are only about 10% of the offset on the master faults.

When the overlap equals the separation (Fig. 4c), a basin develops with the deeper part of the basin along a line joining the ends of the two faults. However, as in Fig. 4b, smaller, less distinct basins develop near the ends of the faults. The greatest depth of the basin is about 14% of the offset on the two faults. The uplift at the ends of the master faults is more than in Fig. 4a but less than Fig. 4b.

When the overlap is zero (Fig. 4d), the deepest part of the basin is exactly perpendicular to the faults, and the greatest depth is about 15% of the offset on the master faults. There is no uplift at the ends of the master faults.

From the vertical displacements, it can be concluded that: (i) a basin will form between *en echelon* right-stepping right-lateral strike-slip faults; (ii) the axis of the deepest part of the basin is on the line joining the ends of the master faults; (iii) the depth and shape of the basin are related to the amount of overlap and the amount of separation between the *en echelon* faults; (iv) there can be small amounts of uplift near the ends of the faults; (v) the amount and shape of the uplift was also a function of the amount of overlap and the separation between the *en echelon* faults; and (vi) the depth of the basin is less than 20% of the offset on the master faults, although this latter conclusion is probably strongly dependent on the elastic properties of the body.

Recalling that the model is for an elastic half-space with infinitesimal displacements on the master faults, the question arises as to the applicability of the models to the earth where the basins are kilometres deep and the master faults have tens of kilometres of offset. Strictly speaking, the models presented above are only valid when describing the initiation of a pull-apart basin. However, once the initial depression forms, it will start to fill with sediments. As movement continues on the master faults, the basin will continue to deepen and fill with sediments. Thus, it seems reasonable to expect that a basin which begins due to an elastic (and hence recoverable) effect will become a permanent feature due to geological processes.

Predicted fault patterns

It is possible to predict fault patterns in the basin from the stresses produced by the *en echelon* faults (Jaeger & Cook, 1969, ch. 4). I will not present a complete analysis of the potential fault pattern as I feel that a simple approach will illustrate the important points. Also, the simplicity of the model probably makes a more detailed analysis unnecessary.

I assume that shear failure will occur at an angle of $\pm 45^\circ$ to the most compressive principal stress axis. Experimental evidence indicates that failure in rock specimens tends to occur at angles between 25–40° to the most compressive principal stress axis (Jaeger & Cook, 1969), so the fault orientations predicted by my assumption could be off by at most 20°. Considering the lack of complexity of the model, this seems to be an acceptable error.

Once again, the tops of the master faults intersect the ground surface unless explicitly stated otherwise.

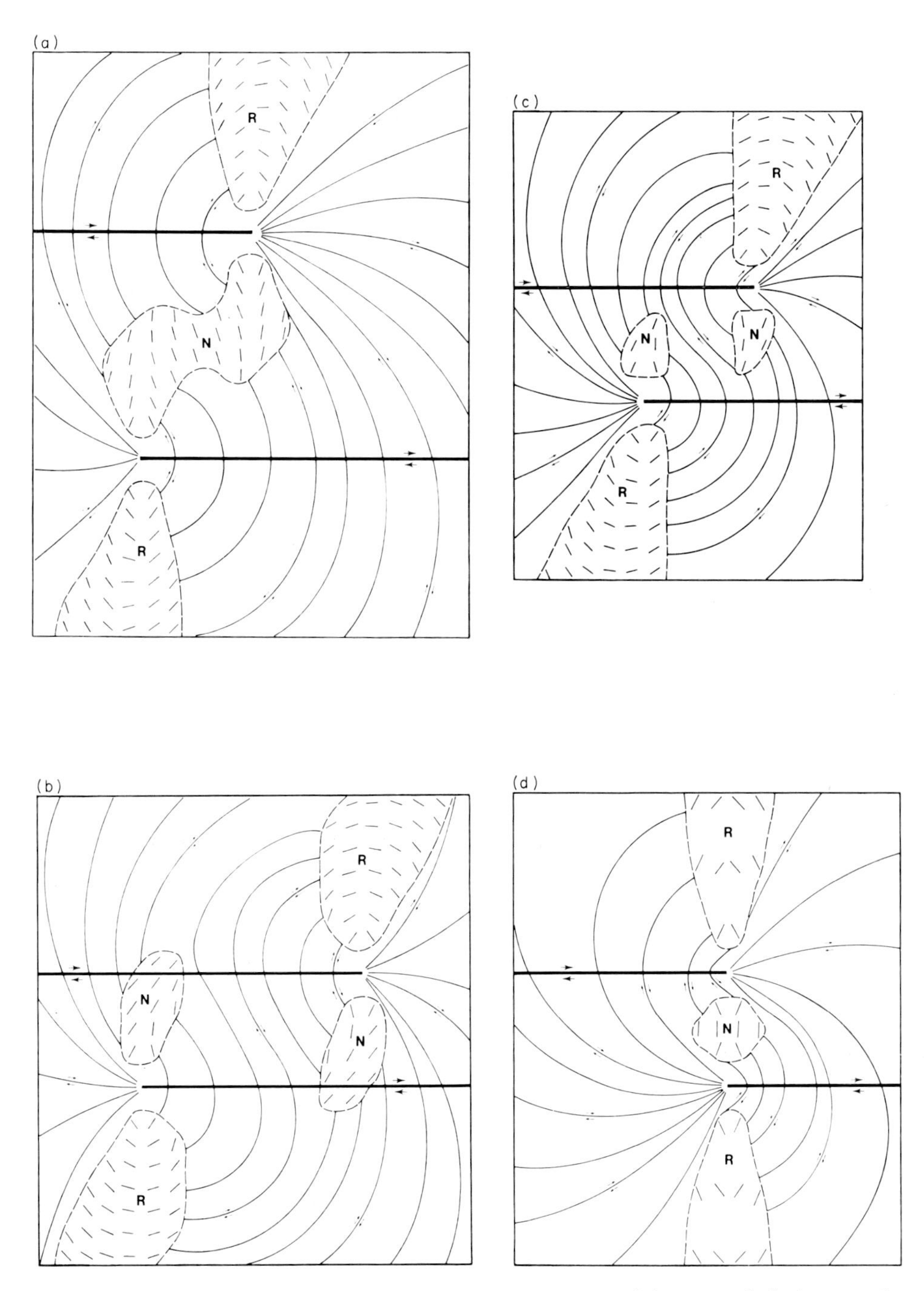

Fig. 5. Potential fault pattern on the ground surface when the tops of the master faults intersect the ground surface. Offset on each master fault is 1 m. Heavy lines show master faults. Possible right-lateral strike-slip fault pattern is shown; the left-lateral fault pattern would be perpendicular to the right-lateral fault pattern. Possible strikes of normal faults shown in area denoted 'N', and possible strikes of reverse faults shown in areas denoted 'R'. Normal and reverse faults would dip 45°.

(a) Separation (20 km) equals twice the overlap; (b) separation (10 km) equals 1/2 the overlap; (c) separation (10 km) equals the overlap; (d) overlap is zero, separation equals 10 km.

When the separation is twice the overlap (Fig. 5a), there is a zone of normal faulting in the central part of the basin and zones of reverse faulting near the ends of the master faults outside of the basin. The model predicts strike-slip faulting in part of the basin. For clarity, only the predicted right-lateral strike-slip faults are shown, but the left-lateral faults would be perpendicular to the right-lateral faults shown. The fault lines drawn are not intended to represent individual faults but rather show schematically the fault patterns which could develop in the model. Thus as in Crowell's model (Fig. 2), there is an extensive zone of normal faulting on the basin sides of the ends of the master faults, and the normal faulting probably dies out in a zone of oblique-slip faulting as one moves along a line perpendicular to one master fault from the end of the fault into the basin. However, there is clearly strike-slip faulting between the zones of normal faulting and the basin sides of the ends of the master faults.

Note that the right-lateral fault trends are not parallel to the master faults. The dislocation model predicts the stress field which is produced by movement on the master faults. In the absence of any other stress field, the resulting stresses would try to remove the initial displacement on the master faults, and this would require left-lateral movement on the master faults. Presumably, the initial right-lateral displacement on the master faults is caused by some regional stress field, but this is not the problem considered in this paper.

When the overlap is twice the separation between the master faults (Fig. 5b), the zone of normal faulting extends from near the end of one master fault across the other master fault, and there is no normal faulting in the centre of the basin. There is strike-slip faulting just at the basin side of the ends of the master faults, and the centre of the basin is dominated by strike-slip faulting. This is in contrast to the fault pattern predicted by Crowell's model (Fig. 2). There are still zones of reverse faulting at the ends of the master faults outside of the basin.

When the overlap equals the separation between the two faults (Fig. 5c), there are zones of normal faulting at the ends of the basin, but the faulting in the centre of the basin is strike-slip. Once again, there is strike-slip faulting between the ends of the master faults and the normal faults toward the basin, and there is a zone of reverse faulting at the ends of the master faults outside of the basin.

When the overlap is zero (Fig. 5d), there is a well developed zone of normal faulting in the centre of the basin which is again separated from the ends of the master faults by zones of strike-slip faulting. There is strike-slip faulting on the flanks of the basin, and zones of reverse faulting are seen at the ends of the master faults outside of the basin.

These models show that the predicted secondary fault patterns in pull-apart basins are dependent on the amount of overlap and the separation between the master faults. In two cases (Fig. 5a and d), normal faulting develops in the centre of the basin and strike-slip develops on the flanks, while in the other two cases (Fig. 5b and c), normal fault zones develop on the flanks of the basin near the ends of the master faults and strike-slip faulting develops in the central part of the basin. In all cases, the faulting on the basin side of the ends of the master faults starts out as strike-slip faulting and changes to normal faulting as one moves into the basin. This suggests that the faulting on the basin side of the ends of *en echelon* faults is generally going to have oblique-slip. The zones of reverse faulting near the ends of the master faults outside of the basin occur in each model. However, it would be unexpected to find the reverse faulting to be well developed in nature because the stresses in these zones will probably not be high enough to cause the rocks to fail.

Once again, the question of applicability of the models can be raised. The fault patterns predicted by the models only apply to the initiation of faulting in the basin. However, once a secondary fault pattern is established in the basin, subsequent movements on the master faults should produce movement on the previously formed secondary fault pattern rather than produce a new fault pattern. This is because a lower stress level is generally required to cause movement on a pre-existing fault than to fracture new rock. Thus the initial secondary fault pattern would be structurally important for a significant part of the basin's history.

Buried faults

The basin shapes discussed in the previous section and the secondary fault patterns discussed here strictly apply only to the basement rocks. As movement continues on the master faults, the basins will deepen and movement will continue on the secondary fault system. The deepening basin will fill with sediments, and the secondary faults will propagate up into the sedimentary fill. If the sedimentation rate is slow enough, the initial secondary fault pattern may propagate through the sediments and be seen at the surface of the basin. If the basin is filled faster than the secondary faults can propagate upward through the sediments, then a tertiary structural pattern may develop in the sediments within the basin. The structural development within the basin would then be very complex and rather difficult to model. In this section a model which is a first approximation to the above problem will be presented.

Consider the case where the overlap is zero (Figs 4d and 5d) and the top edges of the master faults are at a distance below the ground surface equal to the separation. This model is equivalent to assuming that the *en echelon* faults are overlain by a uniformly thick sedimentary layer with the same elastic properties as the semi-infinite

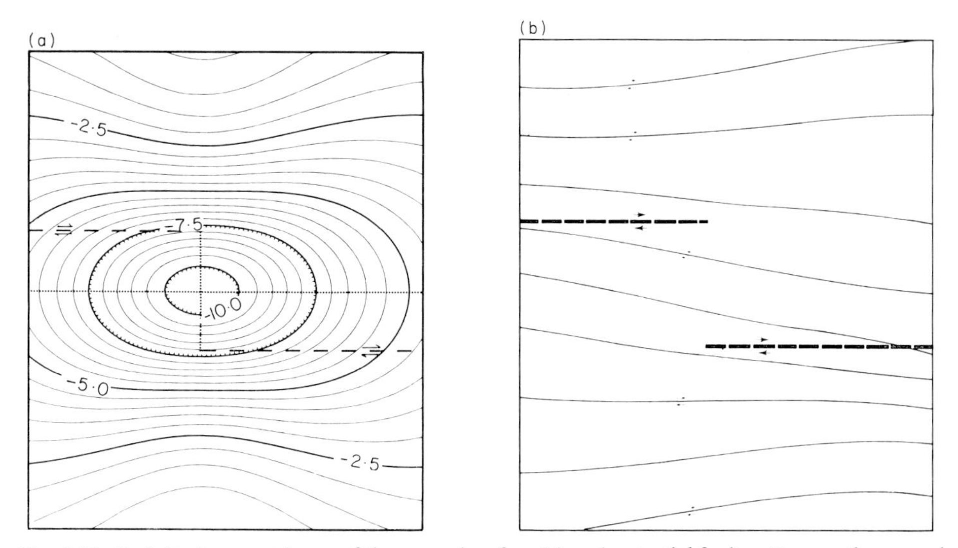

Fig. 6. Vertical displacement in cm of the ground surface (a) and potential fault pattern on the ground surface (b) for 1 m displacement on each master fault when the overlap is zero, the tops of the master faults are 10 km below the ground surface, and the separation is 10 km. Dashed lines show master faults. Negative values are down, and the dotted lines show the axes of the basin. Contour interval 0·5 cm. Only the right-lateral strike-slip pattern is shown. The left-lateral pattern would be perpendicular to the right-lateral pattern. There are no normal or reverse faults.

body. Figure 6a shows contours of the displacement of the ground surface for this model and should be compared with Fig. 4d. Note that the uplifted zones at the ends of the master faults in Fig. 4d are not seen in Fig. 6a. The greatest depth of the basin is about 10% of the offset on the master faults.

Figure 6b shows the predicted fault pattern for this model and should be compared to Fig. 5d. The faulting throughout the basin in Fig. 6b is strike-slip, and there is none of the normal or reverse faulting seen in Fig. 5d. This is important, for it suggests that strike-slip faults should occur in the sediments of a pull-apart basin in the early stages of development before the master faults have propagated to the surface.

The conclusions drawn from this model are also valid for the other models of Figs 4 and 5 with buried master faults. That is, the zones of uplift are not seen and the basins are characterized by strike-slip faulting. Models with varying depths to the tops of the master faults suggest that normal faulting only appears at the ground surface when the fault tops are very near to the ground surface.

It is worth discussing the effects of layering briefly at this time. Previous work on a straight strike-slip fault in a semi-infinite body overlain by a layer which is less rigid than the semi-infinite body (Rodgers, 1976) shows that the layer tends to concentrate the deformation closer to the fault. However, the layer does not affect the predicted fault pattern at the surface. I would expect the same thing to happen for the *en echelon* faults. That is, if the material above the master faults is less rigid than that containing the master faults (e.g. sediments over a crystalline basement), then the basin will be smaller than that shown in Fig. 6 but will still be characterized by strike-slip faulting.

DEVELOPMENT OF PULL-APART BASINS

As the offset across the master faults increases, we might expect the master faults to extend themselves in some direction. If we assume that the master faults extend themselves parallel to their strike, this would have the effect of increasing the overlap while keeping the separation more or less constant. If we start with the configuration of no overlap, then presumably the basin would evolve toward increasing overlap of the master faults. At the same time, a secondary fault system should form in the basin in response to the initial configuration of the master faults. At any later time we would expect the secondary fault system to have the orientation predicted from the current stress field *if* the basin were unfaulted. However, the secondary fault system produced by the initial configuration of the master faults would presumably be present at the later stages of deformation and would thus represent a pre-existing fabric for the basin. Thus the stress fields produced by successive configurations of the master faults would be acting on a pre-existing secondary fault pattern, and it might be expected for movement to continue on the original secondary fault system even though the stress field might be quite different at some later time. We should then be able to use Figs 4 and 5 to suggest the structural evolution of a pull-apart basin.

Figure 7 shows schematically the changes which might be expected if we start with *en echelon* faults with no overlap. The basin shape and secondary fault pattern would be similar to those shown in Figs 4d and 5d. There would be normal faulting in the centre and strike-slip faulting on the flanks of the basin. If the master faults extend themselves along strike, then when the offset on the master faults is about one-half

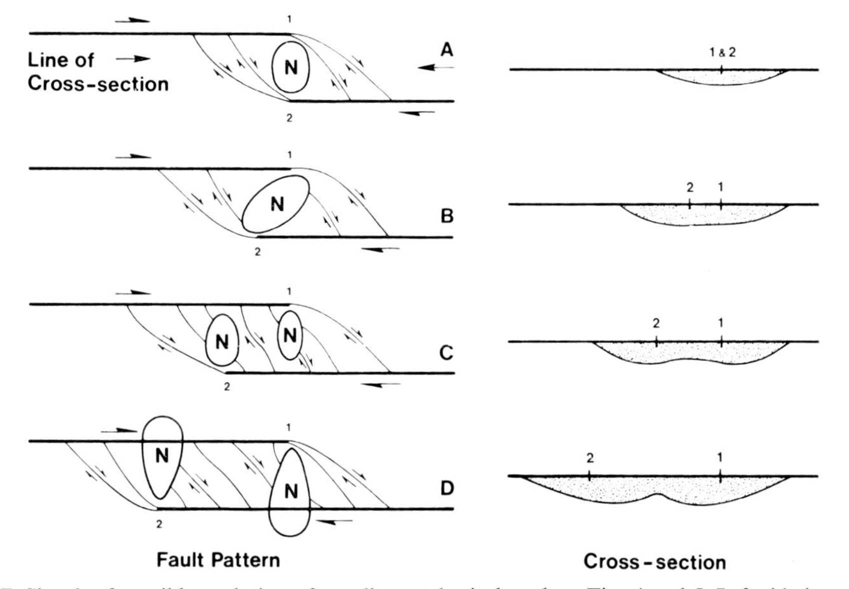

Fig. 7. Sketch of possible evolution of a pull-apart basin based on Figs 4 and 5. Left side is a map view of basement fault development. Numbers refer to locations of ends of master faults. Zones labelled 'N' are zones of normal faulting. Left-lateral strike-slip faults not shown. Right side is a vertical cross-section parallel to master faults through the centre of the basin. Vertical and horizontal scales are arbitrary. Stippled areas are sedimentary fill of basins. Thus cross-sections show suggested evolution of basin shape.

the separation of the faults we have nearly the same situation (Fig. 7B). The zone of normal faulting in the centre of the basin is somewhat larger, and the basin has deepened and broadened. If the basin has been filled by sediments, the secondary fault pattern near the ground surface will probably be dominated by strike-slip faulting such as that shown in Fig. 6.

When the offset across the master faults is about equal to the separation between the faults, some interesting changes occur. Two distinct depocentres begin to develop near the ends of the master faults, and the zones of normal faulting are separated by a zone of strike-slip faults. If the orientations of the secondary faults remain similar to the initial orientations, this means that the sense of offset on some of the secondary faults must change from strike-slip to normal or vice versa as the basin develops. The fault patterns in Fig. 7 still represent the faulting in or near the basement, and, as Fig. 6 suggests, the faulting in the basin fill may still be dominated by strike-slip. Figure 7C suggests that the faulting in the centre of the basin may become strike-slip throughout the section as the basin grows. Figure 7D demonstrates the expected situation when the offset on the master faults is twice their separation. Figure 7C and 7D are similar; and two depocentres are more separated and the zone of strike-slip faulting occupies more of the basin centre in Fig. 7D.

The models suggest that the basement secondary fault patterns will be very complex throughout the history of a pull-apart basin and may include changes in the sense of movement on many of the faults. The fault patterns in the sedimentary fill of the basin should be fairly simple near the ground surface, but there should be a zone within the sediments where the basement and near-surface fault patterns merge. This zone would be structurally complex.

FIELD EXAMPLES

Sharp (1975) suggests that several areas along the San Jacinto fault in California may be pull-apart basins. San Jacinto Valley (Fig. 8, inset) is a deep, narrow valley

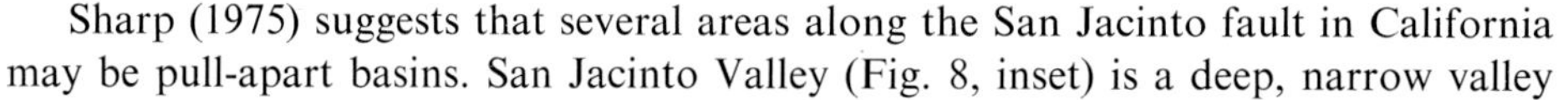

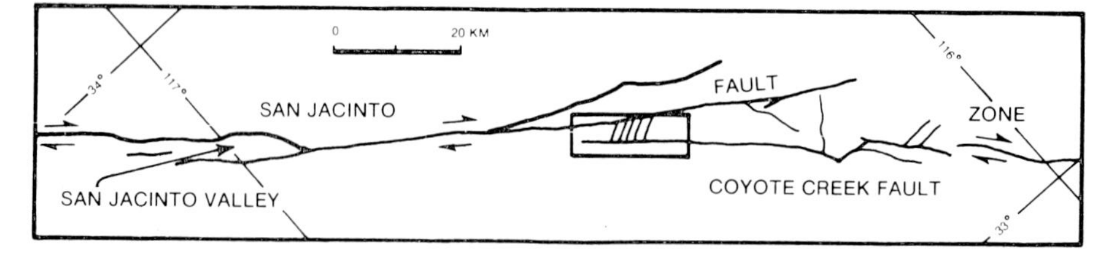

Fig. 8. Generalized fault maps showing pull-apart basins on the San Jacinto fault. Hachures show downthrown sides of normal faults. Barbs are in the upper plates of low-angle reverse faults. Arrows show sense of movement on strike-slip faults. Box in inset shows location of larger part of figure. Inset generalized from Sharp (1975, Fig. 1). Larger figure generalized from a portion of Sharp (1967, plate 1).

between right-stepping right-lateral faults (Lofgren & Rubin, 1975; Lofgren, 1976), and, according to Lofgren & Rubin (1975), in the last 15 000 years the valley has subsided at a rate of about 6 mm/year. The fault pattern in the basement is not exposed in the San Jacinto Valley. However, the basement fault pattern is exposed between the San Jacinto and Coyote Creek faults (Fig. 8). Note that there are several normal faults exposed between the San Jacinto and Coyote Creek faults with the northwestern sides generally downthrown. There appears to be reasonable agreement between the fault patterns seen in Figs 5c and 8.

Other areas that have been suggested as pull-apart basins include the Dead Sea in the Levant fault zone (or Dead Sea Rift; Quennell, 1959) and Death Valley in California (Burchfiel & Stewart, 1966). Certainly both basins are properly located between *en echelon* faults, and both basins are substantially lower than their surroundings (about 1300 m for the Dead Sea and about 3400 m for Death Valley). The structures mapped in Death Valley are not indicative of extension (Hill & Troxel, 1966) as might be expected if Death Valley were a typical Basin and Range structure.

The Salton Trough in southern California lies between NW–SE trending dextral faults, the San Jacinto and Imperial faults on the southwest and the Banning–Mission Creek members of the San Andreas fault zone on the northeast. The trough is filled with about 6100 m of sediment and is characterized by very high temperatures (300°F below about 1300 m) and high seismicity (Hileman, Allen & Nordquist, 1973; Hill, Mowinckel & Peake, 1975). Elders *et al.* (1972) have suggested that there are several small spreading centres in the Salton Trough orientated NE–SW, perpendicular to the bounding faults. However, the fault map of California (Jennings, 1975) does not show any faults in the trough with NE–SW orientation. In addition, the trend of epicentres in the recent earthquake swarms within the trough is nearly N–S, along the line of the Brawley fault. Fault plane solutions from these swarms (Hill *et al.*, 1975) suggest that faulting within the trough is not normal, but is strike-slip and that opening of the spreading centre takes place in a diffuse zone of *en echelon* strike-slip faults which trend parallel to the main bounding faults.

The orientation of the current seismicity together with the lack of mapped faults with the trend of the proposed spreading centres suggests that the Salton Trough may be a complex pull-apart basin. The currently active N–S trending strike-slip faulting is consistent with the fault patterns suggested in the discussion of Fig. 7D. Also, high precision level surveys in the area (Lofgren, 1978) suggest that the northwestern end of the Salton Sea is deepening faster than the southeastern end. Once again, this is consistent with the predictions of Figs 7D and 4b. A spreading centre in the vicinity of Brawley might cause the southeastern end of the Salton Sea to deepen faster than the northwestern end. Of course, none of these data conclusively prove that the Salton Trough is a pull-apart basin, and, if the Trough is a pull-apart basin, it is certainly more complex than the models in this paper. However, the models together with the observational data do clearly allow a pull-apart origin for the Salton Trough.

CONCLUSIONS

Mathematical models have been used to study surface deformation and fault patterns associated with pull-apart basins produced by *en echelon* strike-slip faults. The model studies suggest that the basin shape and the fault patterns are closely related to: (1) the amount of overlap between the faults; (ii) the amount of separation between the faults; (iii) whether or not the faults intersect the ground surface. In particular, strike-slip faults should be well developed in any pull-apart basin, and normal faulting will occur only when the tops of the master strike-slip faults are near the ground surface. In general, the models suggest that pull-apart basins may have forms different from that proposed by Crowell (1974).

The models also suggest that the long-term evolution of a pull-apart basin may be structurally complex, and some of the secondary faults may show different types of movement through time. The faults in the sedimentary fill of the basins may show somewhat simpler patterns than those in the basement.

Examples of pull-apart basins from California show fault patterns which are in reasonable agreement with those predicted by the models. General patterns of basin shape and topographic relief for pull-apart basins in California and the Middle East also agree reasonably well with predictions of the models.

ACKNOWLEDGMENTS

This paper is based on a Cities Service Co. Research Report, and I am grateful to the company for permission to publish the paper. The work benefited from discussions with Rick Groshong. Jack Etter, Rick Groshong, Bill Rizer and Martha Withjack reviewed the manuscript of the company report and made many helpful suggestions. Anonymous reviewers made suggestions which helped to improve the final manuscript. I am grateful to Dr J. C. Crowell and the Society of Economic Paleontologists and Mineralogists for permission to reproduce Fig. 2.

REFERENCES

ALEWINE, R.W. III (1974) *Application of Linear Inversion Theory toward the Estimation of Seismic Source Parameters.* Unpublished Ph.D. Thesis, California Institute of Technology.

BLAKE, M.C., JR, CAMPBELL, R.H., DIBBLEE, T.W., JR, HOWELL, D.G., NILSEN, T.H., NORMARK, W.R., VEDDER, J.C. & SILVER, E.A. (1978) Neogene basin formation in relation to plate-tectonic evolution of San Andreas fault system, California. *Bull. Am. Assoc. Petrol. Geol.* **62,** 344–372.

BURCHFIEL, B.C. & STEWART, J.H. (1966) 'Pull-apart' origin of the central segment of Death Valley, California. *Bull. geol. Soc. Am.* **77,** 439–442.

CHINNERY, M.A. (1961) The deformation of the ground around surface faults. *Bull. seismol. Soc. Am.* **51,** 355–372.

CHINNERY, M.A. (1963) The stress changes that accompany strike-slip faulting. *Bull. seismol. Soc. Am.* **53,** 921–932.

CHINNERY, M.A. & PETRAK, J.A. (1968) The dislocation fault model with a variable discontinuity. *Tectonophysics,* **5,** 513–529.

CLAYTON, L. (1966) Tectonic depressions along the Hope fault, a transcurrent fault in North Canterbury, New Zealand. *N.Z.J. Geol. Geophys.* **9,** 95–104.

COUPLES, G. & STEARNS, D.W. (1978) Analytical solutions applied to structures of the Rocky Mountains foreland on local and regional scales. In: *Laramide Folding Associated with Basement Block Faulting in the Western United States* (Ed. by V. Matthews III). *Mem. geol. Soc. Am.* **151,** 313–335.

CROWELL, J.C. (1974) Origin of late Cenozoic basins in southern California. In: *Tectonics and Sedimentation* (Ed. by W. R. Dickinson). *Spec. Publ. Soc. econ. Paleont. Miner. Tulsa,* **22,** 190–204.

ELDERS, W.A., REX, R.W., MEIDAV, T., ROBINSON, R.T. & BIEHLER, S. (1972) Crustal spreading in southern California. *Science,* **178,** 15–24.

FISHER, M.A., PATTON, W.W., JR, THOR, D.R., HOLMES, M.L., SCOTT, E.W., NELSON, C.H. & WILSON, C.L. (1979) The Norton basin of Alaska. *Oil and Gas Journal,* **77,** 96–98.

FREUND, R. (1965) A model of the structural development of Israel and adjacent areas since Upper Cretaceous times. *Geol. Mag.* **102,** 189–205.

FREUND, R. (1971) The Hope fault, a strike-slip fault in New Zealand. *Bull. N.Z. geol. Surv. (new series)* **86,** 49 pp.

HARDING, T.P. (1976) Predicting productive trends related to wrench faults. *World Oil,* **182,** 64–69.

HILEMAN, J.A., ALLEN, C.R. & NORDQUIST, J.M. (1973) *Seismicity of the southern California region, January* 1, 1932, *to December* 1, 1972. Seismological Laboratory, California Institute of Technology.

HILL, M.L. & TROXEL, B.W. (1966) Tectonics of Death Valley region, California. *Bull. geol. Soc. Am.* **77,** 435–438.

HILL, D.P., MOWINCKEL, P. & PEAKE, L.G. (1975) Earthquakes, active faults and geothermal areas in the Imperial Valley, California. *Science*, **188,** 1306–1308.

JAEGER, J.C. & COOK, N.G.W. (1969) *Fundamentals of Rock Mechanics.* Methuen, London.

JENNINGS, C.W. (1975) Fault map of California, *California Division of Mines and Geology Geologic Data Map Number* 1, Scale 1:750,000.

JOHNSON, A.M. (1970) *Physical Processes in Geology.* Freeman, Cooper & Co., San Francisco. 577 pp.

LOFGREN, B.E. (1976) Land subsidence and aquifer-system compaction in the San Jacinto Valley, Riverside County, California—A progress report. *J. Res. U.S. geol. Surv.* **4,** 9–18.

LOFGREN, B.E. (1978) Salton Trough continues to deepen in Imperial Valley, California (Abstract). *EOS, Trans. Am. geophys. Union.* **59,** 1051.

LOFGREN, B.E. & RUBIN, M. (1975) Radiocarbon dates indicate rates of graben downfaulting, San Jacinto Valley, California. *J. Res. U.S. geol. Surv.* **3,** 45–46.

NORRIS, R.J., CARTER, R.M. & TURNBULL, I.M. (1978) Cainozoic sedimentation in basins adjacent to a major continental transform boundary in southern New Zealand. *J. geol. Soc. Lond.* **135,** 191–205.

QUENNELL, A.M. (1959) Tectonics of the Dead Sea Rift. *Int. geol. Congr.* 1956. **20,** 385–405.

RECHES, Z. & JOHNSON, A.M. (1978) Development of monoclines: Part 11. Theoretical analysis of monoclines. In: *Laramide Folding Associated with Basement Block Faulting in the Western United States* (Ed. by V. Matthews III). *Mem. geol. Soc. Am.* **151,** 273–311.

RODGERS, D.A. (1976) Mechanical analysis of strike slip faults. 11. Dislocation model studies (Abstract). *EOS, Trans. Am. geophys. Union.* **57,** 327.

RODGERS, D.A. (1979) Vertical deformation, stress accumulation, and secondary faulting in the vicinity of the Transverse Ranges of southern California. *Bull. California Div. Mines Geol.* **203,** 74 pp.

SANFORD, A.R. (1959) Analytical and experimental study of simple geological structures. *Bull. geol. Soc. Am.* **70,** 19–51.

SAVAGE, J.C. & HASTIE, L.M. (1969) A dislocation model for the Fairview Peak, Nevada, earthquake. *Bull. seismol. Soc. Am.* **59,** 1937–1948.

SHARP, R.V. (1967) San Jacinto fault zone in the Peninsular Ranges of southern California. *Bull. geol. Soc. Am.* **78,** 705–730.

SHARP, R.V. (1975) En echelon fault patterns of the San Jacinto fault zone. In: *San Andreas fault in southern California* (Ed. by J. C. Crowell) *Spec. Rep. California Div. Mines Geol.* **118,** 147–152.

WILCOX, R.E., HARDING, T.P. & SEELY, D.R. (1973) Basic wrench tectonics. *Bull. Am. Ass. Petrol. Geol.* **57,** 74–96.

Spec. Publ. int. Ass. Sediment. (1980) **4,** 43–62

Basin development along the late Mesozoic and Cainozoic California margin: a plate tectonic margin of subduction, oblique subduction and transform tectonics

D. G. HOWELL, J. K. CROUCH, H. G. GREENE,
D. S. McCULLOCH *and* J. G. VEDDER

U.S. Geological Survey, 345 *Middlefield Road, Menlo Park, California* 94025, *U.S.A.*

ABSTRACT

Along the Californian margin of the North American plate, the configuration and structural stability of late Mesozoic and Cainozoic basins are related to plate kinematics. Three tectonic regimes are recorded; orthogonal high-angle subduction, oblique low-angle subduction, and transform slip. During the first, regionally extensive forearc basins developed; during the second and third, borderland settings formed as a consequence of wrench faulting. In the forearc basins, sedimentological facies constitute regional belts that persist for hundreds of kilometres, with stratigraphic sequences that are 1–15 km thick. Shorelines are relatively straight, shelf facies are broad and well developed, and basin fill is composed of shallow marine, shelf and coalescing submarine-fan facies. Sediment transport in the deeper water facies commonly is parallel to the basin axis. Borderland basins reflect tectonic instability. A principal effect of wrench tectonics is the vertical reciprocation of crustal blocks. Shorelines are generally irregular, and narrow shelves pass abruptly into deep basins. Lithofacies change dramatically along strike, and stratigraphic thicknesses are variable from basin to basin, from tens of metres up to 6 km. Basin-margin facies are marked by unconformities, slump aprons, lithological pinch-outs and submarine canyon channels. Penecontemporaneous slip along the basin-margin faults complicates these lithofacies patterns. Borderland type palaeogeography is most extensively developed in the transform tectonic regime, and therefore the more seaward offshore basins are relatively depleted of terrigenous debris owing to transport barriers.

INTRODUCTION

Sedimentary basin development along the California segment of the North American plate margin can be linked to plate interactions from Cretaceous to Holocene time. Tectonic regimes changed through time in this region. In this paper we shall review the development of sedimentary basins that evolved through three

0141-3600/80/0904-0043$02.00

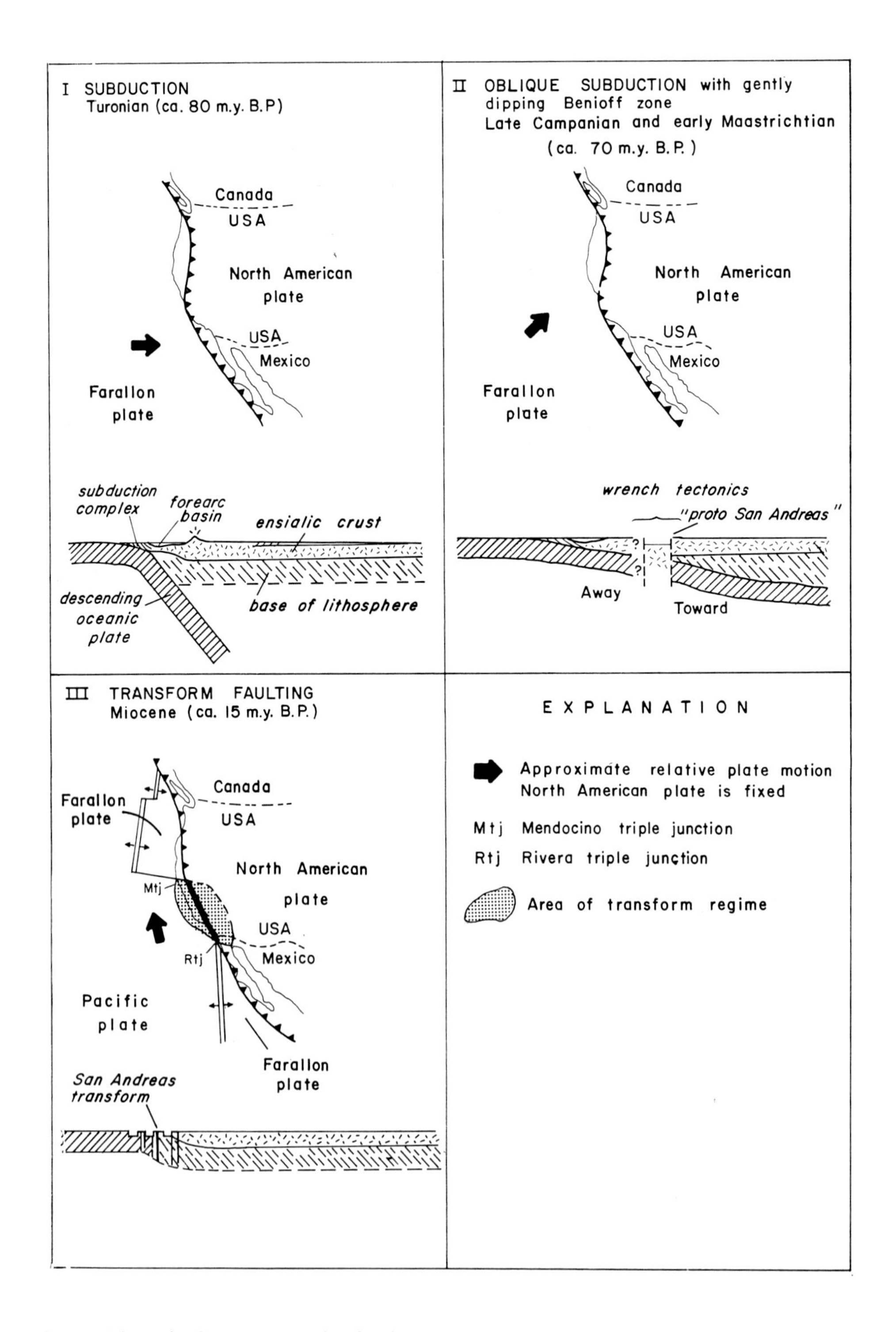

Fig. 1. Schematic plate reconstruction for the California area in the mid-Late Cretaceous and middle and late Tertiary. Figures modified from Dickinson (1979).

contrasting kinematic episodes: subduction, oblique subduction accompanied by strike-slip faulting and transform faulting (Fig. 1).

Subduction and its related processes in western California began in Late Jurassic time. The continental margin that resulted is inferred to have been of Andean type, characterized from east to west by an active continental volcanic arc, a relatively stable and broad forearc basin and a dynamic prism of accretionary material (Hamilton, 1969; Dickinson, 1979). In latest Cretaceous and early Tertiary time, orthogonal convergence changed to oblique convergence (Cooper, Scholl & Marlow, 1976; Coney, 1978). The distribution of Early Cretaceous and early Tertiary volcanism and the associated potassium gradients within the hinterland of the Cordillera imply a flattening of the eastward directed Benioff zone as on Fig. 1 (Lipman, Prostka & Christiansen, 1971; Coney & Reynolds, 1977; Cross & Pilger, 1978; Keith, 1978; Lipman, 1980). Consistent with such a model is the Late Cretaceous and early Tertiary (ca 75–60 m.y. BP) belt of wrench faults along the coast of central California and the development of a proto-San Andreas fault (Suppe, 1970). The total aggregate slip during this early phase of wrench faulting is not known, but it was at least 90 km along the east margin of the Salinian block (proto-San Andreas fault; Graham, 1978) and it may have been an order of magnitude greater along faults west of the Salinian block (Howell & Vedder, 1978). From a sedimentological viewpoint, the principal effect of this faulting was the creation of borderland-like depositional sites caused by the splintering of a piece of continental crust (Salinian block).

Eocene sedimentation patterns adjacent to the Salinian block reflect this early borderland setting (Nilsen & Clarke, 1975). Elsewhere, however, broad forearc sedimentation prevailed as in the California Continental Borderland, where no early phase of wrench tectonics is recognizable (Crouch, 1979). By late Eocene time, high-angle subduction was again well established along the length of the California margin. In late Oligocene time, however, transform tectonic processes began to control sedimentation patterns as the Pacific and North American plates came into contact and the Rivera and Mendocino triple junctions migrated south and north respectively (Atwater & Molnar, 1973). Cumulative slip during the Neogene may be as much as 1000 km on a system of northwest-trending faults that span a wide pliant region of the California margin. The San Andreas fault system encompasses the entire transform regime, 1200 km long (Gulf of California to the Mendocino fracture zone) and at least 500 km wide (Pacific shelf to within the west part of the Basin and Range Province). The present-day master fault in this system is the San Andreas fault.

SUBDUCTION REGIME

Forearc basin

Stratigraphic and sedimentological relations of the Great Valley sequence of central and northern California indicate shelf, slope, and submarine-fan deposition in an elongate forearc basin from Late Jurassic through Cretaceous time (Dickinson, 1971; Ingersoll, Rich & Dickinson, 1977; Ingersoll, 1978, 1979). Although the original continuity of these rocks has been disrupted by late Cainozoic transform faulting, partly equivalent strata are recognized in the Coast Ranges of central California, in

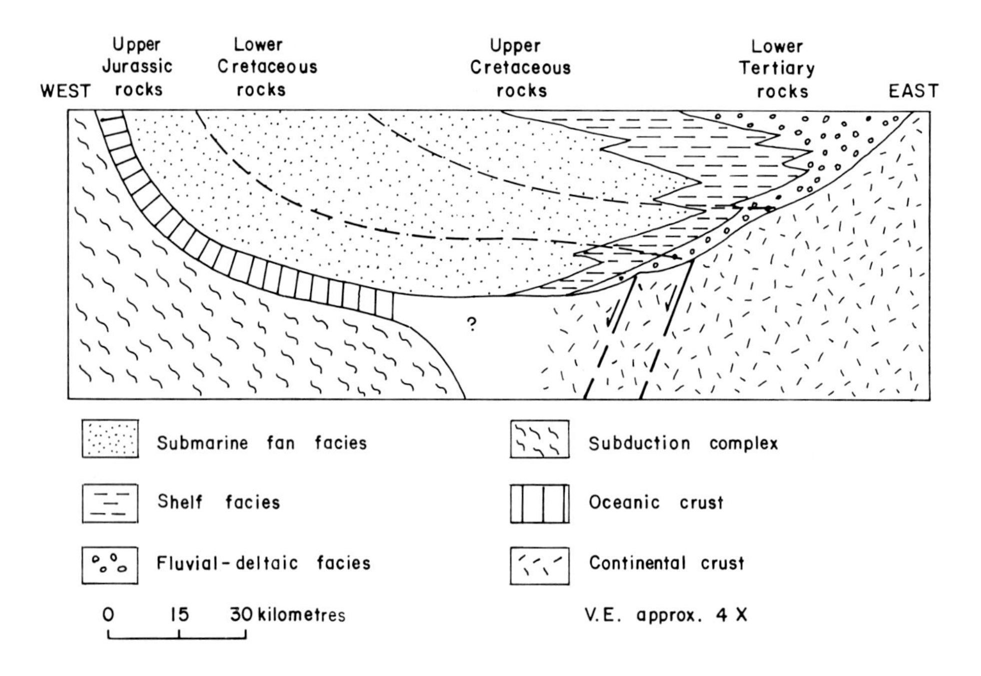

Fig. 2. Schematic cross-section through the forearc basin of the Great Valley sequence showing inferred palaeoenvironmental relations. This large basin system of lithofacies contrasts with similar lithofacies from smaller, steep-sided basins inferred for the Late Cretaceous borderland of the Salinian block (Fig. 4).

southern California along the western flank of the Peninsular Ranges and in the borderland. Lithological and palaeontological characteristics are similar throughout this long belt of late Mesozoic strata. Probably the most obvious criteria that differentiate forearc-basin from wrench-tectonic depositional settings are the regional persistence and great thickness of petrofacies units (Fig. 2). Intrabasin tectonics generally do not alter the basin-wide uniformity of depositional patterns. Gradual widening of the forearc basin (Fig. 3) and progradational and retrogradational cycles (Fig. 2) are uniformly distributed in segments of the forearc basin that are measured in hundreds of kilometres.

Most of the Upper Cretaceous fluvio-deltaic facies form a relatively straight line that can be traced for nearly 1000 km along the western edge of the Sierra Nevada–Peninsular Ranges crystalline terrain. To the west these facies grade into a parallel belt of inferred shelf strata. Within the deeper parts of the forearc basin, thick successions of submarine fan facies prevail. Lithofacies relations suggest that the fans coalesced and that sediment transport was principally toward the south, along the axis of the forearc basin.

In summary, the salient characteristics of the late Mesozoic forearc basin of California and Baja California, Mexico are: (1) dimensional uniformity measured in hundreds of kilometres, (2) stratigraphic, lithological and petrographic uniformity measured in hundreds to thousands of metres of section, (3) relative tectonic stability resulting in regional regressive or transgressive cycles lasting five million years or more.

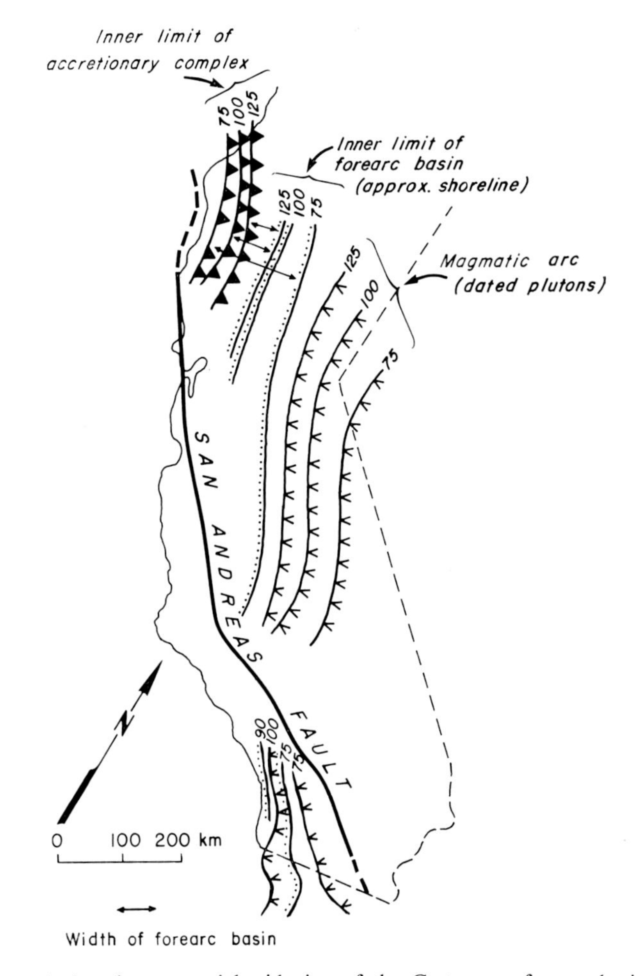

Fig. 3. Diagram relating the sequential widening of the Cretaceous forearc basin of central and northern California as the subduction zone migrated westward and the magmatic zone migrated eastward. The forearc basin is the Great Valley sequence; numbers are ages in m.y. BP. Figure modified from Ingersoll (1978).

OBLIQUE SUBDUCTION

Late Cretaceous and Palaeocene borderland

The Salinian block is an allochthonous ensialic crystalline terrane of chiefly felsic plutonic rocks that range in age from 110 to 79 m.y. BP (Mattinson, Davis & Hopson, 1971; Ehlig & Joseph, 1977; Ross, 1978). This displaced terrain is bounded on both the east and the west by paired belts of the Great Valley sequence (forearc complex) and Franciscan assemblage (accretion complex). Upper Campanian and lower Maastrichtian sedimentary rocks are the oldest strata that lie unconformably on the Salinian basement (Howell *et al.*, 1977). These strata were deposited in local, restricted basins within a tectonically active terrain (Howell & Vedder, 1978). Within each basin the

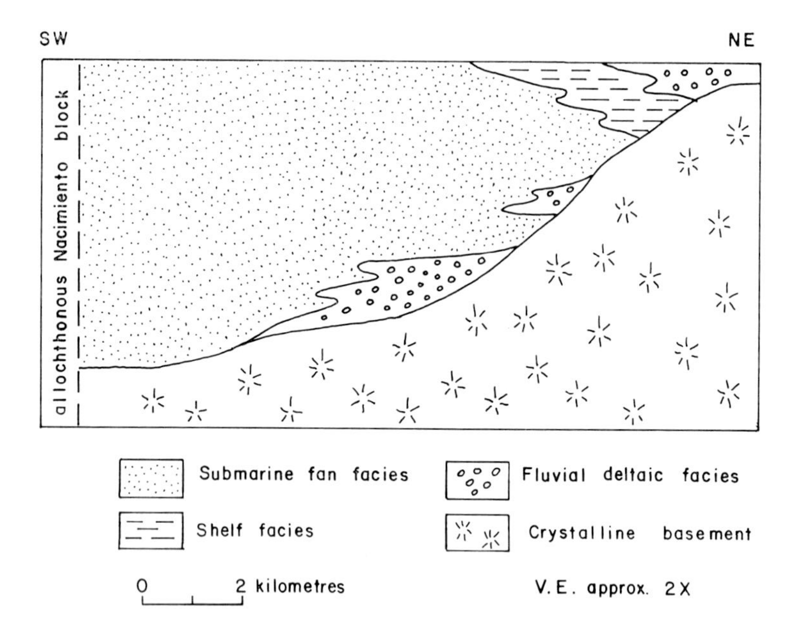

Fig. 4. Schematic cross-section depicting palaeoenvironmental relations inferred from rock units along the Salinian block east of Santa Margarita (or Pozo district). Rocks representing all depositional environments from non-marine to submarine-fan facies abut the crystalline basement. (Note change of scale from Fig. 2.)

Upper Cretaceous and Palaeocene deposits indicate rapid progradational and retrogradational cycles (Figs 4 and 5), and the orientations of non-marine and paralic facies together with the configuration of submarine-fan facies suggest a rugged and scalloped shoreline (Fig. 6). Shelf facies are rarely represented; more typically, fluvio-deltaic facies pass abruptly into proximal submarine-fan facies.

Kinematic analysis of fault displacements within central California indicates a pre-Eocene episode of dextral slip of 100–200 km (Suppe, 1970; Graham, 1978). This period of wrench tectonics may explain the stratigraphic relations and inferred palaeogeography in the Late Cretaceous Salinian terrain (Howell & Vedder, 1978). Assuming that the Late Cretaceous and Palaeocene sedimentological processes of the Salinian block resulted from strike-slip, the following depositional patterns contrast with the patterns displayed by the Great Valley sequence in the forearc basin: (1) proximal facies occur in a highly irregular pattern; (2) shelf facies are rare and submarine-canyon and submarine-fan facies are contiguous with fluvio-deltaic sequences, measurable in terms of tens to hundreds of metres of vertical section; (3) palaeocurrent directions are extremely variable along the margins of the source terrain; and (4) vertical sequences display pronounced bedding and lithological changes within tens to hundreds of metres of section. These Late Cretaceous and Palaeocene sedimentation regimes are much more dynamic than those in any part of a forearc terrain. The inferred plate-tectonic configurations for the Late Cretaceous and Palaeocene margin of California (Coney, 1978) suggests that local strike-slip faulting resulted from oblique subduction, and in this tectonic regime the above patterns are expected.

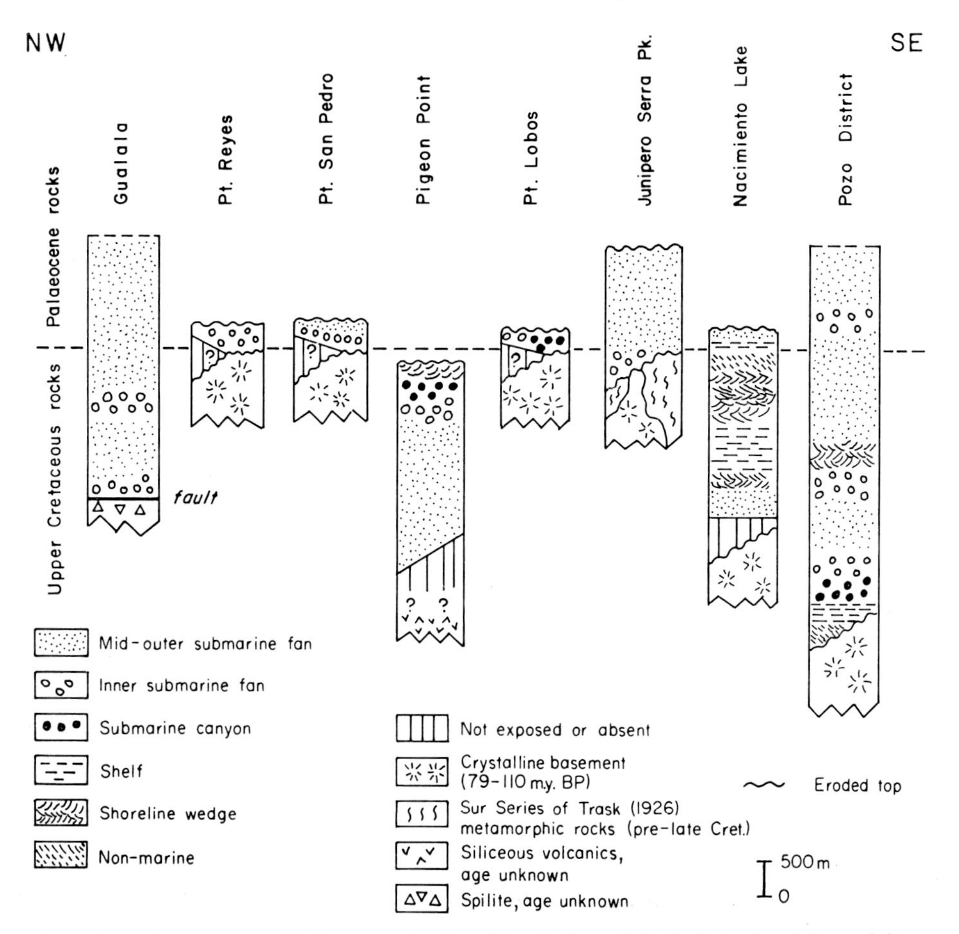

Fig. 5. Late Cretaceous and Palaeocene stratigraphic relations of the Salinian block from eight areas representing five basins (see Fig. 6) of deposition. Vertical facies changes, progradational and retrogradational sequences, and contrasting thicknesses from basin to basin (column to column) are typical.

Eocene palaeogeography

The plate-tectonic mode for central California during most of Eocene time is inferred to have been a return to high-angle subduction; yet within central California a borderland palaeogeography persisted (Nilsen & Clarke, 1975). This borderland setting was largely inherited from the preceding episode of wrench tectonics, and overall sedimentological patterns imply a gradual filling of the older basin topography. By early Oligocene time, regression dominated much of the area. Thus, the coastal instability and dynamic sedimentation that prevailed during the constructional phase was replaced by infilling and levelling of the borderland as it reverted to a more stable forearc basin setting. To the south, in the area of the modern California Continental Borderland, the distribution of lithofacies indicates that a broad, stable forearc basin persisted throughout Eocene time (Howell & Link, 1979). The contrast between this Eocene depositional setting and the areally restricted Late Cretaceous to Eocene depositional episode within the Salinian block emphasizes the problems of correlating stratigraphic units along a structurally dynamic continental margin.

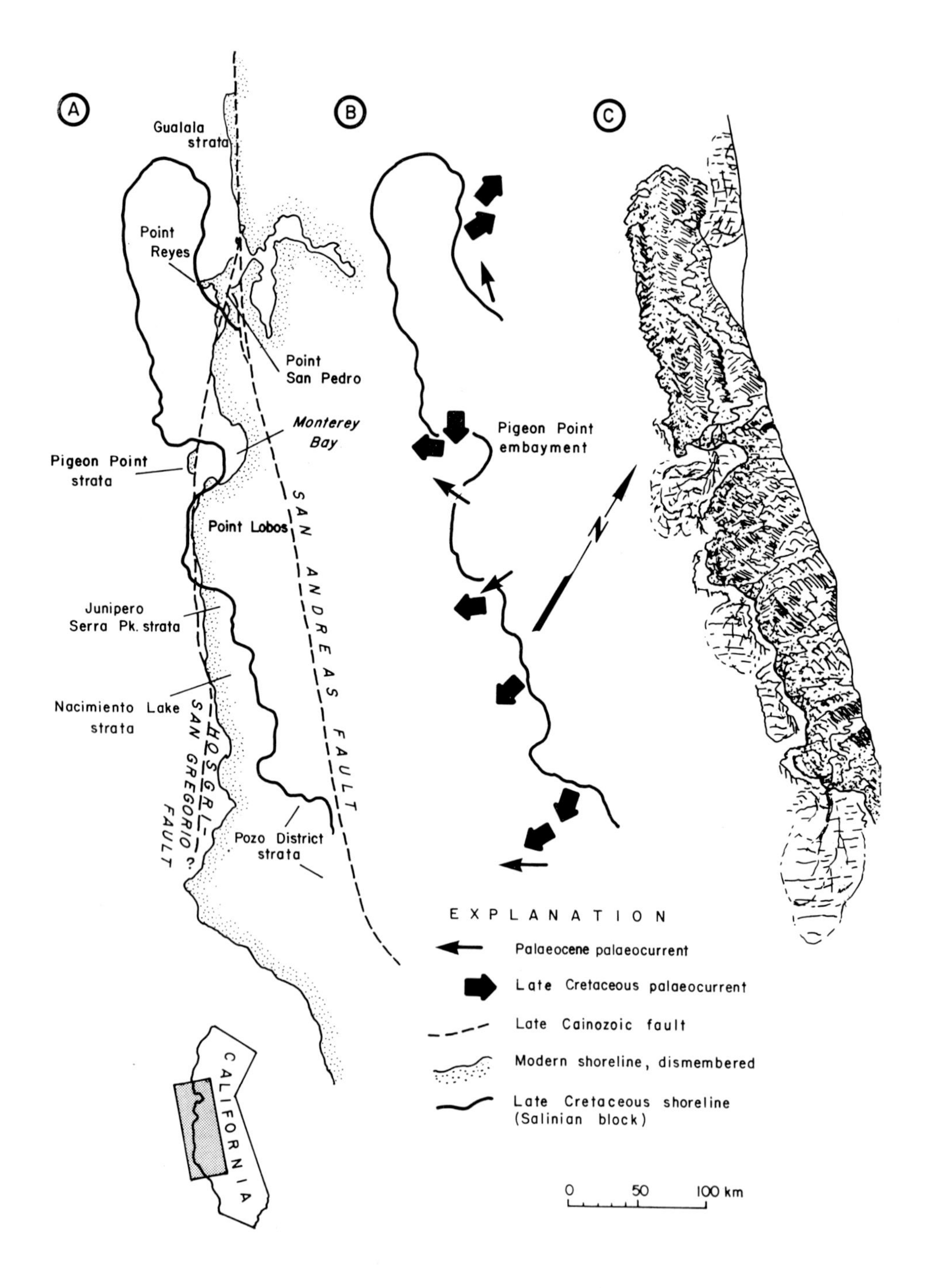

Fig. 6. (A) Locus of Late Cretaceous shoreline superimposed on the modern dismembered shoreline (palinspastic adjustment for 70 km of dextral slip of the San Gregorio fault); (B) generalized palaeocurrent data from Upper Cretaceous and associated Palaeocene strata; (C) generalized latest Cretaceous palaeogeography of the Salinian block.

TRANSFORM REGIME

Oligocene to Holocene borderlands of California

About 30 m.y. BP, transform processes between the Pacific and North American plates introduced an episode of wrench tectonics through California and the neighbouring Pacific shelf that prevails today (Atwater, 1970). The resulting deformation is reflected by the modern topography and bathymetry of *en echelon* linear to lens-shaped ridges and rhomboid-shaped basins. These morphological highs and lows are superposed over contrasting ensialic and ensimatic basement terrains. Structural variations resulting from these different basement types are not pronounced, though downbowed basins appear to be broader where underlain by ensimatic crust.

Within the zone of wrench faulting, vertical tectonism has been the principal factor controlling sedimentation. Summaries of the evolving palaeobathymetry for parts of the Salinian block (Fig. 7) indicate pronounced variations in elevation (Durham, 1974; Vedder, 1975; Graham, 1978). In the Monterey Bay region Neogene wrench faulting and crustal slivering resulted in uplift and depression of smaller structural units within the Salinian block. This local deformation, along with regional tectonism, is responsible for a complicated Tertiary depositional history (Greene & Clark, 1979). The coastline migrated inland as a result of regional subsidence and these episodes were followed by periods of local uplift.

In the California Continental Borderland, middle Miocene schist breccias were shed from subaerial ridges that have since foundered in the region of San Pedro Basin and the Gulf of Santa Catalina (Stuart, 1979). In Miocene and Pliocene time, the Santa Catalina Island platform subsided from subaerial to abyssal depths, and in Pleistocene time it was uplifted and exposed again to subaerial erosion (Vedder, Howell & Forman, 1979). Vertical elevation changes were as much as 1 m per 1000 years for the Santa Catalina Island block (Vedder & Howell, 1980), and local uplift rates along the Ventura coast amount to more than 10 m per 1000 years (Yerkes, Sarna-Wojcicki & Lajoie, 1980). These examples demonstrate a few of the dramatic fluctuations that occur throughout California's broad zone of transform faulting.

Wrench faults express a dynamic regime of compression and tension that accompany lateral displacement. Typically the basins are sags and the ridges are bulges in the crust. Fully developed 'pull-apart' basins, where new oceanic crust develops, are not apparent along the California margin.

The distribution and types of lithological units vary from basin to basin. Acoustic-reflection profiles across the Pacific Shelf of California portray complex structural and depositional patterns and show onlaps and truncations as well as folds and faults. Outcrop and well data, however, allow some generalizations to be made regarding the types of depositional units in the borderland basins (Fig. 8). Typically, thick sedimentary breccias occur as narrow basinward-thinning wedges. Lateral facies of the breccia are commonly diachronous, reflecting different stages of uplift along the adjacent wrench fault zone. Even though the belts of breccia are only a few kilometres wide they are commonly hundreds of metres thick. This basinward pinch-out of breccia units contrasts with the shoreward onlap of fine-grained, deeper water facies. Unconformities are characteristic of basin margins, and instability within the central parts of basins is indicated by thickness changes or lithological variations across structural trends. The margins of basins formed in a wrench-fault realm normally

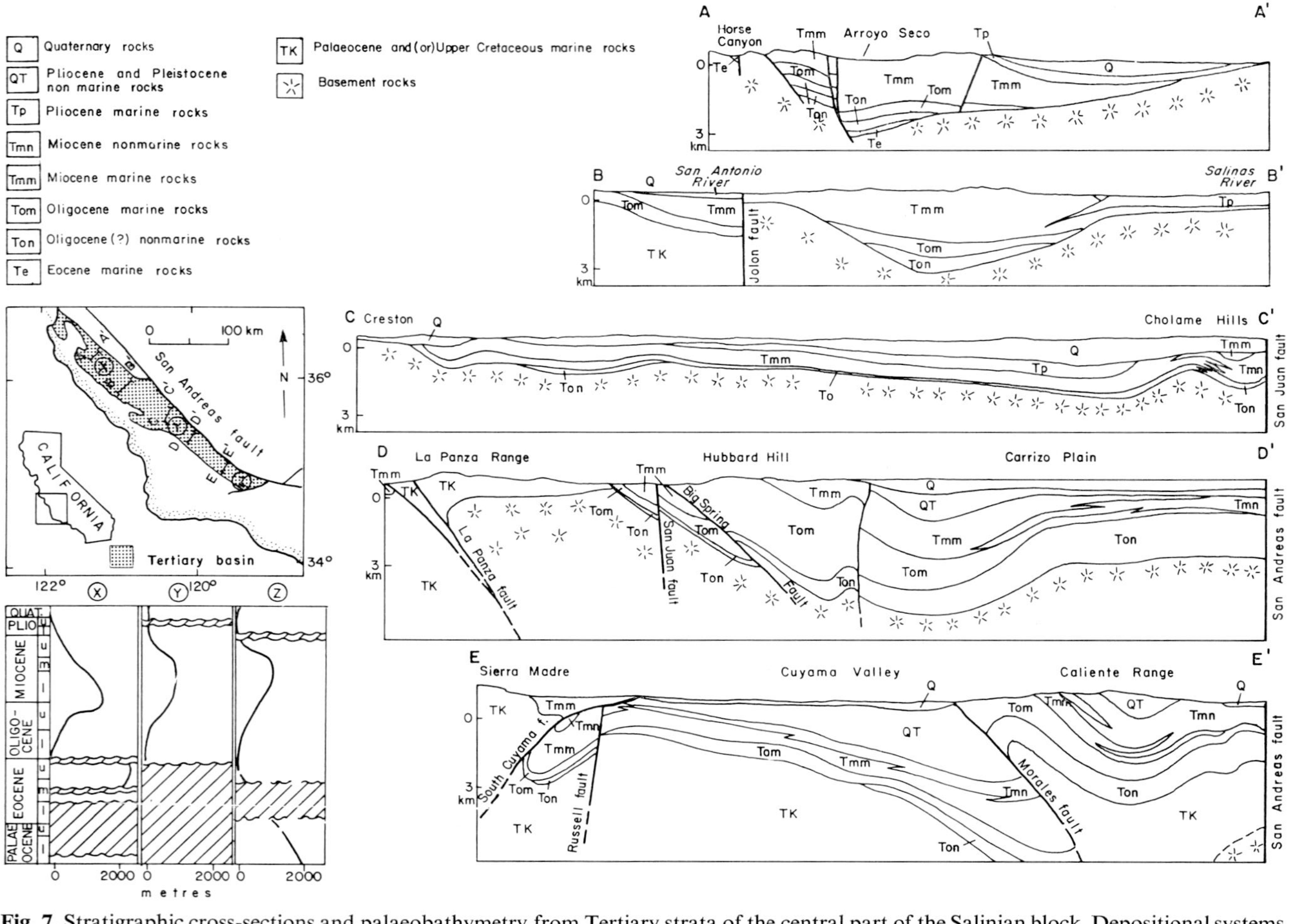

Fig. 7. Stratigraphic cross-sections and palaeobathymetry from Tertiary strata of the central part of the Salinian block. Depositional systems are restricted in area and basin-margin facies reflect spasmodic variations in bathymetry. These relations are typical of wrench-tectonic regions. Data in part from Durham (1974), Vedder (1975), and Graham (1978).

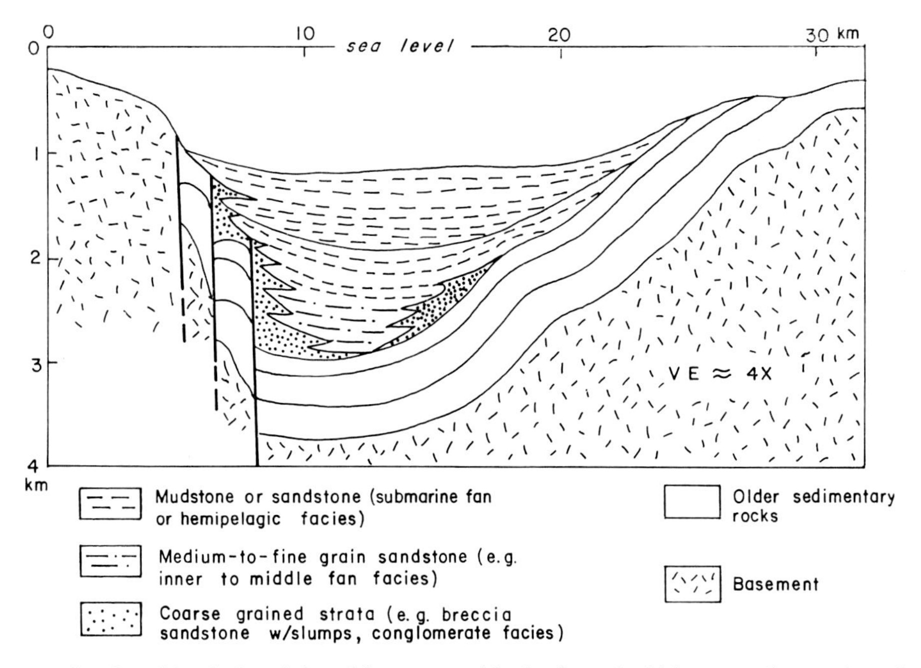

Fig. 8. Stratigraphic relations inferred for structural basins formed within a wrench-tectonic setting of the southern California borderland.

are marked by submarine-canyon facies, slump aprons, and lithological pinch-outs and unconformities. Deposition within basins reflect mass-flow processes where terrigenous source areas are available, and hemipelagic and pelagic deposition in distal, sediment-starved areas.

Late Cainozoic deposits in the Los Angeles and Ventura basins, along the eastern margin of the southern California borderland, locally exceed 6 km in thickness, but correlative strata in more seaward basins rarely exceed 2 km (Vedder *et al.*, 1976; Blake *et al.*, 1978). This rapid attenuation of sediment thickness results from submarine and subaerial barriers which trap sediment shed from the continent. Such barriers also account for wandering palaeocurrent patterns, irregular thicknesses of section and lithological diversity.

Basin formation related to wrench tectonics

With the exception of the western Transverse Ranges, northwest-trending right-slip faults and related *en echelon* folds dominate the structural grain in southern and central California. Development of these wrench-related structures, which began in the Miocene, appears to be genetically linked to late Cainozoic transform motion between the North American and Pacific lithospheric plates (Atwater, 1970). The formation of late Tertiary basins along the central and northern California continental shelf has been attributed to extension due to a shift in relative motion of the Pacific and North American plates (Blake *et al.*, 1978), however, we suggest that these basins formed chiefly in response to wrench tectonics. On the two cross-sections of the offshore Santa Maria Basin (Fig. 9), compression is indicated, particularly in the

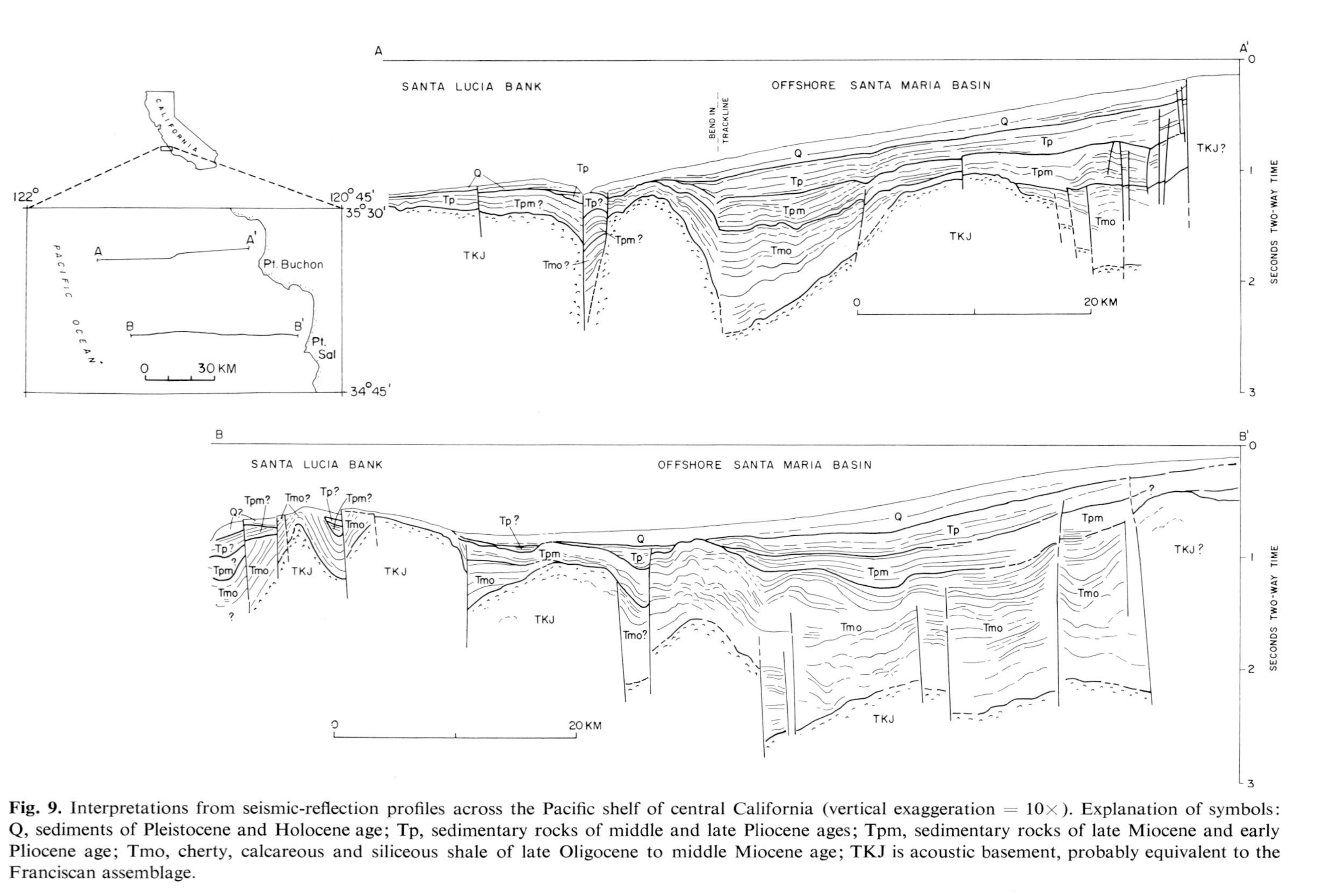

Fig. 9. Interpretations from seismic-reflection profiles across the Pacific shelf of central California (vertical exaggeration = 10×). Explanation of symbols: Q, sediments of Pleistocene and Holocene age; Tp, sedimentary rocks of middle and late Pliocene ages; Tpm, sedimentary rocks of late Miocene and early Pliocene age; Tmo, cherty, calcareous and siliceous shale of late Oligocene to middle Miocene age; TKJ is acoustic basement, probably equivalent to the Franciscan assemblage.

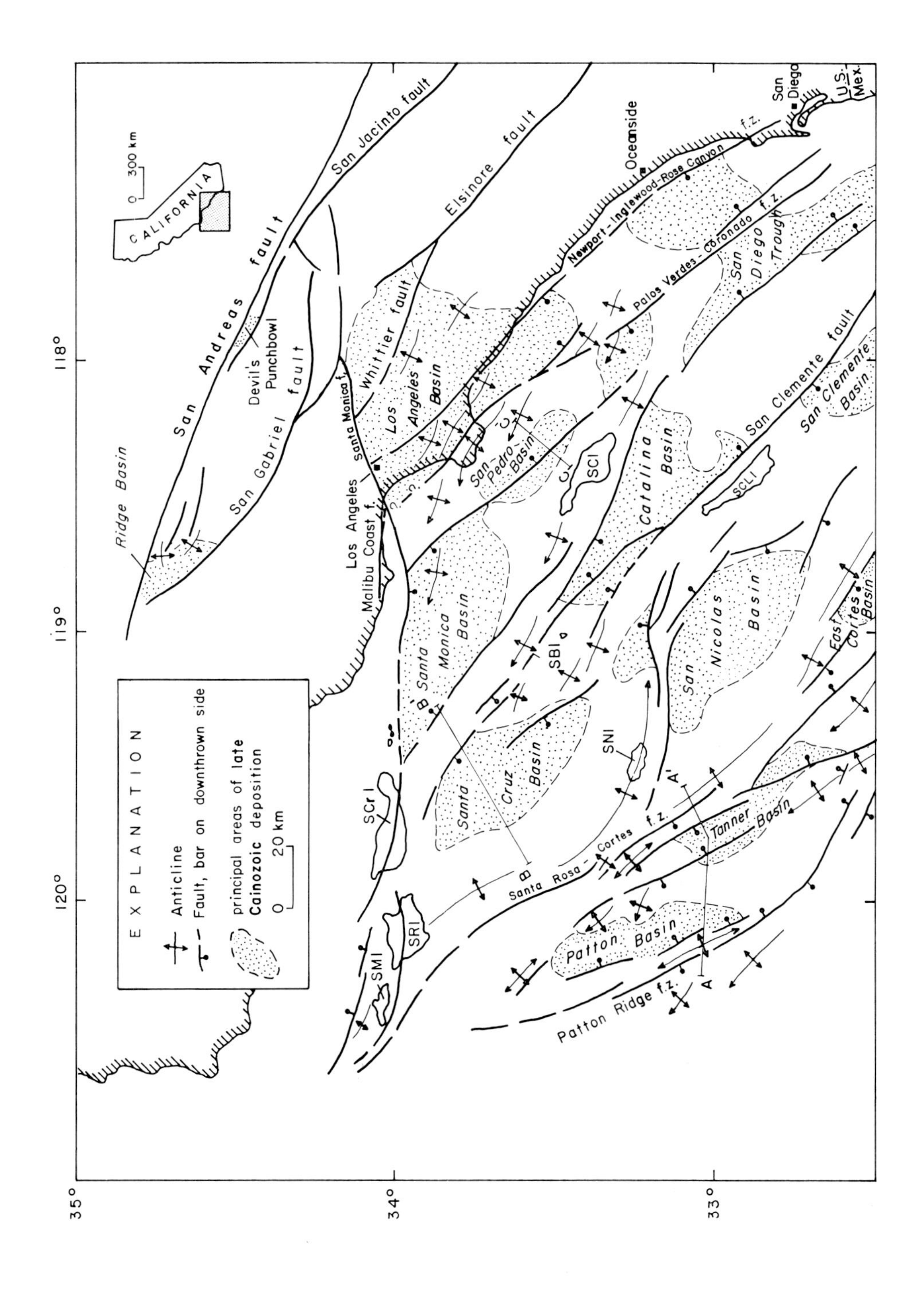

Fig. 10. Selected wrench-tectonic elements related to the development of basins in southern California SMI, San Miguel Island; SRI, Santa Rosa Island; SCrI, Santa Cruz Island; SCI, Santa Catalina Island; SBI, Santa Barbara Island; SNI, San Nicolas Island; SCLI, San Clemente Island. Line A–A′, B–B′, and C–C′ are lines of profiles shown in Fig. 11. Modified from Moore (1969) and Junger (1976).

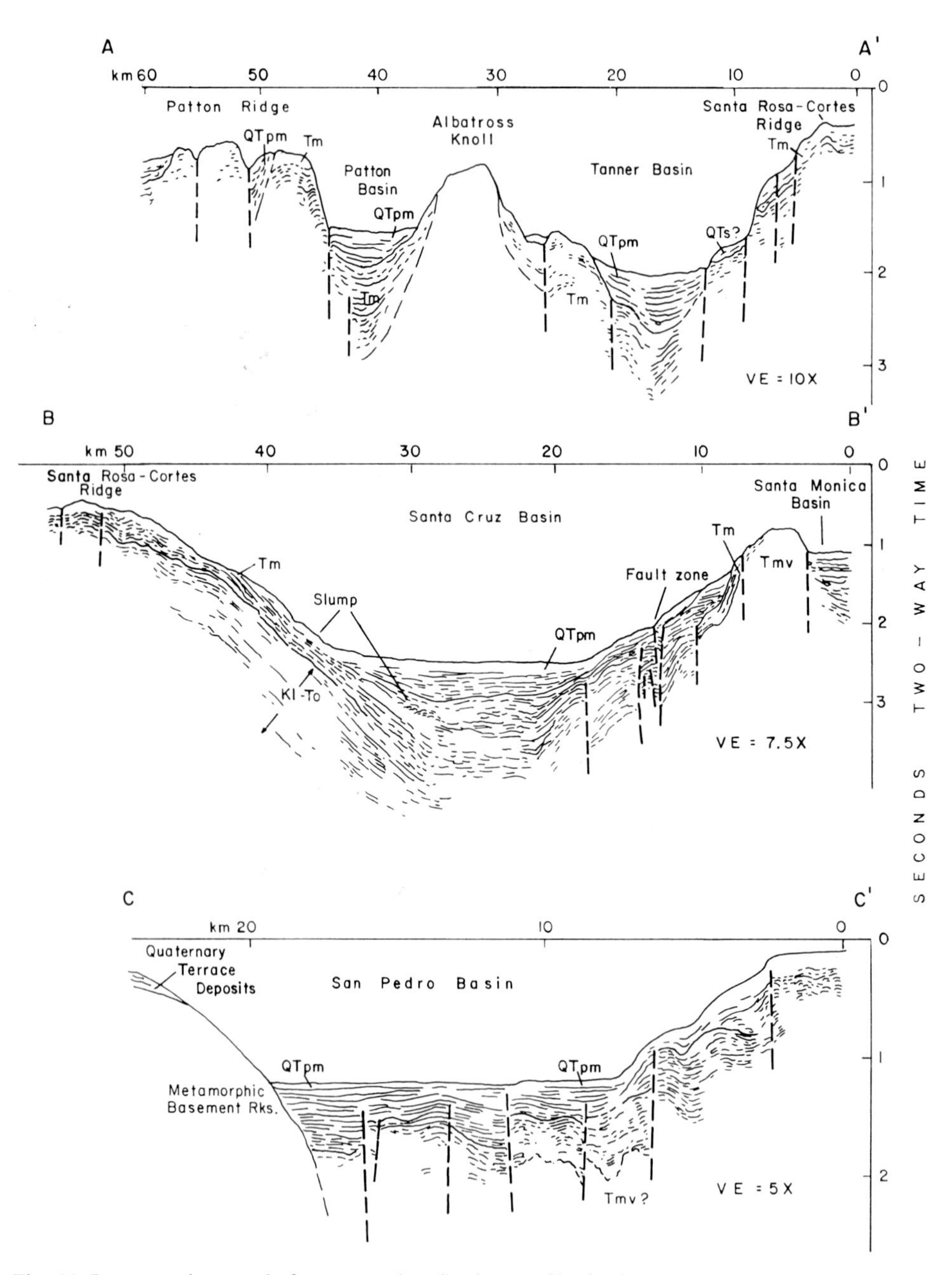

Fig. 11. Interpretations made from acoustic-reflection profiles in the outer borderland. Location of profiles shown in Fig. 10. QTpm, sediments of late Miocene to Holocene age; Tm, sediments of early and middle Miocene age; Tmv, Miocene volcanic rocks; Kl-To, sediments of Late Cretaceous to Oligocene age.

southern part, which is compatible with our alternative basin-forming hypothesis. Sedimentary basins that form by wrench faulting are chiefly the result of crustal extension and sagging as well as folding along braided systems of curving right-slip faults (e.g. Crowell, 1974).

Figure 10 illustrates the structural similarity of the California Continental Borderland to that of onland southern California. Faults such as the San Andreas, San Gabriel, and Newport–Inglewood are well documented right-slip faults (Moody & Hill, 1956; Crowell, 1962; Wilcox, Harding & Seely, 1973). Faults offshore are inferred from bathymetry and geophysical data (Moore, 1969; Greene *et al.*, 1975; Crouch, 1977; Junger & Wagner, 1977; Nardin & Henyey, 1978). Right-slip displacement along these offshore faults is suggested by *en echelon* structures and apparent reversals in relative vertical movement along the strike of the fault zone. Palaeogeographic analyses of Eocene and Miocene sequences also suggest lateral displacement (e.g. Howell *et al.*, 1974). Relatively recent local faulting is indicated throughout much of the region by active seismicity and sea-floor displacements.

All of the sedimentary basins are close to major wrench zones (Fig. 10) and are typically either rhomboid or lens shaped and are fault bounded on at least one flank. *En echelon* folds generally splay into the basins along bordering wrench faults; in some cases, such as along the southwest and western flanks of Santa Cruz and San Nicolas Basins, large-amplitude anticlinoria, instead of faults, form the basin margins.

Where wrench faults diverge, adjacent blocks move away from each other to form basins (Wilcox *et al.*, 1973; Crowell, 1974, 1976), such as Patton Basin, Catalina Basin, and the southern part of Tanner Basin. The sedimentary fill in these basins thickens toward their NNW margins where the fault-bounded block is tipped downward. The gently sloping SE margins of the basins merge with faulted and folded structural highs that apparently were uplifted as a result of compression at the convergent end of the block.

The NW part of Tanner Basin, San Pedro Basin (Fig. 11, Line A–A′, C–C′) and the San Diego Trough may have had a somewhat different structural origin. Judging from interpretations of seismic-reflection profiles, these basins are grabens that formed between parallel strands of bordering wrench zones. On land, a similar geometry and origin is evident for Devil's Punchbowl, which lies between the parallel Punchbowl and San Andreas faults. In contrast to other basins in the southern California region, these fault-bounded basins are narrow and generally have steep flanks. They are characterized by relatively rapid infilling and complex facies relations.

Santa Cruz and San Nicolas Basins form a third distinct basin type. The NE margins of these basins are fault bounded whereas the SW margins are fold bounded (Fig. 11, Line B–B′). Both basins appear to be synclinal troughs that are separated from neighbouring basins by broad, ridge-forming anticlinoria. The margins of the two basins are broader and more gently sloping than the other borderland basins, probably resulting in a generally finer grained, more pelagic basin fill.

Basins onshore that formed as a consequence of wrench faulting are known to have complex structural and depositional histories. In Los Angeles basin for example, Miocene faulting, folding and volcanism occurred contemporaneously with rapid subsidence and infilling of the basin (Yerkes *et al.*, 1965; Crowell, 1974). Within the basin, facies relationships are complex and change abruptly. Wedges of coarse clastics containing glaucophane schist detritus were rapidly shed into the basin from adjoining fault scarps to the west, and these deposits interfinger with and are interbedded with a host of lithofacies reflecting shallow to deep-water deposition (Woodford *et al.*, 1954).

Complexities similar to those in the Los Angeles basin presumably are present at many places in the borderland. For example, Patton and Tanner Basins both lie within a block bounded by right-lateral wrench zones (Figs 10 and 11). Extensive faulting, folding and volcanism as well as uplift and erosion of pre-basin (lower middle and lower Miocene) strata have occurred along the basin margins (Crouch, 1977). Younger strata lap onto these older structures and are much less deformed; however, unconformities within the basin strata together with faulting and uplift of basin sediments along the margins attest to renewed periods of tectonism during basin filling. Albatross Knoll (Fig. 10, Line A–A′), which separates the two basins, is a local basement block that consists of metamorphic and volcanic rocks. The flat-topped crest of the knoll,

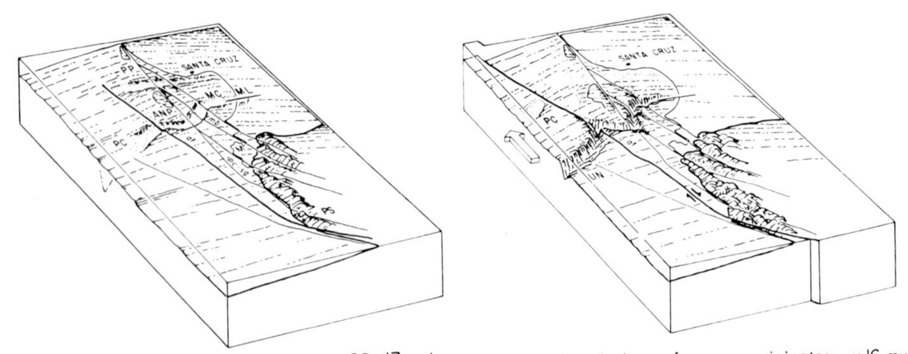

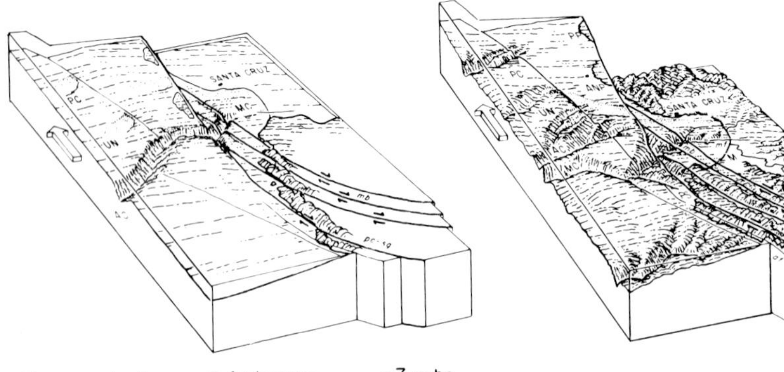

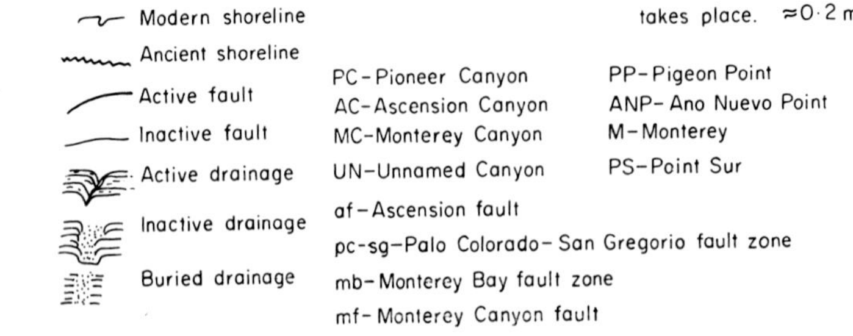

Fig. 12. Sequential development and splintering of a submarine canyon in the Monterey Bay region of California. Resultant bathymetry displays a number of headless canyons northwest of canyon-forming area (data from H. G. Greene, drawn by Tau Rho Alpha).

which at present lies at a water depth of about 670 m, appears to have had a history of late Miocene subaerial erosion followed by subsidence and westward tilting.

Submarine-canyon facies are common along the fault-bounded margins of basins within a wrench-tectonic regime. Subsequent offset of these lithofacies may create complicated rock distribution patterns that do not reflect original sediment dispersal systems. For example, a palinspastic reconstruction of the Monterey Bay region (Greene, 1977) suggests that the headless canyons north of Monterey Bay (Pioneer Canyon, an unnamed canyon, and Ascension Canyon) once were connected to Monterey Canyon, which heads only a few metres offshore and a few kilometres north of the mouth of the Salinas River (Fig. 12). Right-slip on the Ascension fault displaced northward the lower part of Monterey Canyon, which had eroded subaerially during an early Miocene highstand (Greene & Clark, 1979). The displaced canyon, inactive and buried during the time of its displacement, was rejuvenated in Pleistocene time as Pioneer Canyon, 110 km northwest of the present-day Monterey Canyon.

The unnamed headless canyon between Pioneer and Ascension canyons formed in the same manner as did Pioneer Canyon, in middle and late Miocene time. Between 14 and 10 m.y. ago other offshore fault zones (the Palo Colorado–San Gregorio and Monterey Bay fault zones) within the Monterey Bay region became active, and the evolution of the present configuration of Monterey Bay coastline began. Ascension Canyon may have begun to develop as a seaward channel to Monterey Canyon by progressive offset along the Palo Colorado–San Gregorio fault zone. Submarine fans developed at the mouth of the submarine canyons, and the fans, too, have been displaced northward.

SUMMARY

Basin configurations and sedimentological relations in California reflect contrasting tectonic processes along a segment of the west margin of the North American plate. During periods of high-angle, orthogonal subduction, broad forearc-basin deposition prevailed; but during periods of oblique, low-angle subduction or transform faulting, borderland settings evolved with the regional extent of the ridge and basin topography being greatest in the transform regime. Wrench tectonics controlled the development of the various borderlands, and sedimentological patterns reflect components of vertical displacements attendant with lateral slip. This alternating up and down movement of crustal blocks bordering the basins produced unconformities, lithological pinch outs, and slump aprons at a much smaller scale than those developed in the larger depositional systems that accompanied orthogonal subduction.

REFERENCES

Atwater, T.M. (1970) Implications of plate tectonics for the Cenozoic tectonic evolution of western North America. *Bull. geol. Soc. Am.* **81**, 3513–3536.

Atwater, T. & Molnar, P. (1973) Relative motion of the Pacific and North American plates deduced from sea-floor spreading in the Atlantic, Indian and South Pacific Oceans. In: *Proceedings of the Conference on Tectonic Problems of the San Andreas Fault System* (Ed. by R. L. Kovach and A. Nur). *Stanford Univ Publ. Geol. Sci.* **13**, 136–148.

BLAKE, M.C. JR, CAMPBELL, R.H., DIBBLEE, T.W. JR, HOWELL, D.G., NILSEN, T.H., NORMARK, W.R., VEDDER, J.C. & SILVER, E.A. (1978) Neogene basin formation in relation to plate-tectonic evolution of San Andreas fault system, California. *Bull. Am. Ass. Petrol. Geol.* **62,** 344–372.

CONEY, P.J. (1978) Mesozoic Cordilleran plate tectonics. In: *Cenezoic Tectonics and Regional Geophysics of the Western Cordillera* (Ed. by R. B. Smith and G. P. Eaton). *Mem. geol. Soc. Am.* **152,** 344–372.

CONEY, P.J. & REYNOLDS, S.J. (1977) Cordilleran Benioff zones. *Nature,* **270,** 403–406.

COOPER, A.K., SCHOLL, D.W. & MARLOW, M.S. (1976) Mesozoic magnetic lineations in the Bering Sea marginal Basin. *J. geophys. Res.* **81,** 1916–1934.

CROSS, T.A. & PILGER, R.H. JR (1978) Constraints on absolute motion and plate interaction inferred from Cenozoic igneous activity in the western U.S. *Am. J. Sci.* **278,** 865–902.

CROUCH, J.K. (1977) Structure of the outer California Borderland and a possible analogue in the region between the San Andreas and San Gabriel faults. *Abstr. Progm. geol. Soc. Am.* **9,** 407.

CROUCH, J.K. (1979) Neogene tectonic evolution of the California Continental Borderland and western Transverse Ranges. *Bull. geol. Soc. Am.* **90,** 338–345.

CROWELL, J.C. (1962) Displacement along the San Andreas fault, California. *Spec. Pap. geol. Soc. Am.* **71,** 61 pp.

CROWELL, J.C. (1974) The Orocopia Thrust, southeastern California (abs.): *Abstr. Progm. geol. Soc. Am.* **6,** 159.

CROWELL, J.C. (1976) Implications of crustal stretching and shortening of coastal Ventura basin, California, In: *Aspects of the Geologic History of the California Continental Borderland* (Ed. by D. G. Howell). *Misc. Publ. Am. Ass. Petrol. Geol. Pac. Sect.* **24,** 365–382.

DICKINSON, W.R. (1971) Clastic sedimentary sequences deposited in shelf, slope, and trough settings, between magmatic arcs and associated trenches. *Pacif. Geol.* **3,** 15–30.

DICKINSON, W.R. (1979) Cenozoic plate tectonic setting of the Cordilleran region in the U.S. In: *Cenozoic Paleogeography of the Western U.S. Pacific Coast Paleogeography Symposium,* 3 (Ed. by J. M. Armentrout, M. R. Cole and H. TerBest). *Spec. Publ. Soc. econ. Paleont. Miner., Pacif. Sect.* 1–13.

DURHAM, D.L. (1974) Geology of the southern Salinas Valley area, California. *Prof. Pap. U.S. geol. Surv.* **819,** 111 pp.

EHLIG, P.L. & JOSEPH, S.E. (1977) Polka dot granite and correlation of La Panza quartz monzonite with Cretaceous batholithic rocks north of Salton Trough. In: *Cretaceous Geology of the California Coast Ranges West of the San Andreas Fault* (Ed. by D. G. Howell, J. G. Vedder and K. McDougall). *Soc. econ. Paleont. Miner. Pacif. Sect., Pacific Coast Palaeogeography Field Guide,* **2,** 91–96.

GRAHAM, S.A. (1978) Role of Salinian Block in evolution of San Andreas fault system, California. *Bull. Am. Assoc. Petrol. Geol.* **62,** 2214–2232.

GREENE, H.G. (1977) Geology of the Monterey Bay Region. *Open-File Rep. U.S. geol. Surv.* **77–718.** 347 pp.

GREENE, H.G. & CLARK, J.C. (1979) Neogene Paleogeography of the Monterey Bay Area, California. In: *Cenozoic Paleogeography of the Western U.S. Pacific Coast Paleogeography Symposium,* 3 (Ed. by J. M. Armentrout, M. R. Cole and H. TerBest). *Spec. Publ. Soc. econ. Paleont. Miner., Pacif. Sect.* 277–295.

GREENE, H.G., CLARKE, S.H., FIELD, M.E., WAGNER, H.C. & LINKER, F. (1975) Preliminary report on the environmental geology of selected areas of the southern California Continental Borderland. *Open-File Rep. U.S. geol. Surv.* **75–596.** 70 pp.

HAMILTON, W. (1969) Mesozoic California and the underflow of Pacific mantle. *Bull. geol. Soc. Am.* **80,** 2409–2430.

HOWELL, D.G. & LINK, M.H. (1979) Eocene conglomerate sedimentology and basin analysis, San Diego and the southern California Borderland. *J. sedim. Petrol.* **49,** 517–540.

HOWELL, D.G. & VEDDER, J.G. (1978) Late Cretaceous paleogeography of the Salinian block, California. In: *Mesozoic paleogeography of the western U.S. Pacific Coast Paleogeography Symposium,* 2 (Ed. by D. G. Howell and K. A. McDougall). *Spec. Publ. Soc. econ. Paleont. Miner. Pacif. Sect.* 107–116.

HOWELL, D.G., STUART, C.J., PLATT, J.P. & HILL, D.J. (1974) Possible strike-slip faulting in the southern California Borderland. *Geology,* **2,** 93–98.

Howell, D.G., Vedder, J.G., McLean, H., Joyce, J.M., Clarke, S.H. & Smith, G. (1977) Review of Cretaceous geology, Salinian and Nacimiento Blocks, Coast Ranges of central California. In: *Cretaceous Geology of the California Coast Ranges, West of the San Andreas Fault* (Ed. by D. G. Howell, J. G. Vedder & K. A. McDougall). *Soc. econ. Paleont. Miner. Pacif. Sect., Pacific Coast Paleogeography Field Guide* **2,** 1–46.

Ingersoll, R.V. (1978) Submarine fan facies of the Upper Cretaceous Great Valley sequence, northern and central California. *Sedim. Geol.* **21,** 205–230.

Ingersoll, R.V. (1979) Evolution of the Late Cretaceous fore-arc basin, northern and central California. *Bull. geol. Soc. Am.* **90,** 813–826.

Ingersoll, R.V., Rich, E.I. & Dickinson, W.R. (1977) Great Valley Sequence, Sacramento Valley field trip guide. *Geol. Soc. Am., Cordilleran Sect.* 73 pp.

Junger, A. (1976) Tectonics of the southern California borderland. In: *Aspects of the Geologic History of the California Continental Borderland* (Ed. by D. G. Howell). *Misc. Publ. Am. Ass. Petrol. Geol. Pacif. Sect.* **24,** 486–498.

Junger, A. & Wagner, H.E. (1977) Geology of the Santa Monica and San Pedro Basins, California Continental Borderland. *U.S. geol. Surv. Misc. Field Studies Map*, MF-820, scale 1:250,000.

Keith, S.B. (1978) Paleosubduction geometries inferred from Cretaceous and Tertiary magmatic patterns in western North America. *Geology*, **6,** 516–521.

Lipman, P.W. (1980) Cenozoic volcanism in the western U.S.: implications for continental tectonics. *Monogr. Am. geophys. Union.* (In press.)

Lipman, P.W., Prostka, H.J. & Christiansen, R.L. (1971) Evolving subduction zones in the western United States, as interpreted from igneous rocks. *Science*, **174,** 821–825.

Mattinson, J.M., Davis, T.E. & Hopson, C.A. (1971) U-Pb studies in the Salinian block of California. *Carnegie Inst. Wash. Yearbook*, **70,** 248–251, illust. sketch map.

Moody, J.D. & Hill, M.J. (1956) Wrench-fault tectonics. *Bull. geol. Soc. Am.* **67,** 1207–1246.

Moore, D.G. (1969) Reflection profiling studies of the California Continental Borderlands. *Spec. Pap. geol. Soc. Am.* **107,** 138 pp.

Nardin, T.R. & Henyey, T.L. (1978) Plio-Pleistocene diastrophism in the area of the Santa Monica and San Pedro shelves, California Continental Borderland. *Bull. Am. Ass. Petrol. Geol.* **62,** 247–272.

Nilsen, T.H. & Clarke, S.H. Jr (1975) Sedimentation and tectonics in the early Tertiary continental borderland of central California. *Prof. Pap. U.S. geol. Surv.* **925.** 64 pp.

Ross, D.C. (1978) The Salinian Block: a mesozoic granite orphan in the California Coast Ranges. In: *Mesozoic Paleogeography of the Western U.S. Pacific Coast Paleogeography symposium*, 2 (Ed. by D. G. Howell and K. A. McDougall). *Spec. Publ. Soc. econ. Paleont. Miner., Pacif. Sect.* 509–522.

Stuart, C.J. (1979) Middle Miocene Paleogeography of coastal southern California and the California Borderland: evidence from schist-bearing Sedimentary Rocks. In: *Cenozoic Paleogeography of the Western U.S. Pacific Coast Paleogeography Symposium*, 3 (Ed. by J. M. Armentrout, M. R. Cole and H. TerBest). *Spec. Publ. Soc. econ. Paleont. Miner., Pacific. Sect.* 29–44.

Suppe, J. (1970) Offset of late Mesozoic basement terranes by the San Andreas fault system. *Bull. geol. Soc. Am.* **81,** 3253–3258.

Trask, P.D. (1926) Geology of Point Sur quadrangle, California. *Sci. Bull. Calif. Univ. Publ. Dept Geol.* **16,** 119–186.

Vedder, J.G. (1975) Juxtaposed Tertiary strata along the San Andreas fault in the Temblor and Caliente Ranges, California. In: *San Andreas Fault in Southern California* (Ed. by J. C. Crowell). *Spec. Rep. Calif. Div. Mines Geol.* **118,** 234–240.

Vedder, J.G., Greene, H.G., Scott, E.W. & Taylor, J.C. (1976) A summary report of the regional geology, petroleum potential, environmental geology and technology for exploration and development in the area of proposed Lease Sale 48, California Borderlands. *Open-File Rep. U.S. geol. Surv.* **76–787.** 43 pp.

Vedder, J.G. & Howell, D.G. (1980) Topographic evolution of the southern California borderland during late Cenozoic time. In: *The California Islands: Proceedings of a Multidisciplinary Symposium* (Ed. by D. M. Power). Santa Barbara Museum of Natural History, Santa Barbara, California. (In press.)

VEDDER, J.G., HOWELL, D.G. & FORMAN, J.A. (1979) Miocene rocks and their relation to older strata and basement, Santa Catalina Island, California. In: *Cenozoic Paleogeography of the Western U.S. Pacific Coast Paleogeography Symposium*, 3 (Ed. by J. M. Armentrout, M. R. Cole and H. TerBest). *Spec. Publ. Soc. econ. Paleont. Miner. Pacif. Sect.* 239–256.

WILCOX, R.E., HARDING, T.P. & SEELY, D.R. (1973) Basic Wrench Tectonics. *Bull. Am. Ass. Petrol. Geol.* **57,** 74–96.

WOODFORD, A.O., SCHOELLHAMER, J.E., VEDDER, J.G. & YERKES, R.F. (1954) Geology of the Los Angeles basin: *Bull. Calif. Div. Mines Geol.* **170,** 65–81.

YERKES, R.F., MCCULLOH, T.H., SCHOELLHAMER, J.E. & VEDDER, J.G. (1965) Geology of the Los Angeles basin, California—An introduction. *Prof. Pap. U.S. geol. Surv.* **420–A.** 57 pp.

YERKES, R.F., SARNA-WOJCICKI, A.M. & LAJOIE, K.A. (1980) Geology and Quaternary deformation of the Ventura area, Ventura County, California. *Prof. Pap. U.S. geol. Survey.* (In press.)

Spec. Publ. int. Ass. Sediment. (1980) **4,** 63–78

Evolution of a strike-slip fault-controlled basin, Upper Old Red Sandstone, Scotland

B. J. BLUCK

Department of Geology, University of Glasgow, Glasgow G12 8QQ, Scotland

ABSTRACT

The Upper Old Red Sandstone of Midland Scotland is an upward fining, upward maturing, red bed sequence which exceeds 3 km in thickness. It begins with polymict conglomerates of alluvial fan origin, passes upward into sub-lithic arenite and breccia-conglomerates deposited by braided streams and alluvial fans, and ends with quartz conglomerates, quartz arenites, and caliche beds. This sequence was laid down in two embayments: the main Midland Valley Basin with an axial flow to the NE and E and the Kintyre Basin with an axial flow to the SW. In both basins, older rocks are overlapped by younger as the area of sedimentation extended. The Midland Valley Basin was initiated in the Clyde area by sinistral movement on the Highland Boundary Fault—a fault with a long and diverse history. This major fracture is either offset or has a sharp bend in the Firth of Clyde region, so that the ground to the south of it broke into fault blocks. As each fault-bounded block developed, an upward fining conglomerate sequence formed on the downthrow side and the lithostratigraphy shows a progression in the genesis of conglomerate wedges towards the SW. This type of basin contrasts with many other fault controlled basins in its comparatively thin sequence; in having palaeocurrents for the coarse sediments whose orientation is similar to those of the fine (rather than perpendicular as is the usual case); in having an upward maturing sequence; and in having a systematic progression in the development of fault controlled conglomerate wedges.

INTRODUCTION

Upper Old Red Sandstone rocks, sporadically occurring over much of Scotland, are studied in detail in the Midland Valley.

Structural setting

The Midland Valley of Scotland lies between two fold belts: the Southern Highlands to the north and the Southern Uplands to the south. The Highland rocks, belonging to the Dalradian Supergroup, were folded and metamorphosed in Arenig or pre-Arenig times (Dewey & Pankhurst, 1970) and lie in a complex synform whose

0141-3600/80/0904-0063$02.00

axis is parallel to a major fault zone, the Highland Boundary Fault Zone. The Dalradian Supergroup comprises a wide range of lithologies, variably metamorphosed and originally deposited in a fault-bounded basin of late Precambrian–Cambrian age (Harris *et al.*, 1978).

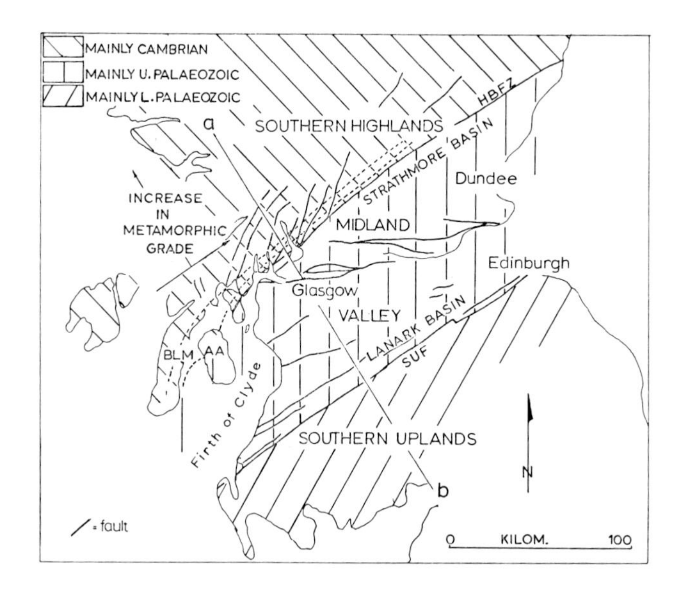

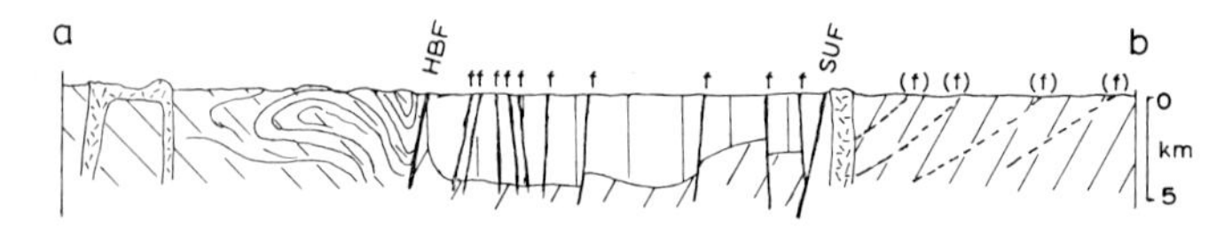

Fig. 1. Major structural divisions of central and southern Scotland. HBFZ, Highland Boundary Fault Zone; SUF, Southern Uplands Fault; AA, Aberfoyle Anticline (synform); BLM, Ben Ledi Monoform; (f) faults (conjectural). After the Tectonic Map of Great Britain and Northern Ireland, I.G.S., 1966.

The southern fold belt, forming the Southern Uplands, comprises mainly greywacke, siltstone and shale of Ordovician and Silurian age which rest on an Arenig foundation of spilite, black shale and radiolarian chert. These rocks are complexly folded and faulted but have suffered little metamorphism and McKerrow, Leggett & Eales (1977) have proposed that they are part of an accretionary prism formed on the landward edge of a northerly subducting plate. The Southern Uplands fold belt is bounded on its northern margin by the Southern Uplands Fault (Fig. 1).

The Midland Valley comprises rocks mainly of Upper Palaeozoic age which are now disposed in a rather complex syncline, subjected to much faulting. Within the Midland Valley there are fairly extensive exposures of Old Red Sandstone rocks. The bounding faults, the Highland Boundary and Southern Uplands Faults, in being major *en echelon* fractures, are typical of old faults reactivated through a younger cover (Wilcox, Harding & Seely, 1973). The Highland Boundary Fault has evidence of this older history in cherts, spilite, black shale, sheared gabbro and serpentinite which are exposed along the Fault zone. This sedimentary–igneous rock association ranges in age from Upper Cambrian to Lower Ordovician (Downie *et al.*, 1971). Longman, Bluck & van Breemen (1979) have suggested that these rocks are part of an ophiolite assemblage, formed in a marginal sea which closed along the Highland Boundary Fault. This fault was also active in Devonian times when there was a massive downthrow of several kilometres to the SE. It continued activity into post-Lower Carboniferous times at least, with a comparatively minor NW throw. The Midland Valley, in general, continued to be structurally active up to and beyond Permo-Triassic times, the evidence for this activity being particularly strong in the Firth of Clyde (McLean & Deegan, 1978).

Because of this post Old Red Sandstone fault activity the structure of the Old Red Sandstone rocks in the Firth of Clyde region is, in places, quite complex.

The Old Red Sandstone

The Old Red Sandstone rocks, roughly of Devonian age, have accumulated in ground bounded to the north and south by regions of considerable crustal instability. In the Southern Uplands, deposition of the Old Red Sandstone closely followed the orogeny but in the Highlands there was a considerable time interval between the two. However, both these regions contributed to the Old Red Sandstone sediments which accumulated in the subsiding Midland Valley. The Lower Old Red Sandstone is a conglomerate–lithic arenite–volcanic rock association which accumulated in at least two basins; the Strathmore Basin to the north and the Lanark Basin to the south, separated from each other partly by a volcanic chain. The Highlands contributed much sediment to the Strathmore Basin and the Southern Uplands to the Lanark Basin so that each basin has a belt of conglomerate lying adjacent to the major fault which forms one of its margins. Both had axial flow to the SW parallel to the bounding faults and the structural grain of the adjacent source rocks. The Strathmore Basin has a sediment fill 8 km thick in the NE, which thins partly by overlap to the SW where in the region of the Firth of Clyde it rests with prominent unconformity on the underlying Dalradian metamorphic rocks (George, 1960; Armstrong & Paterson, 1970; Bluck, 1978). The nature of the structural control on the sedimentation in this basin is being investigated and is likely to involve strike-slip movement. During Lower Old Red Sandstone times there was much plutonic activity in the whole Scottish region (see Pidgeon & Aftalion, 1978).

UPPER OLD RED SANDSTONE

The Upper Old Red Sandstone rests unconformably on the Lower and, despite the absence of a detailed biostratigraphy, there seems good reason to believe that most, if not all, the Middle Old Red Sandstone (Middle Devonian) is missing.

The Upper Old Red Sandstone is thinner, finer grained and generally more mature than the Lower. It comprises a large-scale sedimentary sequence which, in general terms, begins with alluvial fans and ends with coastal sediments. There are four major lithological divisions with a fifth appearing locally at the base in the Northern Firth of Clyde (Fig. 2). This lowest unit is a sandstone of possible aeolian and fluvial overbank origin. The main divisions are: (2) the basal conglomerate, (3) pebbly sandstone, (4) breccia and conglomerate, and (5) sandstone-with-calcrete. The sequence is thickest and most complete in the Northern Firth of Clyde region (Fig. 3).

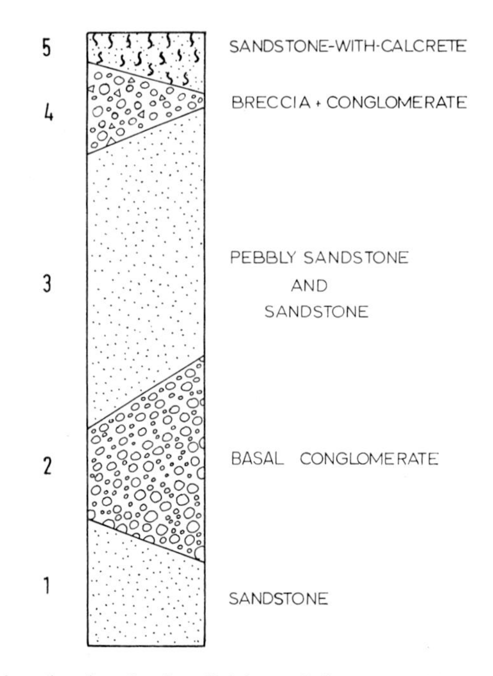

Fig. 2. Composite section showing the five divisions of the Upper Old Red Sandstone in the Firth of Clyde region.

The basal conglomerate division

Conglomerates are coarser grained and comprise more of the succession in the region of the Firth of Clyde than in any other area of the Midland Valley. In the Firth of Clyde region the basal conglomerates (2, Fig. 2) range in thickness from 250 to 2000 m and, with the exception of those in Kintyre, were deposited by palaeocurrents flowing to the NE, roughly parallel with the Highland Boundary Fault (Figs 4 and 5). In this way they are in marked contrast with the deposits of the Lower Old Red Sandstone where the axial flow is confined to the sandstones and a lateral flow typifies the conglomerates. Taken as a whole there is no systematic change in the maximum grain size of the basal conglomerates as they are traced in the palaeocurrent direction (Fig. 5), and the basal conglomerates remain in a facies of debris flows and proximal braided streams along the whole basin length. In addition, the composition of the clasts in the conglomerate differs considerably, indicating quite a variation in provenance (Fig. 5). Taken together these data demonstrate that there is no single dispersal system in the Clyde region, but rather a number of dispersal systems (shadows) each with its own lateral and vertical grain size and facies change. The rocks on the east side of the Firth of Clyde in the region from Wemyss Bay to Galston illustrate the stratigraphic disposition of these conglomerates (Fig. 3). At Wemyss Bay the basal conglomerate (the Skelmorlie Formation, Bluck, 1978, Fig. 8) with a mixed metamorphic and igneous provenance rests on the lower sandstone, but at Portencross there is another conglomerate, also proximal, but of different provenance (the Hunterston Formation).

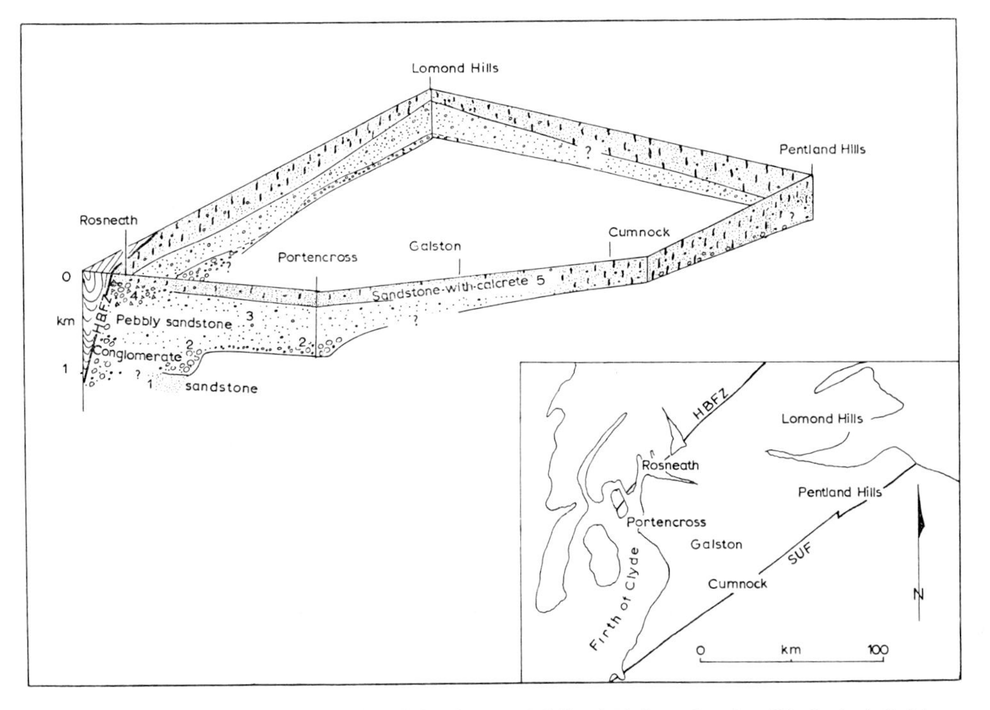

Fig. 3. Section through Upper Old Red Sandstone, Midland Valley, showing lithological divisions 1, etc. as indicated in Fig. 2. HBFZ, Highland Boundary Fault Zone; SUF, Southern Uplands Fault. The boundary between divisions is not always easily determined, so some margin of error is to be expected. For location of Wemyss Bay, see Fig. 5.

Since the conglomerate at Wemyss Bay (the Skelmorlie Formation) has the same palaeocurrent direction as the one at Portencross (the Hunterston Formation), the two conglomerates cannot have formed in the same basin of deposition. The lower sandstone and its overlying conglomerate have been overlapped by the younger Hunterston Formation which at Portencross rests unconformably on Lower Old Red Sandstone. This SW younging of the conglomerates, and their wedge-like shape is confirmed by the presence of fine conglomerate with a composition similar to the Hunterston Formation occurring in sandstones which are thought to overlie the Skelmorlie Formation of Wemyss Bay. Further to the SE, near Galston (Fig. 3) in a succession only 200–300 m thick, sandstone of the Upper Old Red Sandstone rests on volcanic rocks of the Lower Old Red Sandstone, thus demonstrating the overlap of the Hunterston Formation. A similar stratigraphy is emerging from a study of the sequences on the western side of the Firth of Clyde, suggesting therefore that the discrete dispersal shadows are really wedges of conglomerate each of which gets younger in a SW direction. Four dispersal shadows have been identified in the Firth of Clyde region, and each comprises a sequence which begins with coarse conglomerates and passes upwards into sandstones (e.g. Fig. 6). The coarse conglomerates, in being mainly debris flows often with clasts exceeding 50 cm in diameter, require steep slopes in both drainage and depositional area. Their thickness requires these slopes to have initially persisted for some time, to be later pared back or diminished when sandstones replaced conglomerates in the sequence (Bluck, 1967). This type of sequence

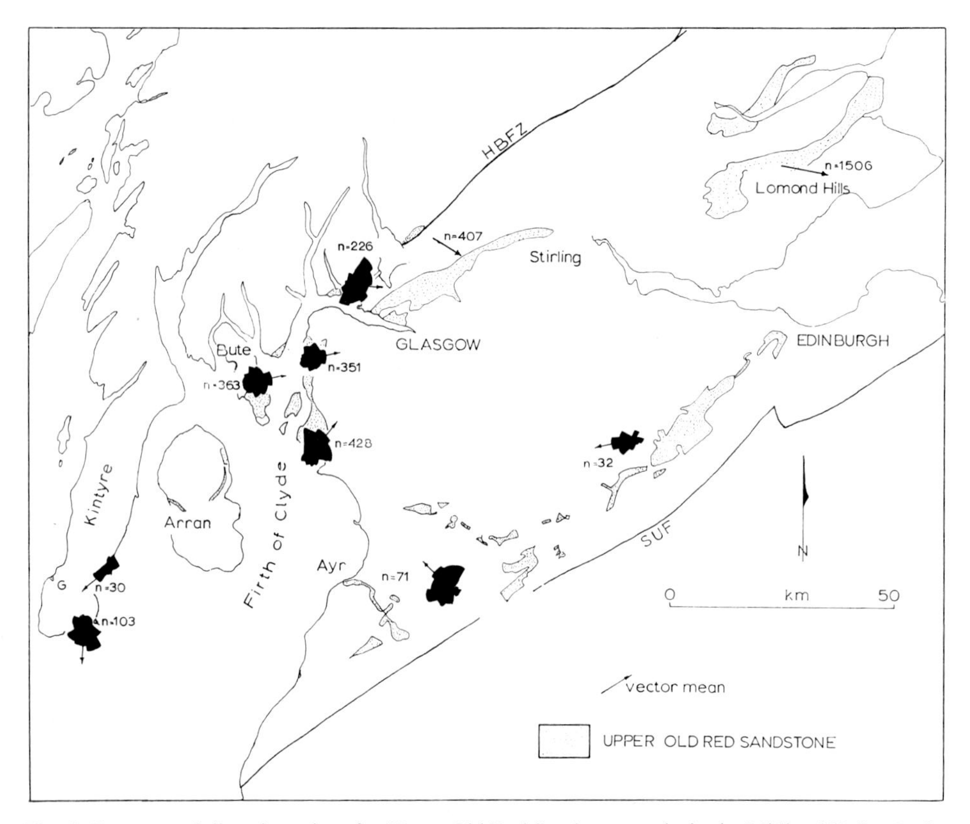

Fig. 4. Cross stratal dip orientations for Upper Old Red Sandstone rocks in the Midland Valley Basin. Rose diagrams give total cross stratal readings for selected areas; arrows, vector mean; single arrows record data from other workers: Stirling area—Read & Johnson (1967); Lomond Hills area—Chisholm & Dean (1974). *n*, number of observations; G, Galdrings.

requires early rapid uplift or subsidence which is best achieved by faulting. These postulated faults would need to trend normal to the palaeocurrent flow, get younger or produce younger sequences to the SW and produce a thickening succession to the NE. The presence and orientation of contemporaneous faults in Upper Old Red Sandstone sediments is supported by the presence and orientation of structures indicative of soft sediment faulting—which include sandstone and conglomerate dykes, and growth faults of various kinds, some of which have soft sediment deformation alongside them (see Fig. 7).

The location, source and stratigraphic position of the fault bounded basins are given in Fig. 8, where it is postulated that the Hunterston Formation belongs to the same fault basin as the sequence in Bute. The geometry of each basin fill is only now beginning to emerge as mapping proceeds. It is clear that each sequence generated at one of the postulated faults thickens to the NW, i.e. towards the Highland Boundary Fault. In the Rosneath area, for example, the basal conglomerate is overlapped by a finer one at Ardmore Point, where the fine conglomerate rests unconformably on Lower Old Red Sandstone rocks. And further to the SE, this fine conglomerate is overlapped by sandstones which rest on the Old Red Sandstone rocks at Killearn (roughly the line c–d in Fig. 8D).

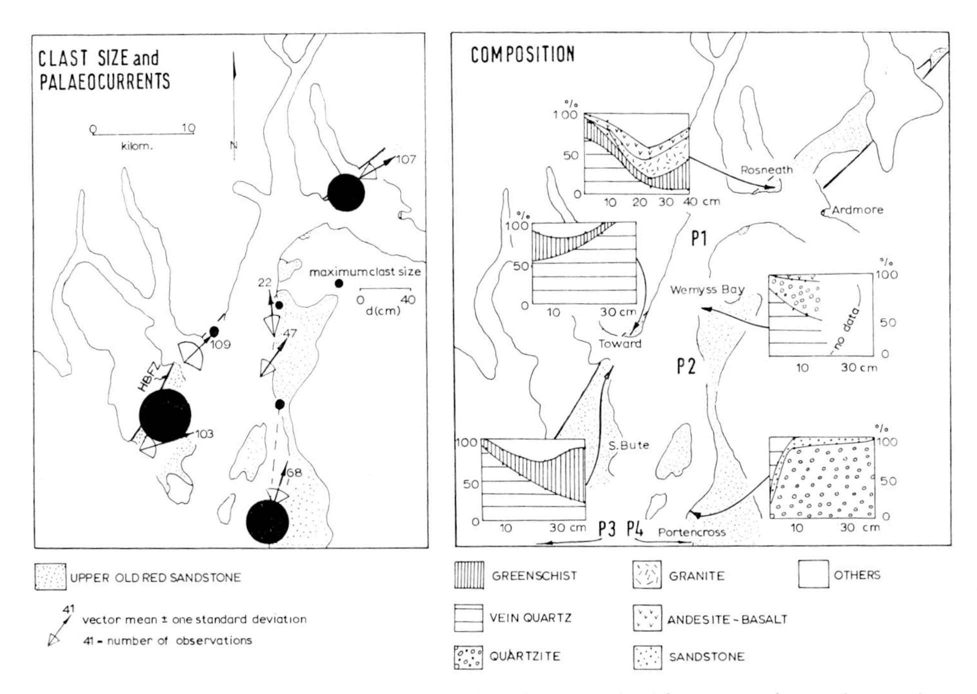

Fig. 5. Data on basal conglomerate. Maximum clast size determined from averaging ten largest clasts from 1 m^2, from ten coarsest parts of section. Composition data obtained from > 300 counts per size grade, and up to six different size grades per locality. P_1 etc. refers to source area; d, clast diameter in centimetres.

The type of faulting envisaged here resembles the faulting seen during basement extension (Kingma, 1958, Fig. 2; Crowell, 1974, Fig. 5), where a whole basin area undergoes subsidence, but the basement, in being divided into fault bounded blocks, provides source areas on the (relative) upthrown fault margins. The asymmetrical nature of the basin cross-section (Rosneath–Killearn, roughly c–d in Fig. 8D) and its elongation parallel with the master fracture (Highland Boundary Fault, Figs 8 and 12) have analogies in pull-apart basins of strike-slip origin. Mild crustal extension in the Clyde area is supported by the presence of Upper Old Red Sandstone andesitic volcanic rocks on Arran. Clasts of andesite in conglomerates as far away as Wemyss Bay (Skelmorlie Formation of Bluck, 1978) and Rosneath, may indicate contemporaneous volcanic activity over a wider area.

Fault-bounded basins of the kind postulated for these Upper Old Red Sandstone rocks (Fig. 8) are thought to have been produced by a sinistral shift on the Highland Boundary Fault, and this suggestion is supported by the orientation of folding in the beds of the topmost Lower Old Red Sandstone and Lower Upper Old Red Sandstone (see Fig. 7).

The sequence further to the SW, in Kintyre, has a dispersal to the SW, so defining a basin boundary zone to the NE in the region west of Arran (Figs 4 and 12).

The Pebbly Sandstone division

The unit is made up of alluvia, mostly of braided stream origin, along with a variety of overbank sediment and local but thick sequences of lacustrine beds (Fig. 6).

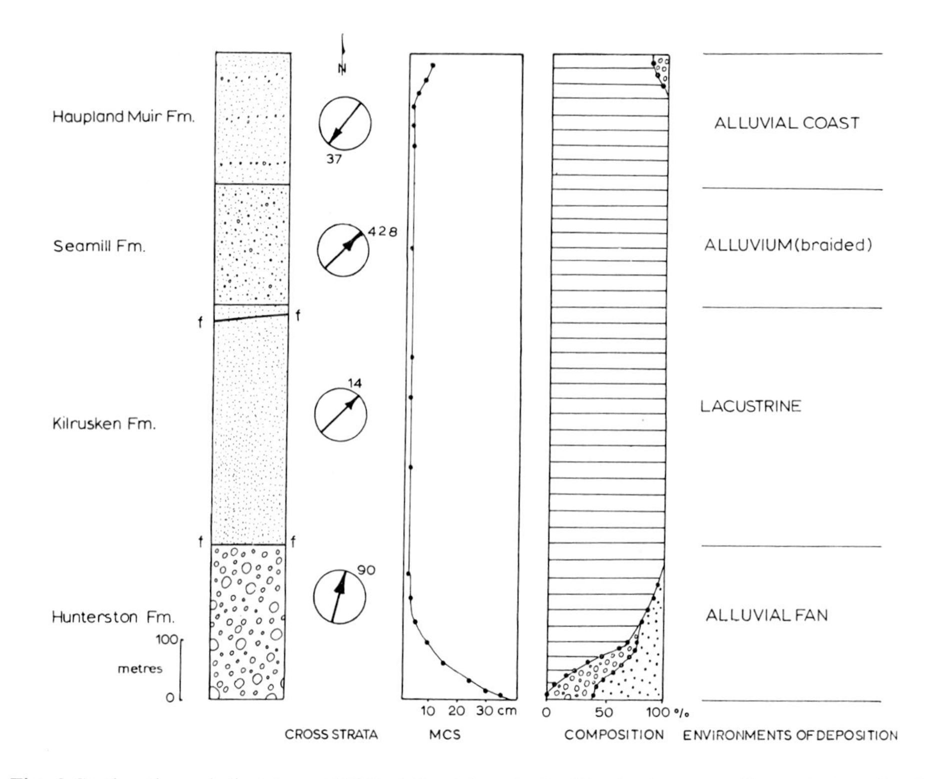

Fig. 6. Section through the Upper Old Red Sandstone in the Hunterston area. Cross strata refers to cross stratal dip orientation, arrow = vector mean, numbers refer to number of observations, MCS, maximum clast size in centimetres, being the average of the ten largest clasts from 1 m². Composition ornament as for Fig. 5.

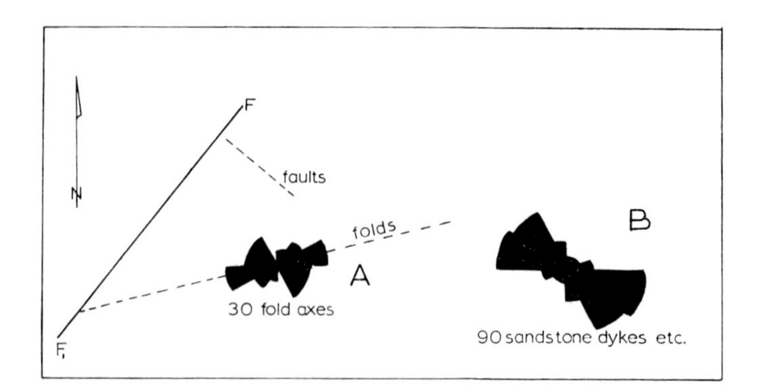

Fig. 7. Orientation of fold axes (from the upper part of Lower Old Red Sandstone and lower part of Upper Old Red Sandstone) and orientation of dykes, rotated back to horizontal (Upper Old Red Sandstone) in the northern Firth of Clyde. Not all folds are *en echelon*, and most have an amplitude of <1 km. F–F_1 is a sinistral fault with the same orientation as the Highland Boundary Fault. 'Folds' refer to the orientation of the *en echelon* folds found by Wilcox *et al.* (1973) to be generated by a sinistral fault with the orientation F–F_1. 'Faults' refer to orientation of active faults controlling sedimentation at the time; note their trend in relation to sandstone dykes.

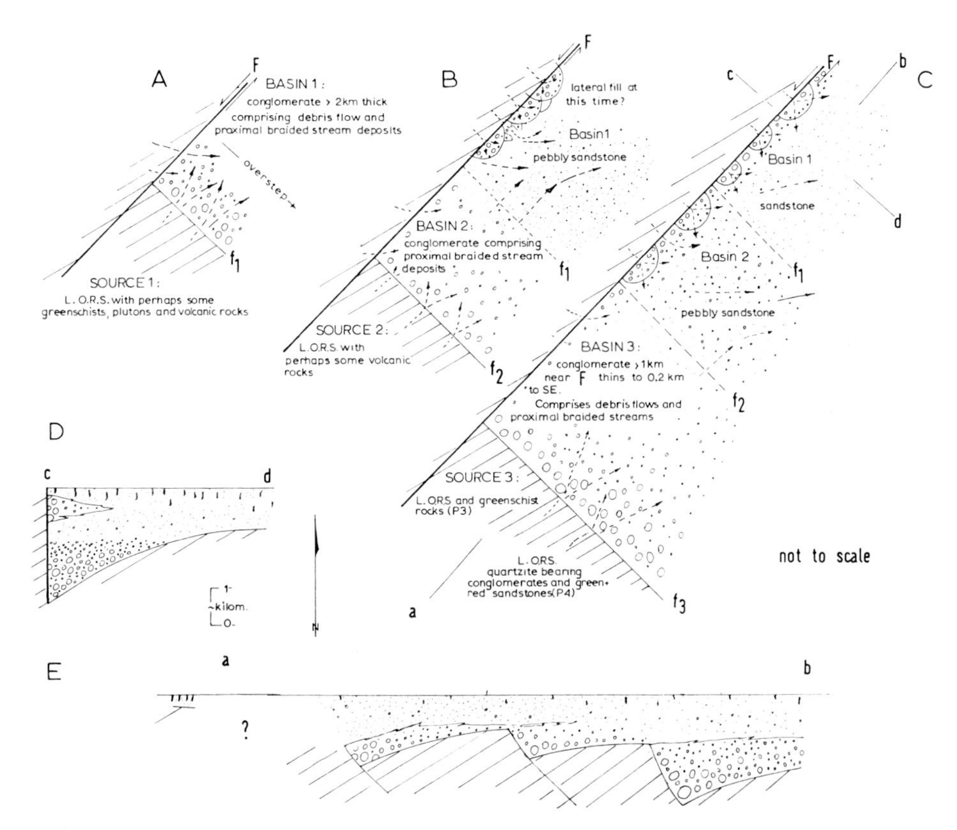

Fig. 8. Synoptic description of sedimentary and stratigraphic details of the Upper Old Red Sandstone rocks, together with a diagrammatic explanation of them. A, B, C shows progressive evolution of basins as the master fracture (F) (= Highland Boundary Fault) sinistrally moves to extend the basement on its SE side. This extension results in faults f_1, f_2, f_3 bounding three basins and giving rise to three sources within the general downthrown margin SE of F. In C, source 3 has two provenances P_3, P_4, referred to in Fig. 5. D and E are sections through C.

The sandstones are thickest in the northern Firth of Clyde, thinning both to the east and, partly by overlap, to the south. These sandstones are thought to be partial equivalents of the conglomerates but overlap them, clearly indicating an expansion in the basin of sedimentation (see Fig. 3). The dispersal pattern differs from that of the conglomerates in being more easterly (see Fig. 12 and Bluck 1978, Fig. 12). In Southern Kintyre the sandstones thicken to the SW, in the palaeocurrent direction.

Most of the sandstones associated with this division are sub-lithic arenites (Fig. 9). Conglomerates and pebbly sandstones are a minor component of this division and they contain many clasts of vein quartz, which often include patches of low grade metamorphic rock. This composition is fairly uniform in the Clyde and Kintyre areas, perhaps indicating the dominance of the Dalradian source. Highland Dalradian-type rock clasts are also found in the Stirling district (Read, in Francis *et al.*, 1970, p. 99).

The Breccia and Conglomerate division

A second major rudaceous division exists in the upper part of the Upper Old Red Sandstone sequence (Fig. 2). This division contrasts with the lower in being

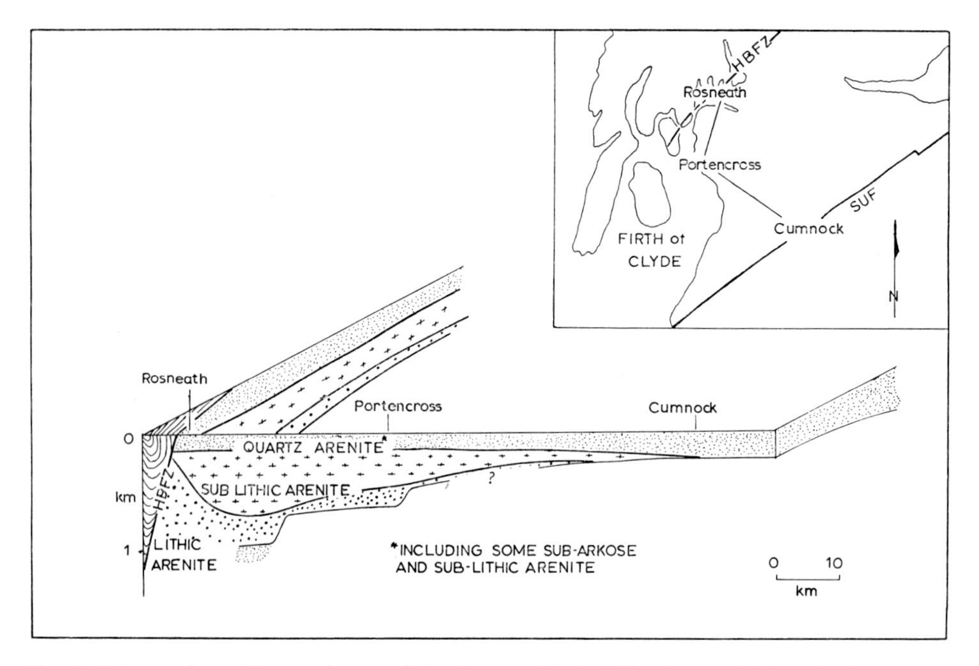

Fig. 9. Petrography of the sandstone of the Upper Old Red Sandstone divisions in the western Midland Valley.

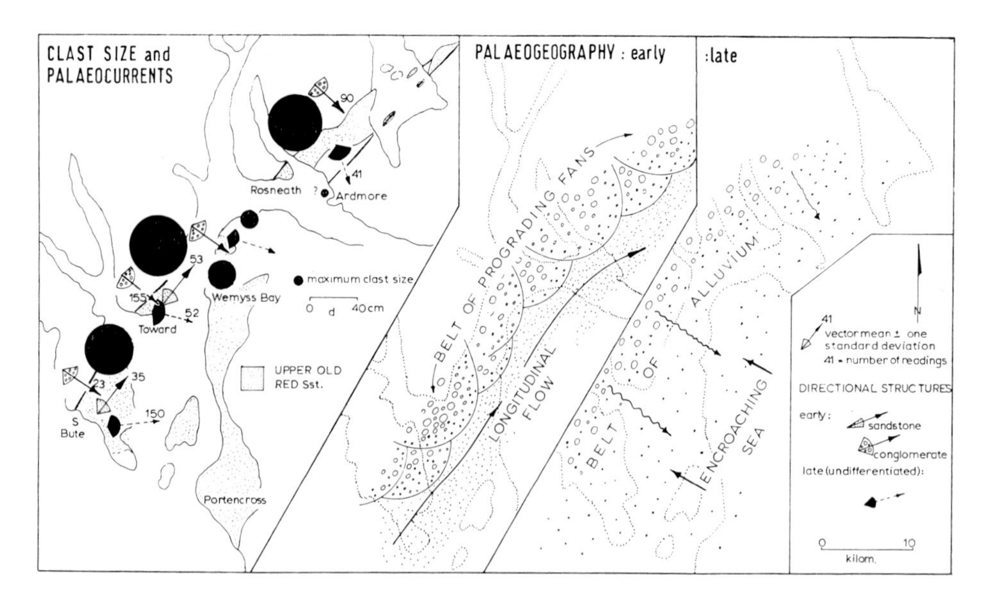

Fig. 10. Data for and palaeogeography of the breccia and conglomerate division. Maximum clast size data determined as in Fig. 5; d, clast diameter in centimetres.

restricted to the NW margin of the basin, i.e. along the Highland Boundary Fault; in comprising mainly angular greenschist clasts of local Dalradian type; and in having a SE dispersal. The conglomerate fines (Fig. 10) and appears to thin to the SE. It was laid down by debris flows and proximal braided streams which, in the region of the

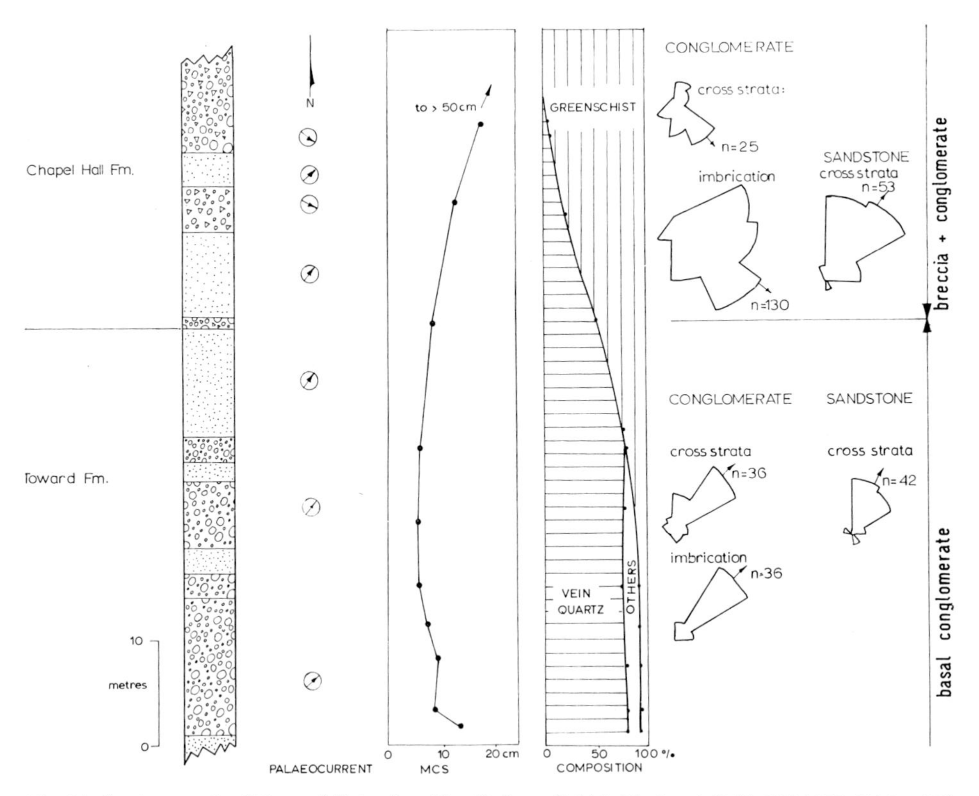

Fig. 11. Section north of Toward Point (see Fig. 5), from British National Grid: NS 1367 6716 to NS 1395 6843 (top). Arrow, vector mean of cross strata and for imbrication dip (+180°). MCS refers to maximum clast size, being the average of the ten largest clasts from 1 m^2. Rose diagrams: arrow, vector means, *n*, number of readings.

Highland Boundary Fault, were capable of transporting clasts over 1 m in diameter. These data suggest this division to be the product of a series of alluvial fans concentrated over and SE of the Highland Boundary Fault and having local drainage in the adjacent Dalradian rocks (Fig. 10).

The relationship between the upper breccia and the basal conglomerates is clearly seen in a critical section in Cowal, just north of Toward Point (Fig. 11). The section shows basal conglomerates with a uniform NE dispersal passing upward into a sequence where sandstones of a similar dispersal interstratify with breccias which are dispersed to the SE. The breccias coarsen upward, and the NE dispersed sandstones are replaced by ones which show the SE dispersal of the breccias (Fig. 11). The sequence is overlain by sandstone-with-calcrete.

The sandstone-with-calcrete division

The topmost division comprises an alternation of sandstone, conglomerate, argillite and calcretes. The conglomerates are usually thin and fine, often part of the coarse member of upward fining cycles. The calcretes, forming the top of the upward fining cycles, are sometimes quite thick (> 9 m) and some can be traced laterally for > 5 km. The calcretes sometimes show classic soil profiles (Burgess, 1960). The sandstone-with-calcrete division is a sheet-like unit over much of the Midland

Valley (Fig. 3) but in the northern region of the Firth of Clyde thins to the SW from ca 300 m north of Glasgow to a few tens of metres in Arran. In the southern part of the Midland Valley it overlaps the older divisions (Fig. 3) to rest on rocks of the Lower Old Red Sandstone.

The sandstones associated with this part of the sequence vary greatly. Some are thin sheet-like bodies associated with the upward fining units, others are thick, poorly cemented units typified by very large-scale cross-stratification which may be aeolian; yet others have burrows resembling *Skolithos* and sandstones of this kind have been interpreted by Chisholm & Dean (1974) as being marine. The latter type of sandstones, although present in the west of the Midland Valley, are more common in the east; and with the presence of marine Devonian in the central North Sea (Ziegler, 1975, p. 133) it seems likely that marine incursions into the Midland Valley Basin came in from this easterly direction.

Sandstone-with-calcrete is also found in Southern Kintyre where it occurs at the top of the sequence; but when traced northwards it oversteps the underlying sandstone and conglomerate divisions to rest directly on the Dalradian rocks at Galdrings (McCallien, 1927) (see Fig. 4 for locality). This northerly overstep of the sandstone-with-calcrete division confirms the NE source predicted by the palaeocurrents in Kintyre.

The conglomerates in this division are mainly quartz and quartzite bearing, but greywacke of the Southern Uplands appears in the southerly exposures. The associated sandstones are quartz arenites (Fig. 9) and this, together with the presence of calcrete in the thinner, more proximal parts of the succession, confirms the current view that calcretes mark periods of tectonic stability.

PALAEOGEOGRAPHY AND STRUCTURAL FRAMEWORK

The Upper Old Red Sandstone was deposited in two embayments: the Midland Valley Basin (stratigraphically a basin but palaeogeographically an embayment, see Bluck, 1978, p. 272) which opened to the east and northeast; and the Kintyre embayment, the shape of which is not known but which opened to the southwest. Both have palaeocurrent orientations parallel to the Highland Boundary Fault and to the other structures in the older rocks flanking the Midland Valley (see Figs 4 and 12).

The upland zone separating the Midland Valley and Kintyre embayments probably had a NNW–SSE orientation (Fig. 12) which corresponds to a prominent lineament in the Precambrian rocks to the north (Bowes, 1976, p. 369). The origin of the Upper Old Red Sandstone embayments may well be partly related to the intersection of this Precambrian lineament with the Highland Boundary Fault, causing the latter to be offset in this region. When the Highland Boundary Fault moved sinistrally the Clyde part of the Midland Valley, and possibly the Kintyre, embayments were formed.

During the deposition of subsequent rocks, the sediments in the fault basin were overstepped, so that a basin formed which covered the greater part of the Midland Valley, even into the Southern Uplands, and into the Highlands (see George, 1960). Evidently, by late Upper Old Red Sandstone time, the basin of deposition had greatly expanded and a more uniform lithofacies spread across it. But coarse proximal sediments persisted in the region of the Highland Boundary Fault with major fans

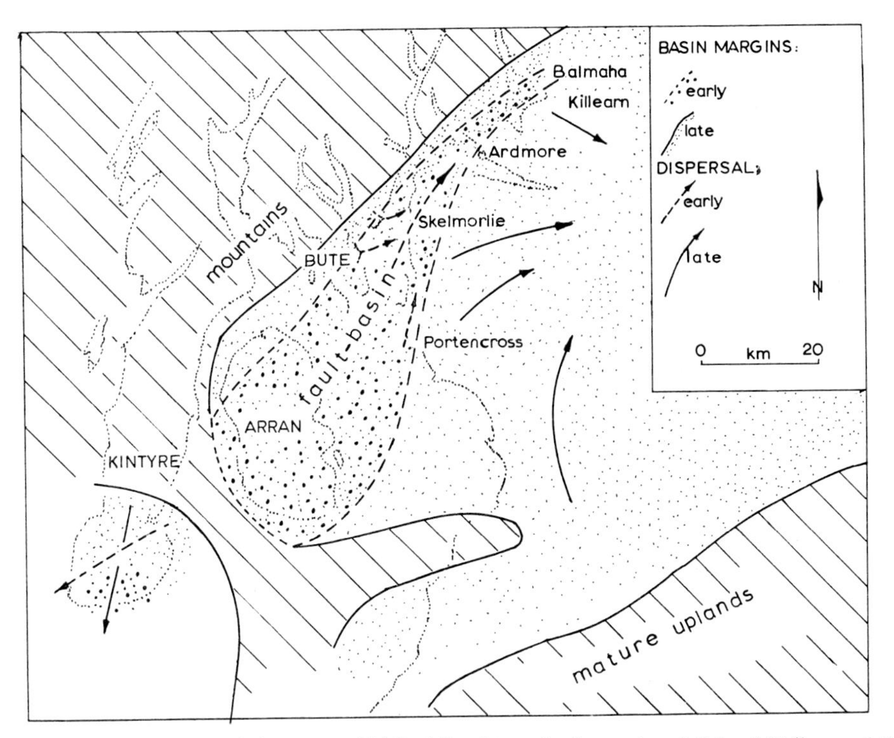

Fig. 12. Palaeogeography of the Upper Old Red Sandstone in the western Midland Valley, and the dispersal directions for the basal conglomerate (early) and the pebbly sandstone (late) divisions (see Bluck 1978, Fig. 12).

prograding towards the SE. This breccia development may be the direct result of uplift in the Highlands. Alternatively, it may result from a decrease in the volume of sediment reaching subsiding basins which, because of source recession, were becoming more distal through time (see Fig. 8). In the context of the whole Midland Valley basin, the sedimentation in the fault controlled Firth of Clyde area is seen as a local expression of a more regional tectonic event. Much of this later Upper Old Red Sandstone sediment is dispersed from the SW, and in a SW direction there is little room for a large source area (see Fig. 12). This problem of imbalance between source and basin may be partly explained by the diachronism in faulting and lithofacies (Fig. 13); partly by having a more extensive source area in the region of the southern Firth of Clyde; and partly by having an extensive source to the NW, but with the rivers turning to the east when entering the basin of deposition. The latter case has been seen to occur in the South Atlas Fault region (Bluck, 1978). Upward in the Upper Old Red Sandstone sequence palaeocurrent dip directions change from NE to E, and more Dalradian clasts appear in the rocks, suggesting the Southern Highlands to be a more potent source as time proceeds.

The early Upper Old Red Sandstone fault controlled basin contrasts with other basins for which a fault control origin has been proposed and for which there are sedimentary details available.

(1) The sequence is relatively thin, although it may be as much as 3–4 km thick at the very downstream end of the Clyde fault basin (i.e. near Loch Lomond; cf. Qureshi, 1970).

(2) The palaeocurrent orientations for both the coarse and fine sediments are almost the same. In other basins Füchtbauer (1967), Wilson (1971), Steel & Wilson (1975), and Steel (1976) record palaeocurrents in conglomerates almost perpendicular to those of the intertonguing finer sediments (see Fig. 13).

(3) The basal conglomerates, with some exceptions along the Highland Boundary Fault, record a simple upward fining sequence, accompanied by a change from mass-flow to braided stream deposition. In some cases the sequence is fine–coarse–fine, with the initial fine part of the sequence being very thin compared with the latter (see Fig. 13A). This type of sequence is postulated to have occurred when, after a period of rapid initial uplift, the fault block was worn down and back. The other kind of basin (Wilson, 1971; Steel & Wilson 1975, Fig. 2; Steel, 1976) may show more complex profiles (see Fig. 13B) due to fluctuations in fan and axial flood-plain growth. These fluctuations may be brought about by changing rates of uplift or the changing topography of a source which was laterally moving past some imaginary point in the basin.

(4) The whole sequence, including the sandstones, shows an upward maturing which reflects the upward change to stability. Within this regional upward change there are the more local changes in conglomerate units discussed in 3. Little is known about the changes in maturity in the laterally filling basins (Fig. 13B); they may have been initially variable, then finally became mature.

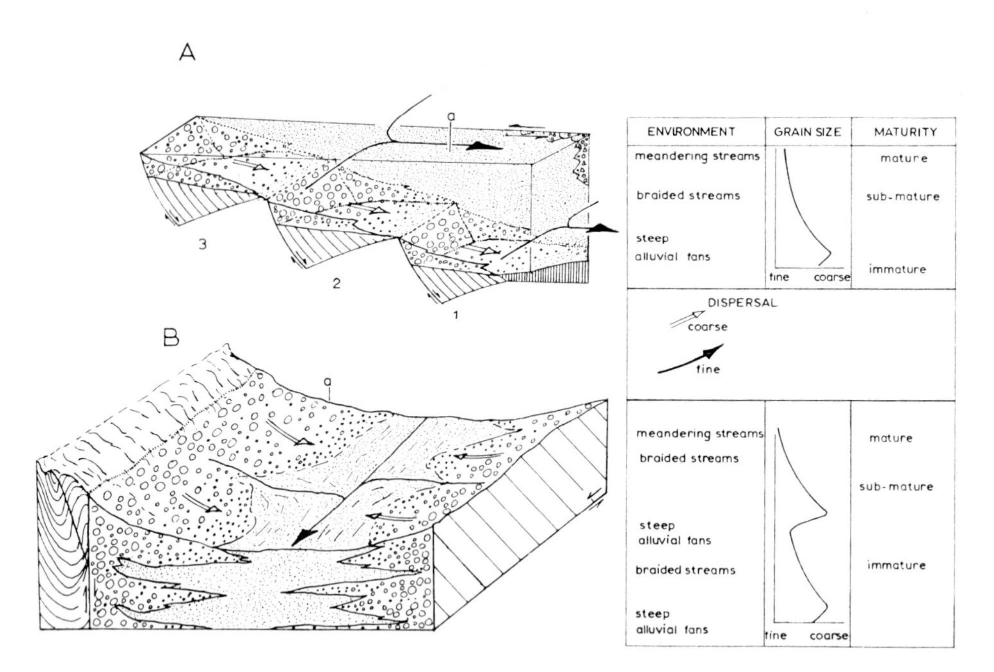

Fig. 13. Some important differences between two types of fault controlled basin. A, the basin, as postulated to occur in the Upper Old Red Sandstone of the Midland Valley, comprises a block-faulted basement caused by extension due to a sinistral shift on the main bounding fault at a. B, laterally or vertically shifting source area to a fault bound basin which has axial and lateral sediment dispersal. a, the section in which the vertical variations in environment, grain size and maturity are likely to occur.

CONCLUSIONS

The Upper Old Red Sandstone Midland Valley Basin differs from most late orogenic basins in having conglomerates and sandstones with roughly the same direction of dispersal. Other basins of this kind usually have a coarse lateral and a fine longitudinal fill. This characteristic, together with stratigraphic evidence for wedge-like conglomerates which young in the direction of source and evidence of fault activity contemporaneous with sedimentation all suggest alluvial deposition in a pull-apart basin.

The basin which developed in Firth of Clyde in particular formed as a consequence of a sinistral shift on the Highland Boundary Fault. This fault, with a long history of movement, may have been initially an Ordovician zone of thrusting, and a Lower Old Red Sandstone zone of wrench faulting.

The basin development in the Clyde is only a local expression of a wider structural event which encompasses most of Midland Scotland and the adjacent blocks. The sediment filled the basin so that a more extensive, more uniform lithofacies developed, which, because of the tectonic stability, was subject to a great deal of reworking and maturing. The basin sequence matures upward from lithic arenites to quartz arenites and from polymict conglomerates to quartz conglomerates. There is also evidence for a change in environment from steep-sloped alluvial fans to braided streams, meandering streams with laterally extensive calcretes, and eventually to a littoral environment.

ACKNOWLEDGMENTS

I thank Drs Harold Reading and Peter Friend for their reading of the manuscript and their perceptive comments.

REFERENCES

Armstrong, M. & Paterson, I.B. (1970) The Lower Old Red Sandstone of the Strathmore region *Rep. Inst. geol. Sci.* No. 70/12. 24 pp.

Bluck, B.J. (1967) Deposition of some Upper Old Red Sandstone conglomerates in the Clyde area: a study in the significance of bedding. *Scott. J. Geol.* **3,** 139–167.

Bluck, B.J. (1978) Sedimentation in a late orogenic basin: the Old Red Sandstone of the Midland Valley of Scotland. In: *Crustal Evolution in Northwest Britain and Adjacent Regions* (Ed. by D. R. Bowes and B. E. Leake). *Geol. J. Spec. Issue*, **10,** 249–278.

Bowes, D.R. (1976) Tectonics in the Baltic Shield area in the period 2000–1500 million years ago. *Acta. Geol. Pol.* **26,** 354–376.

Burgess, I.C. (1960) Fossil soils in the Upper Old Red Sandstone of South Ayrshire. *Trans. geol. Soc. Glasgow*, **24,** 138–153.

Crowell, J.C. (1974) Sedimentation along the San Andreas Fault, California. In: *Modern and Ancient Geosynclinal Sedimentation* (Ed. by R. H. Dott, Jr, and R. H. Shaver). *Spec. Publ. Soc. econ. Paleont. Miner., Tulsa,* **19,** 292–303.

Chisholm, J.I. & Dean, J.M. (1974) The Upper Old Red Sandstone of Fife and Kinross: a fluviatile sequence with evidence of marine incursion. *Scott. J. Geol.* **10,** 1–30.

Dewey, J.F. & Pankhurst, R.J. (1970) The evolution of the Scottish Caledonides in relation to the isotopic age pattern. *Trans. R. Soc. Edin.* **68,** 361–389.

DOWNIE, C., LISTER, T.R., HARRIS, A.L. & FETTES, D.J. (1971) A palynological investigation of the Dalradian rocks of Scotland. *Rep. Inst. geol. Sci.* No. 71/9. 30 pp.

FRANCIS, E.H., FORSYTH, I.H., READ, W.A. & ARMSTRONG, M. (1970) The geology of the Stirling District. *Mem. geol. Surv. U.K.*

FÜCHTBAUER, H. (1967) Die Sandsteine in der Molasse nördlich der Alpen. *Geol. Rdsch.* **56,** 266–300.

GEORGE, T.N. (1960) The stratigraphical evolution of the Midland Valley. *Trans. geol. Soc. Glasgow,* **24,** 32–107.

HARRIS, A.L., BALDWIN, C.T., BRADBURY, H.J., JOHNSON, H.D. & SMITH, R.A. (1978) Ensialic basin sedimentation: the Dalradian Supergroup. In: *Crustal Evolution in Northwest Britain and Adjacent Regions* (Ed. by D. R. Bowes and B. E. Leake). *Geol. J. Spec. Issue,* **10,** 115–138.

KINGMA, J.T. (1958) Possible origin of piercement structures, local unconformities, and secondary basins in the Eastern Geosyncline New Zealand. *N.Z. J. Geol. Geophys.* **1,** 269–274.

LONGMAN, C.D., BLUCK, B.J. & VAN BREEMEN, O. (1979) Ordovician conglomerates and evolution of Midland Valley. *Nature,* **280,** 578–581.

McCALLIEN, W.J. (1927) A preliminary account of the post-Dalradian geology of Kintyre. *Trans. geol. Soc. Glasgow.* **18,** 40–126.

McKERROW, W.S., LEGGETT, J.K. & EALES, M.H. (1977) Imbricate thrust model of the Southern Uplands of Scotland. *Nature,* **267,** 237–239.

McLEAN, A.C. & DEEGAN, C.E. (1978) A synthesis of the solid geology of the Firth of Clyde region. In: *The Solid Geology of the Clyde Sheet* (55 *N*/6 *W*) (Ed. by A. C. McLean and C. E. Deegan). *Rep. Inst. geol. Sci.* No. 78/9, 95–114.

PIDGEON, R.T. & AFTALION, M. (1978) Cogenetic and inherited zircon U–Pb system in granites: Palaeozoic granites of Scotland and England. In: *Crustal Evolution in Northwest Britain and Adjacent Regions* (Ed. by D. R. Bowes and B. E. Leake). *Geol. J. Spec. Issue,* **10,** 183–220.

QURESHI, I.R. (1970) A gravity survey in the region of the Highland Boundary Fault in Scotland. *Q. J. geol. Soc. Lond.* **125,** 481–502.

READ, W.A. & JOHNSON, S.H.H. (1967) The sedimentology of sandstone formations within the Upper Old Red Sandstone and lowest Calciferous Sandstone Measures west of Stirling, Scotland. *Scott. J. Geol.* **3,** 242–269.

STEEL, R.J. (1976) Devonian basins of western Norway: sedimentary response to tectonism and to varying tectonic context. *Tectonophysics,* **36,** 207–224.

STEEL, R.J. & WILSON, A.C. (1975) Sedimentation and tectonism (? Permo-Triassic) on the margin of the North Minch basin, Lewis. *J. geol. Soc. Lond.* **131,** 183–202.

WILCOX, R.E., HARDING, T.P. & SEELY, D.R. (1973) Basic wrench tectonics. *Bull. Am. Ass. Petrol. Geol.* **57,** 74–96.

WILSON, A.C. (1971) *Lower Devonian sedimentation in the north-west Midland Valley of Scotland.* Unpublished Ph.D. Thesis, University of Glasgow.

ZEIGLER, P.A. (1975) North Sea basin history in the tectonic framework of North Western Europe. In: *Petroleum and the Continental Shelf of North West Europe,* 1, *Geology* (Ed. by A. W. Woodland), pp. 165–187. Applied Science Publishers, Barking, Essex.

Spec. Publ. int. Ass. Sediment. (1980) **4**, 79–103

Late Caledonian (Devonian) basin formation, western Norway: signs of strike-slip tectonics during infilling

RON STEEL *and* TOR GUNNAR GLOPPEN*

Geological Institute (A), University of Bergen, Norway

ABSTRACT

Some of the Devonian, late orogenic basins of western Norway are likely to have developed along strike-slip faults generated in response to the latest stages of Iapetus Ocean closure and continental collision. General characteristics of these basins such as elongate shape, high sedimentation rates, extremely thick sedimentary piles compared to basin area, rapid lateral facies variations and scarce igneous/metamorphic activity are consistent with a hypothesis of strike-slip origin. More specific features such as mappable proximal onlap of successive basin-fill segments, significant basin asymmetry with regard to both sediment facies and sequence thickness and dominant longitudinal mode of basin infill are all indicative of such a mode of origin while mis-matches between fan materials and geology of adjacent drainage area, skewed fan-body geometry and the direction of onlap with respect to sediment transport all suggest a dextral sense of fault movement. Sedimentary facies draping the strike-slip basin margin (north) are conglomerates, mainly of sediment gravity flow origin deposited on small alluvial fans and fan deltas, which consistently interfinger and mix with floodbasin and lacustrine fines from the basin axis, fluvial/lacustrine delta system. Conglomerates draping the 'normal' basin margin (south) are of both mass flow and stream flow facies and formed on larger alluvial fans. The sedimentary pile is permeated by cyclothems of dominantly upward coarsening character (irrespective of facies) and, in places, demonstrably basin-wide. Sequences of the order of 5–20 m in thickness are thought to have been the response to largely vertical basin floor movements, while superimposed sequences some 100–200 m thick (the onlapping segments) are likely to have been deposited between intervals of mainly strike-slip movement. Data from one area, after some assumptions about basin segment geometry, suggest an average horizontal component of movement some two to three times greater than the vertical one. The difficulties of establishing a set of criteria which can be used to distinguish strike/oblique-slip generated sedimentation from rift sedimentation are discussed.

* Present address: Statoil, Stavanger, Norway.

0141-3600/80/0904-0079$02.00

INTRODUCTION

A series of small, late Caledonian (largely early–middle Devonian) basins is present along coastal Norway between Bergen and Trondheim (see summary by Nilsen, 1973). The best known group, lying between Sognefjord and Nordfjord (Fig. 1), are characterized by great stratigraphic thicknesses (up to 25 km) despite their small size (<2000 km^2), by alluvial/lacustrine infills and by a prominent basin-wide cyclicity of inferred tectonic origin (Steel, 1976; Steel *et al.*, 1977). It is suggested that some of these basins have developed in response to dextral (E–W) strike-slip faulting, which was generated in connection with Iapetus Ocean closure, continental collision and major sinistral translation along the Arctic–North Atlantic Caledonides. The latter movement appears to have involved crustal translation of major proportions (Harland, 1973; Morris, 1976; Phillips, Stillman & Murphy, 1976; Ziegler, 1978). Important elements of the system were the Great Glen Fault of Scotland and the Cabot Fault and related fault systems of New Brunswick and Newfoundland (Haworth, 1974; Morris, 1976). It has recently been argued that the Upper Old Red Sandstone of Midland Scotland accumulated in strike-slip basins (Bluck, 1978).

Steel (1976) previously suggested that the origin of Hornelen Basin in western Norway was related to strike-slip tectonics, but a documentation of the case is made here for the first time. It is suggested that Kvamshesten Basin, also mainly sand-filled, is of similar origin, whereas Solund and possibly Haasteinen Basins (conglomerate-filled) resulted from dominantly normal faulting (Fig. 1).

The main objects of the present contribution are: (1) to present and discuss criteria used to postulate a strike-slip origin for some of the Norwegian late Caledonian basins, and, (2) to describe briefly the sedimentation and sedimentary sequences in an attempt to identify at which level, if any, fault movements can be detected in the sedimentary pile, and whether there exist aspects of the sequences which are peculiarly a result of the lateral as opposed to merely the vertical components of basin floor movement.

The fault belts: evidence of alternating extension and contraction

The Devonian basins in western Norway all appear to have experienced some late or post-Devonian eastward thrusting, so that although their western ends rest unconformably on Cambro-Silurian greywacke, shale, basalt, gabbro and granodiorite, their eastern parts now lie on Precambrian quartzite, schist and gneiss (Fig. 1). The northern and southern margins of the basins are presently fault-bounded and there are also a number of important east–west faults in the intervening basement blocks (Bryhni, 1964). None of these faults is known to be of strike-slip character, but this would be difficult to demonstrate because Caledonian structural trends are also east–west in this region (Fig. 3). However, the present high-standing character of most of the basins (e.g. Fig. 2) and their position in the zone of convergence of some of the faults (Fig. 1) suggests tipped wedges due to converging strike-slip faults and associated thrusting. It can then be speculatively argued that there were phases of both transtension (terminology of Harland, 1971) (with consequent downwarp and sedimentation) and of transpression (contraction, thrusting, uplift) along some of these fault belts. This is a characteristic situation along strike-slip fault zones with braided or curved fault

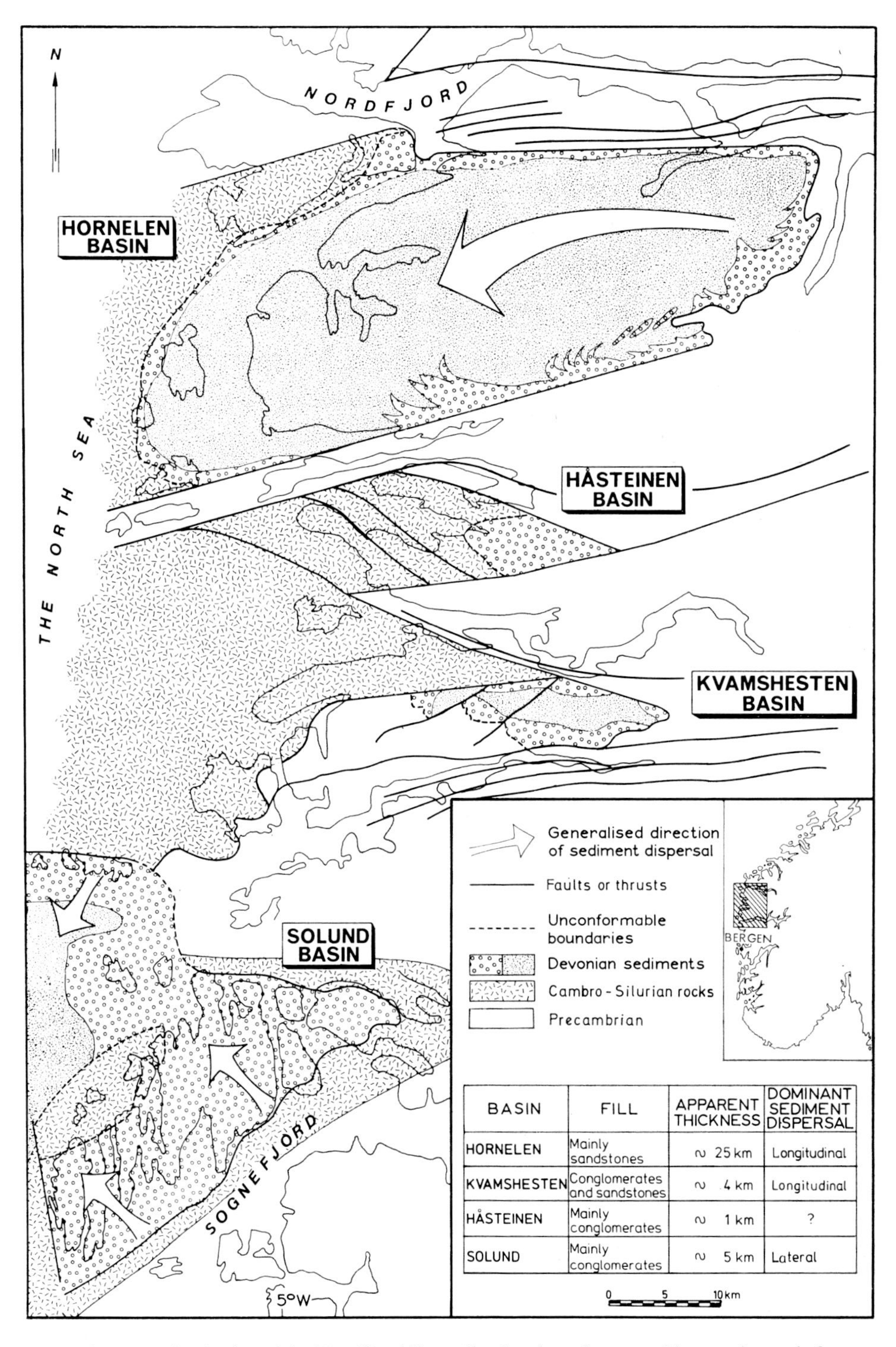

BASIN	FILL	APPARENT THICKNESS	DOMINANT SEDIMENT DISPERSAL
HORNELEN	Mainly sandstones	~ 25 km	Longitudinal
KVAMSHESTEN	Conglomerates and sandstones	~ 4 km	Longitudinal
HÅSTEINEN	Mainly conglomerates	~ 1 km	?
SOLUND	Mainly conglomerates	~ 5 km	Lateral

Fig. 1. The Devonian basins of the Nordfjord/Sognefjord region of western Norway (tectonic features after Bryhni, 1964; Kildal, 1970; Ramberg *et al.*, 1977; Solund palaeocurrents after Nilsen, 1968). Note the contrasts in sediment thickness, grain size and sediment dispersal direction between Hornelen and Solund Basins.

Fig. 2. Oblique aerial view, northwestwards, across Kvamshesten Basin. This high-standing basin is thrust along its eastern edges and appears to lie in a zone of convergence of major faults (see Fig. 1).

Fig. 3. View along the faulted, northern margin of Hornelen Basin. The marginal belt of conglomerates (diagonally across the photograph) is tilted to near vertical while the sediments along the basin axis are dipping more gently eastwards. There are several subparallel faults in the basement (near distance left) which run parallel to the Caledonian 'grain' in the area (see map in Fig. 11B).

Fig. 4. Aerial view of Hornelen Basin. The basin is bounded by faults and thrusts (north, south and east) and by an unconformity (western islands). The 25 km thick succession, generally dipping gently eastwards, is organized into prominent, upward coarsening sequences standing out now as step-like topography. The basin is some 70 km × 20 km.

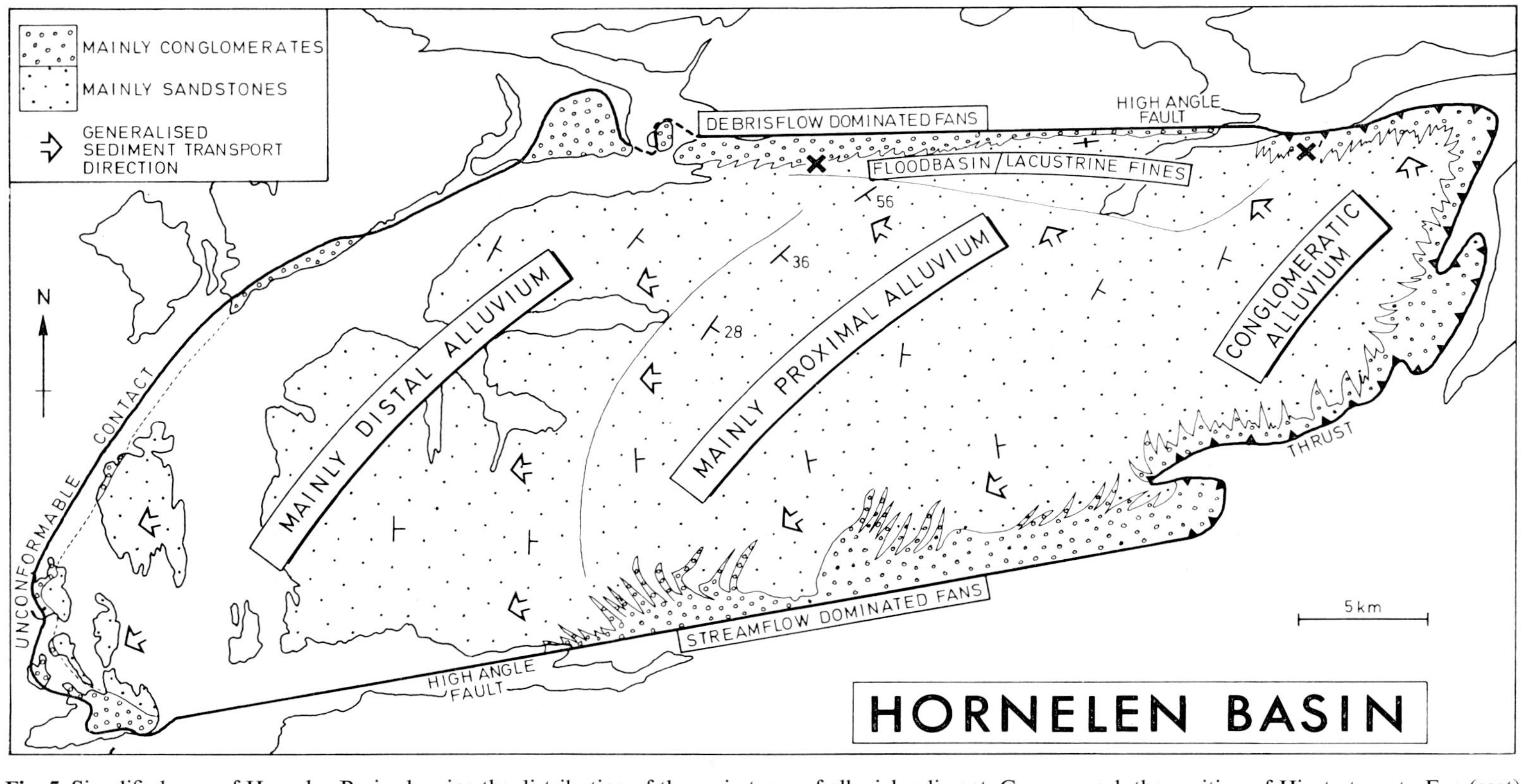

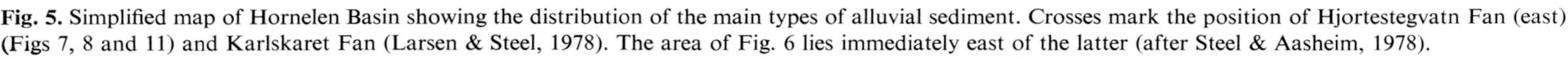

Fig. 5. Simplified map of Hornelen Basin showing the distribution of the main types of alluvial sediment. Crosses mark the position of Hjortestegvatn Fan (east) (Figs 7, 8 and 11) and Karlskaret Fan (Larsen & Steel, 1978). The area of Fig. 6 lies immediately east of the latter (after Steel & Aasheim, 1978).

traces (Crowell, 1974.) There are a number of other features which are consistent with looking at both the Devonian deposition and the late Devonian deformation in terms of early extensional and late contractile phases along the same strike-slip system.

(1) The present synclinal form of both Hornelen and Kvamshesten Basins (near-vertical upturned, northern limbs (Fig. 3)) can be accounted for by crustal contraction as adjacent blocks converged during strike-slip motion. In addition to the prominent thrusting at the eastern margins, there has probably also been local thrusting of Devonian sedimentary rocks out across the northern margins (e.g. Fig. 3).

(2) The dominant westward direction of sediment transport (Fig. 1) together with the eastward migrating locus of sedimentation (and therefore eastward migrating area of uplift) suggest dextral slip along the northern marginal fault (Steel *et al.*, 1977) during the early extensional phase. The same sense of strike-slip movement, though now with a component of contraction across the blocks, can be deduced for the later history, resulting in uplift and thrusting of the basins.

PATTERNS OF SEDIMENTATION IN HORNELEN BASIN

Prior to a discussion of the criteria by which strike-slip faulting is recognized to have been important *during* the infilling of some of these basins, some of the general aspects of sedimentation and of sedimentary sequences are outlined here (see Steel *et al.*, 1977; Steel & Aasheim, 1978 for details). As noted by Reading (this volume), there may be little direct evidence of the strike-slip component of movements in strike-slip basin successions; indications of extension and normal components of faulting appear to dominate. This is also the case here, although we suspect that certain aspects of the all-pervading cyclicity (Fig. 4), particularly its internal configuration and geometry on the 100–200 m scale, point to intermittent strike-slip.

The vertical component of movement on the floor of a small basin will often be recognized after an analysis of the resultant vertical sedimentary succession. In Hornelen Basin the key to repeated vertical movement is a repeated, basinwide, upward coarsening motif; a record of widespread retreat (base level drop) and subsequent progradation (response) of the alluvial and lacustrine systems. Any additional horizontal component of movement, however, will be recognized rather in the geometric relationship of one increment of infill (sequence) to the next. Offset stacking of sedimentary bodies may well be hinted at in the time trend of a vertical sequence but will usually be ambiguous without supportive three-dimensional control.

The discussion below centres on Hornelen Basin because its history of development is best documented, but it is emphasized that Kvamshesten Basin (Bryhni & Skjerlie, 1975) bears a striking resemblance to Hornelen and is likely to have had a similar origin. Some of the contrasts between these two basins and the conglomeratic Solund Basin have been described by Steel (1976).

Infill across the strike-slip margin

The northern fault margin of Hornelen Basin is draped by a belt of conglomerate which tongues out within several kilometres, largely into fine-grained sediments of floodbasin and lacustrine origin (Figs 5 and 6). Mapping of the well exposed conglomerate succession has demonstrated that it is cyclically organized into a series

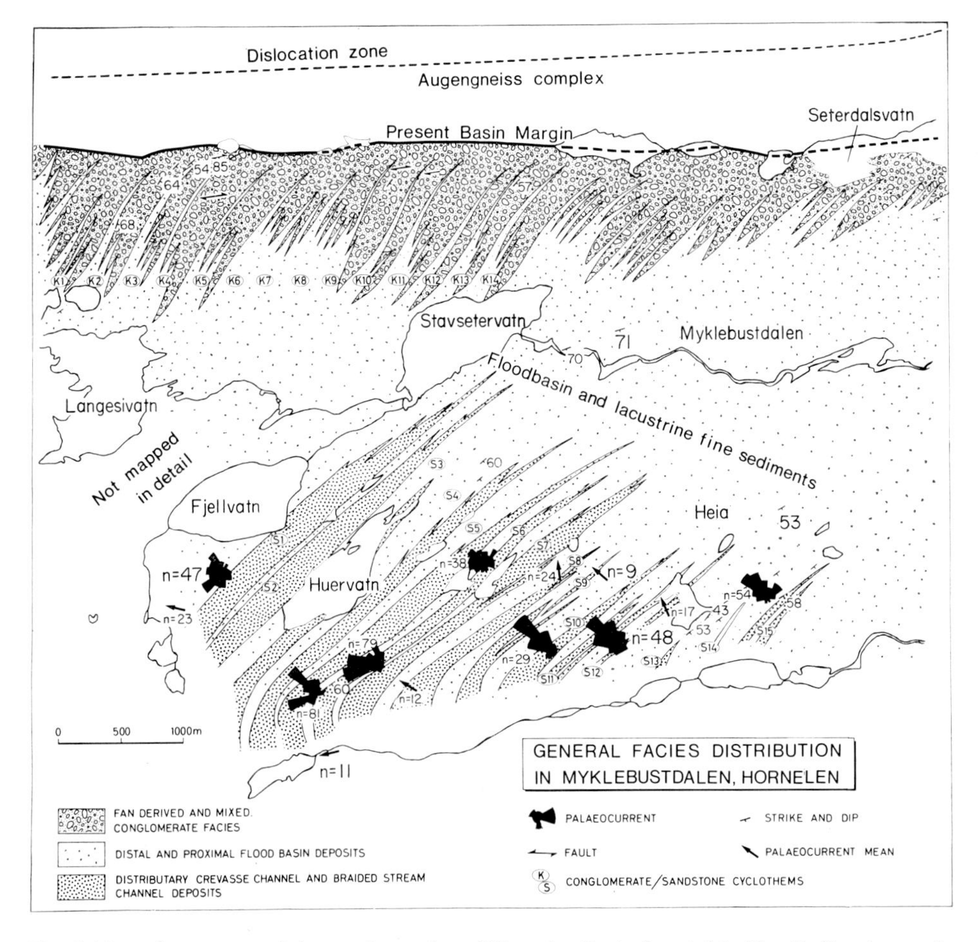

Fig. 6. Map of a segment of the northern edge of Hornelen Basin (located in Fig. 5). Conglomeratic fans interfinger with a belt of floodbasin/lacustrine fines which in turn pass southwards into the coarser, fluvial sediments of the axial region (from V. Larsen, unpublished data).

of upward coarsening and thickening wedges of alluvial fan (when interfingering with subaerial floodbasin/floodplain deposits) or fan delta (interfingering with lacustrine deposits) origin. The segment of the northern margin shown in Fig. 6 contains more than twenty-five such fans. In general, the fans along this margin are dominated by conglomeratic and sandy sediment gravity flows. Details of such deposits together with their organization and relationship to the impinging floodbasins have been documented by Larsen & Steel (1978). Fan deltas have been recognized by identification of beach gravels in the fan sequences (Gloppen, 1978), by the presence of debris flows with anomalous (high) bed thickness/maximum particle size ratios or matrix content across the lower fan reaches, and by conglomerates showing textural inversion due to rigorous mixing, slumping and sliding into the adjacent lacustrine fines. Internal details of an alluvial fan which interfingers with floodbasin sequences along this margin, are shown in Fig. 7. The subaerial nature of this fan (Hjortestegvatnet) is suggested by the dominance of streamflow and sheetflood conglomerates around the toe reaches, in contrast to the sheetlike, graded, subaqueous sediment gravity flows on the toe of

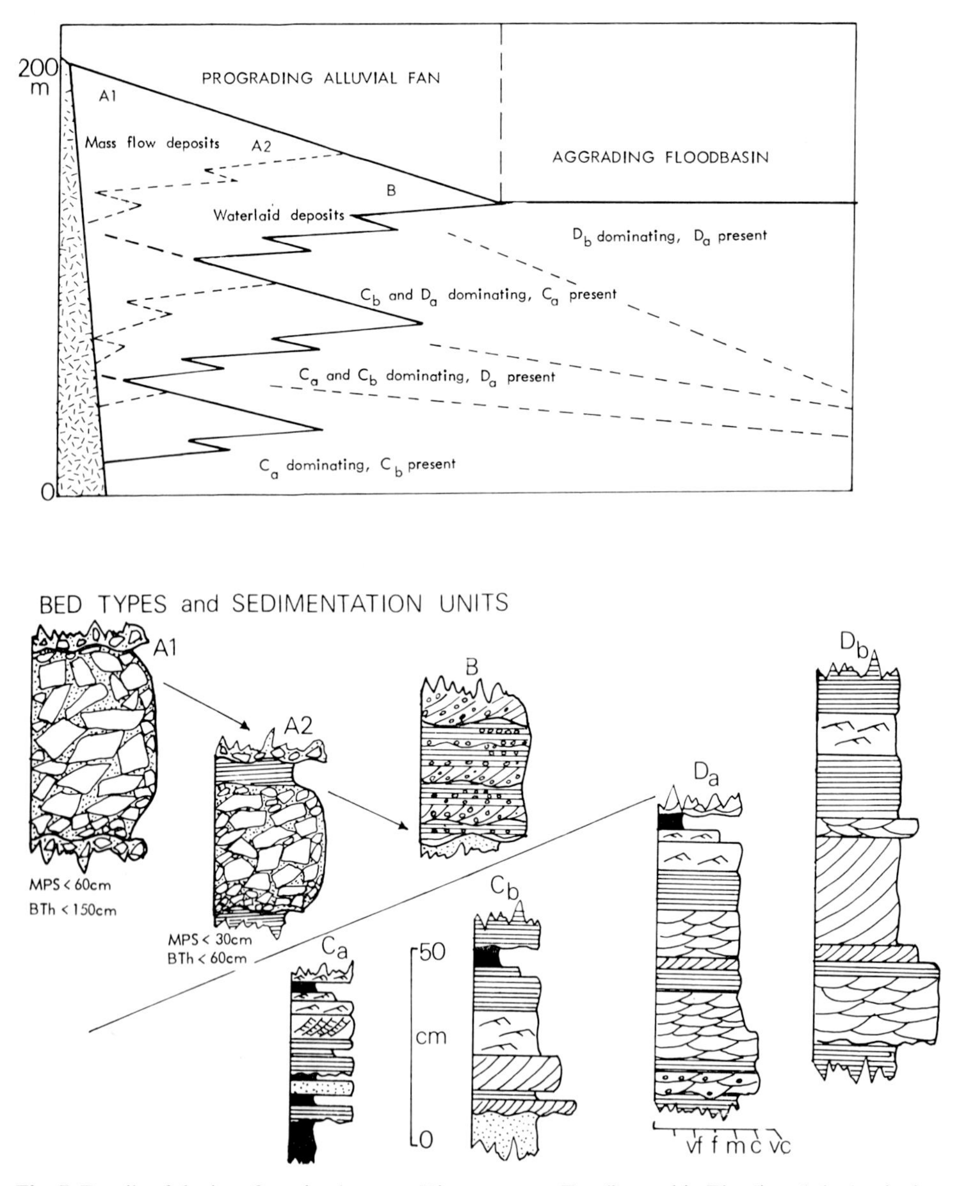

Fig. 7. Details of the interfingering between Hjortestegvatn Fan (located in Fig. 5) and the impinging floodbasin deposits, together with a summary of bed types and sedimentation units (after Gloppen, 1978).

the fan delta described by Larsen & Steel (1978). When mapped, it is clear that the upwards coarsening in the Hjortestegvatnet conglomerate body (~200 m thick) is exactly laterally equivalent to a corresponding upwards coarsening in the coarse sandstone of the axial fluvial system farther south (Fig. 8), despite the fact that the conglomerate and sandstone derive from different dispersal systems. In addition a belt of intervening mudstone, siltstone and fine sandstone (floodbasin) wedge out northward, southward and upward (Fig. 8). It is evident that the upward coarsening motif is basinwide (independent of facies), i.e. an important argument for its tectonic origin

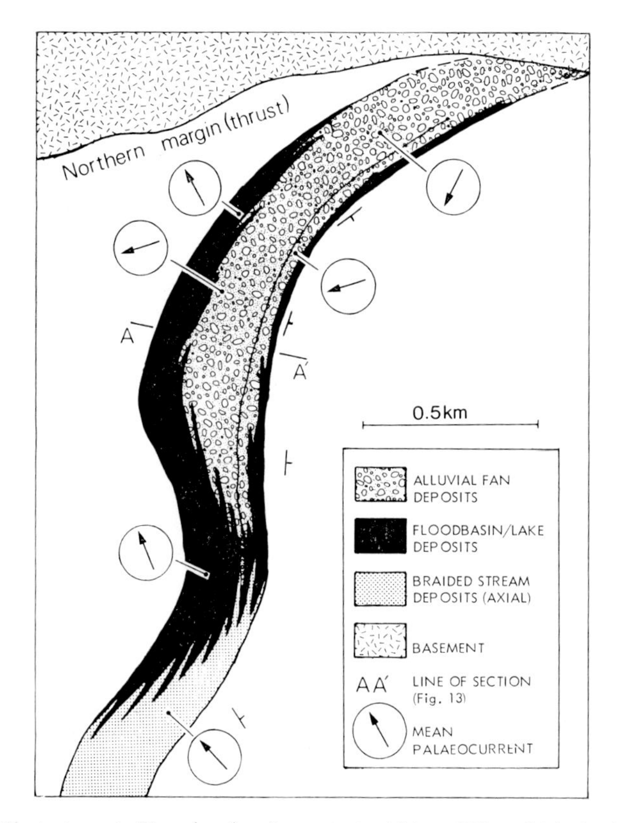

Fig. 8. Map of Hjortestegvatn Fan showing its context within a 200 m thick, basinwide cyclothem. Smaller scale, upward coarsening sequences can be mapped within this major unit. (After Gloppen, 1978). AA′ is line of section for Fig. 13.

see Steel *et al.*, 1977). A single phase of basin floor subsidence produced simultaneous progradation of the marginal fans and progradation (westward) of the axial fluvial system. This pattern can be identified along most of the northern margin, although in places the floodbasin/lacustrine sequence is mappable as a continuous belt (Fig. 6) which separates the coarser axial sandstones from the marginal fanglomerates.

The marginal fanglomerate bodies are further organized into a series of smaller scale (10–25 m) upward coarsening sequences (Figs 8 and 13). These are likely to reflect fan lobe progradation and shifting in the first instance, but may well have been ultimately controlled by fault movements along the margin. Tectonic control also at this level of organization is suggested by the uniformity of these cycles (see also Steel *et al.*, 1977, their Fig. 5), by the fact that lacustrine or floodbasin fine-grained sediments also interfinger at this level (Figs 7 and 8) and, most important, cycles of a similar magnitude are prominent in the sandy axial system (i.e. it is likely that these cycles are also extremely laterally persistent) (see also Fig. 8).

It has previously been argued (Steel *et al.*, 1977) that the small-scale sequences were the most immediate response to basin floor subsidence (i.e. primarily vertical movement) against the northern fault, while the large-scale cycles (or more precisely the boundaries between such cycles) reflect intervals of dominantly lateral movement of basin floor with respect to source areas. It is suggested that the upward fining

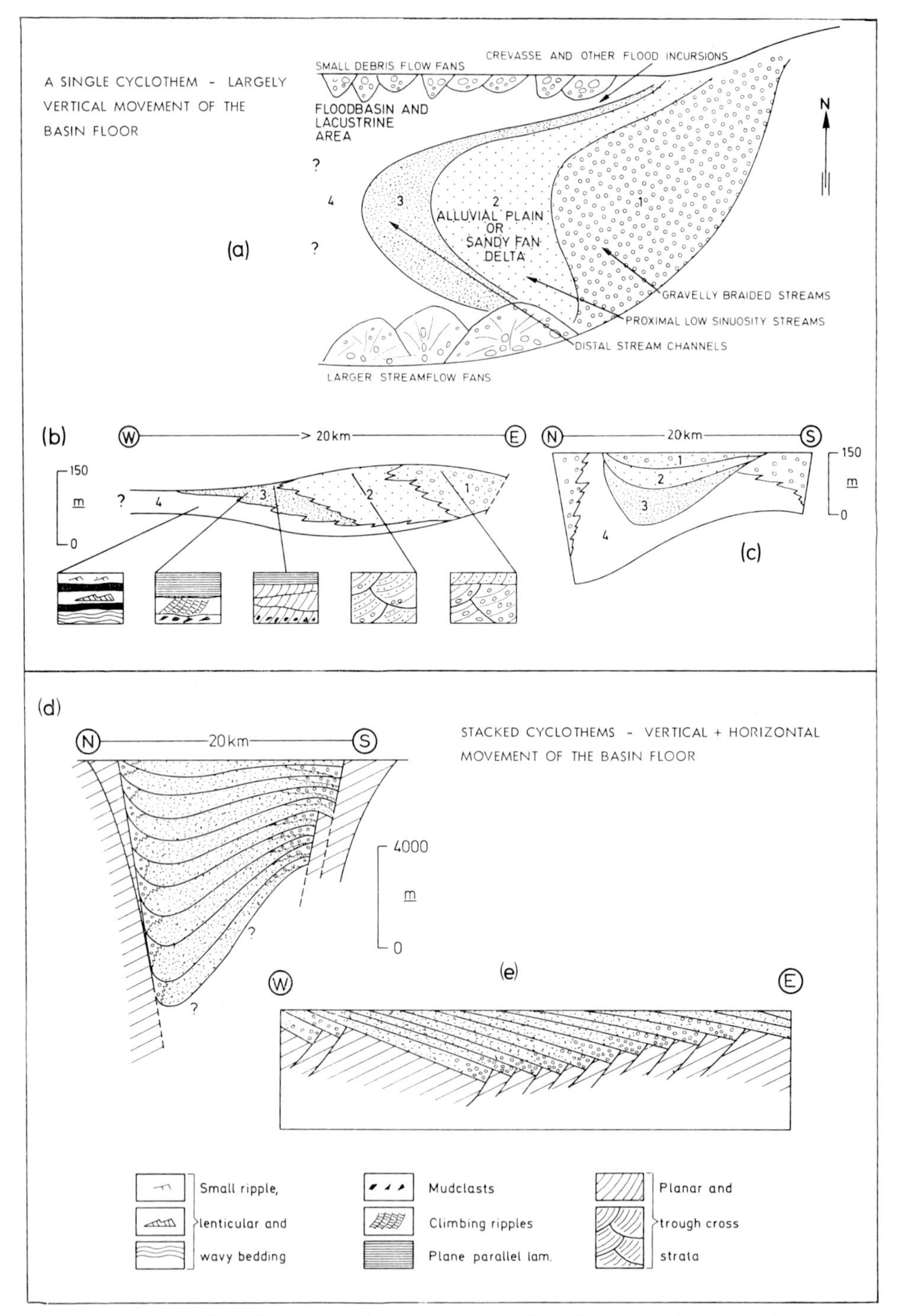

Fig. 9. Summary of the palaeogeography (a), and internal configuration of individual cyclothems (b and c), together with simplified north-south and east-west sections (d and e) showing basin asymmetry, cyclic organization of the infill and onlap of basement by the Devonian succession. The palaeogeography is necessarily simplified and most typical of the period represented by the uppermost two-thirds of the basin succession.

segment sometimes capping the large sequences (e.g. the upper 60 min the fan sequence Fig. 13) may well reflect a skewness in the uppermost portion of the fan body which in turn can be interpreted in terms of a gradual lateral translation of the basin edge (as discussed below). It is worth noting that this upward fining segment is also organized into smaller, upward coarsening units.

Infill across the 'normal' margin

The southern margin of Hornelen Basin also had a series of fans banked against it (Figs 5 and 9), although they often differ considerably from those of the northern margin. The fans here are of larger radius, have a higher component of streamflow (e.g. cross-stratified, clast-supported) conglomerate than those in the north, commonly have trough cross-stratified, sandy braided stream deposits on their lower reaches rather than a fine-grained lacustrine drape, and are enveloped by cycles commonly less than half the thickness of those in the north (see Steel, 1976 for details). All of these features point to less rapid subsidence along the southern margin, lower slopes, more fluvial processes on the fans, greater sediment yield from larger drainage areas and the likelihood that the fans here were sometimes important contributors of material to the axial fluvial system.

Sedimentation along the basin axis

The axial system of Hornelen Basin accounts for up to 90% of sediment in the basin, although this system may not be independent of the fans along the southern margin in that the southern and eastern edges of the basin were continuous (Fig. 9a). The axial system is dominated by fluvial deposits and was earlier referred to as an alluvial plain system (Steel, 1976) in which fluvial conglomerate was deposited proximally, sandy braided stream deposits in the middle reaches, and finer sediments distally. It was a basin of inland drainage in which sediments became finer grained, sedimentary structures changed significantly and sandstone bodies thinned out in a westward direction (see Steel *et al.*, 1977, for details). Later work on the finest-grained sediments showed that these are typically organized into flood-generated units of sedimentation (each typically less than 20 cm thick), which were deposited in lacustrine or floodbasin belts (Fig. 6).

The difficulty in always making a clear distinction between 'lacustrine' and 'flood basin' facies arises because (1) the area of fine-grained sedimentation was clearly dominated by flood-influx rather than sedimentation from suspension; and (2) the identifiably subaqueous units often apparently accumulated in water shallow enough to dry up, as shown when the flood units are organized into upward coarsening (shoaling), lake-fill sequences (<5 m). Those flood units showing clear evidence of having accumulated in standing water are typically sheet-like, massive or graded (often inversely at the base) sandstone beds, and are separated from each other by mudstone conspicuously lacking in mudcracks. The variety of subaqueous flows has been described by Larsen & Steel (1977) and Steel & Aasheim (1978).

A curious feature of these fine-grained successions is their great persistence in time (thickness) within tracts of very restricted width; for example, a vertical succession of more than 10 km of such deposits is sandwiched in a thin belt between debris flow fans of the northern margin and coarser sandstone of the axial fluvial deposits (Fig. 5). We now consider that the axial system is more properly designated as a lacustrine fan delta system. This is consistent with the lobate shape of the axial sand bodies (the marginal fine-grained sediments accumulated between these bodies and the fans; Fig. 6), the uniform but relatively rapid decrease in grain size westward (from conglomerates to mudstone in ~20 km), and the prominent prograding tendency to which the axial system was subject (as demonstrated by the upwards coarsening on individual sandstone bodies; Fig. 4). Details of processes and subenvironments within this axial fan delta system have been documented by Steel & Aasheim (1978).

As already noted, a repetition of alluvial cyclothems is especially obvious along the axial zone of Hornelen Basin (Fig. 4). The geometry and internal make-up of individual bodies together with a suggestion as to how some 200 of these have been stacked in the basin is shown in Fig. 9. Those aspects of the cyclicity relevant to the question of strike-slip faulting during sedimentation are discussed further below.

EVIDENCE OF SYN-DEPOSITIONAL STRIKE-SLIP MOVEMENT

Some of the general characteristics of strike-slip basins, for example high sedimentation rates, thick sedimentary piles, rapid lateral facies variations and scarce igneous/metamorphic activity, have been discussed by Crowell (1974) and Mitchell & Reading (1978). Examination of Hornelen Basin has provided a number of other criteria which may be additionally helpful in identifying ancient strike-slip basins. These fall into two categories: the first group relates to basin scale and can be applied only if a large part of the original basin can still be identified; the other can be applied to small portions of the succession.

Basin scale criteria

Migrating locus of sedimentation

A unique aspect of some basins which develop during strike-slip fault movement is the tendency for the depocentre to migrate parallel with the fault through time. This is particularly the case when movement occurs against a restrictive double bend, causing a lateral movement of both the area of relief and the area of downwarp (Crowell, 1974). This situation can be identified both directly and indirectly in Hornelen Basin.

(a) There is a mappable, progressive overlap of specific facies from cyclothem to cyclothem in an eastwards direction. This is seen most clearly in the Grøndalen area where successively younger fanglomerate bodies, dispersed from the southeastern edge of the basin, can be seen to migrate eastwards in discrete steps (Fig. 10). The setting of each fanglomerate body in Fig. 10 is such that it grades northeastwards into sandy braided alluvium of the (westwards dispersed) axial system and westwards into fine floodbasin deposits. The implicit eastwards onlap of each cyclothem onto basement, speculatively suggested across step faults in Fig. 9e, cannot be observed because of later tilting and faulting around the basin edges. The overlap in Grøndalen was first recognized by Bryhni (1964) who also realized its significance with respect to basin migration although he did not elaborate on a tectonic mechanism.

(b) Indirectly, depocentre migration is suggested by the 25 km stratigraphic

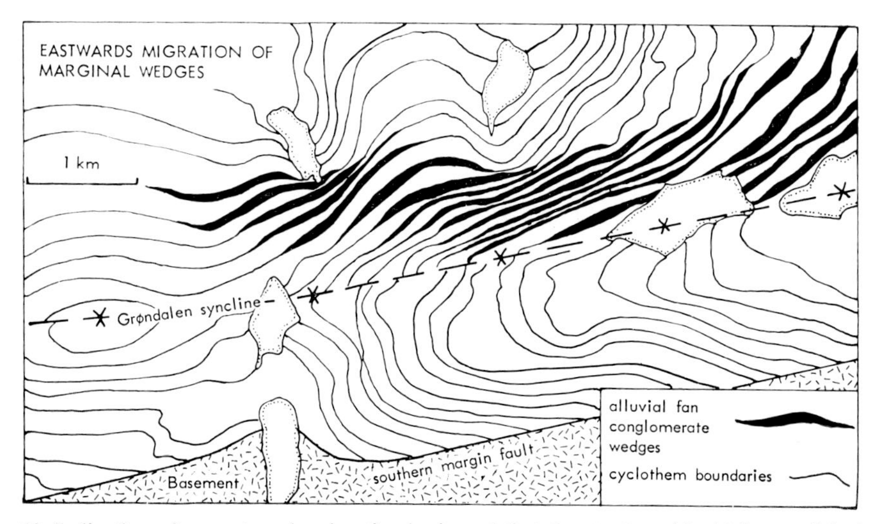

Fig. 10. Indication of an eastwards migrating basin and that the stratigraphic thickness of the basin infill is much greater than the basin depth. Systematic eastwards migration of successive alluvial fan conglomeratic wedges in an area near the southeastern margin. These conglomerates also crop out along parts of the area south of the syncline but have been omitted there for clarity.

thickness of the basin succession, suspiciously high considering that the degree of metamorphism at the base of the pile is no greater than that at the top. A simple but important consequence of depocentre migration is the production of stratigraphic thicknesses much greater than the true basin depth at any point. In the model discussed below it is suggested that the true vertical depth of Hornelen Basin probably does not exceed 8 km. This type of argument has also been used by Crowell (1975) to explain the 11 km Mio-Pliocene succession in the Ridge Basin along a palaeo-strand of the San Andreas Fault in southern California.

Basin asymmetry

The fault zones along which strike-slip basins develop commonly have a greater degree of movement along one edge than along the other. This leads to an asymmetry, with the basin floor having an important component of slope toward the major edge. In addition to the obvious asymmetry in the *thickness* of the basin pile which will result, there will be an important *facies* asymmetry. In Hornelen Basin the thickness of strata along the northern edge is commonly twice that along the southern edge (Fig. 9d), but more important, the slope of the basin floor is also responsible for the differing fan geometries and facies on opposite margins (Fig. 9a and d). Small, mass flow dominated fans are particularly important along the deeper edge. The creation and persistence of the belt of fine-grained lacustrine and floodbasin deposits draping the northerly fanglomerates is another consequence of this gradient (Fig. 9c). The lacustrine facies is characteristic of present-day strike-slip fault zones on land (Reading, 1980). In ancient strike-slip basins the persistence of this facies in time despite its restricted lateral extent is often most impressive. In the Pliocene Ridge Basin of southern California this facies is some 9000 m thick, though it rarely attains a width greater than 3 km (Link & Osborne, 1978), while in Hornelen Basin more than 10 000 m of lacustrine and floodbasin deposits are largely confined to a tract less than 4 km wide

(Figs 5, 6 and 8). An additional consequence of this consistent lapping of fine-grained sediment around the steep fans bordering the most unstable side of the basin is the production of a diamictite facies (caused by mixing and textural inversion) on the feather edge between the fans and the fine-grained belt (Larsen & Steel, 1978).

Dominance of longitudinal fill

Because of the lateral persistence and the width of strike-slip fault zones it could be expected that important drainage systems often develop parallel to them. In addition, because both uplift and subsidence are commonly created along different segments of the same zone due, for example, to curvature or to braiding of the fault pattern, a dominance of longitudinal drainage will be encouraged. This is conspicuously the case in Hornelen Basin where the lateral infilling appears to have been minimal. There is a real contrast between Hornelen and Solund basins in this sense, as discussed elsewhere (Steel, 1976).

Criteria relating to the strike-slip margin

Pebble mis-matches

This criterion, a mis-match between clast types in the conglomerate draping the basin margin and the adjacent 'basement' rocks, may often be most telling (e.g. Crowell, 1952), particularly if it can be used together with criterion 1. If the marginal alluvial fan bodies can be mapped and their radius and area estimated, then crude estimates of the drainage areas can be calculated and plotted on the adjacent basement. A simple check can then be made as to whether there are pebble types which cannot easily be matched with the geology of the drainage area. In some cases the true source area of distinctive pebble types may be recognized and the actual amount of lateral displacement calculated. More often, especially where a number of possible sources exist or where the structural grain of the adjacent basement parallels the basin margin, it can only be stated that some lateral displacement has occurred. An important limitation of this type of evidence is that it may not be possible to distinguish between lateral movement resulting from later tectonics and that from syn-depositional movements.

A number of lithological discrepancies are present along the northern margin of Hornelen Basin (Fig. 11A). Fanglomerate near the western end commonly contains considerable amounts of augengneiss, a rock type which presently crops out farther east in the basement. Fanglomerate at the eastern end of this margin is rich in gabbro clasts whose source cannot now be identified at all. The method is illustrated in more detail with reference to Hjortestegvatn Fan (already discussed in the previous section) in Fig. 11B. The mapped fan has a minimum radius of 1·5 km assuming the fan shape was semicircular. A range of calculated fan areas (A_f), in the case of the fan having been more elongate, are given in Table 1 together with the corresponding drainage basin areas (A_d). The latter were calculated from the empirical relationship $A_f = cA_d^n$ (Bull, 1964; Denny, 1965). Values of $c(= 0{\cdot}15)$ and $n(= 0{\cdot}9)$ were used from fans along the eastern side of Death Valley (Fig. 12a) (Hooke, 1972), because they are thought to be similar in many respects (size, processes, climate, tectonics) to Hjortestegvatn Fan. The maximum and minimum drainage area values from Table 1 are plotted on

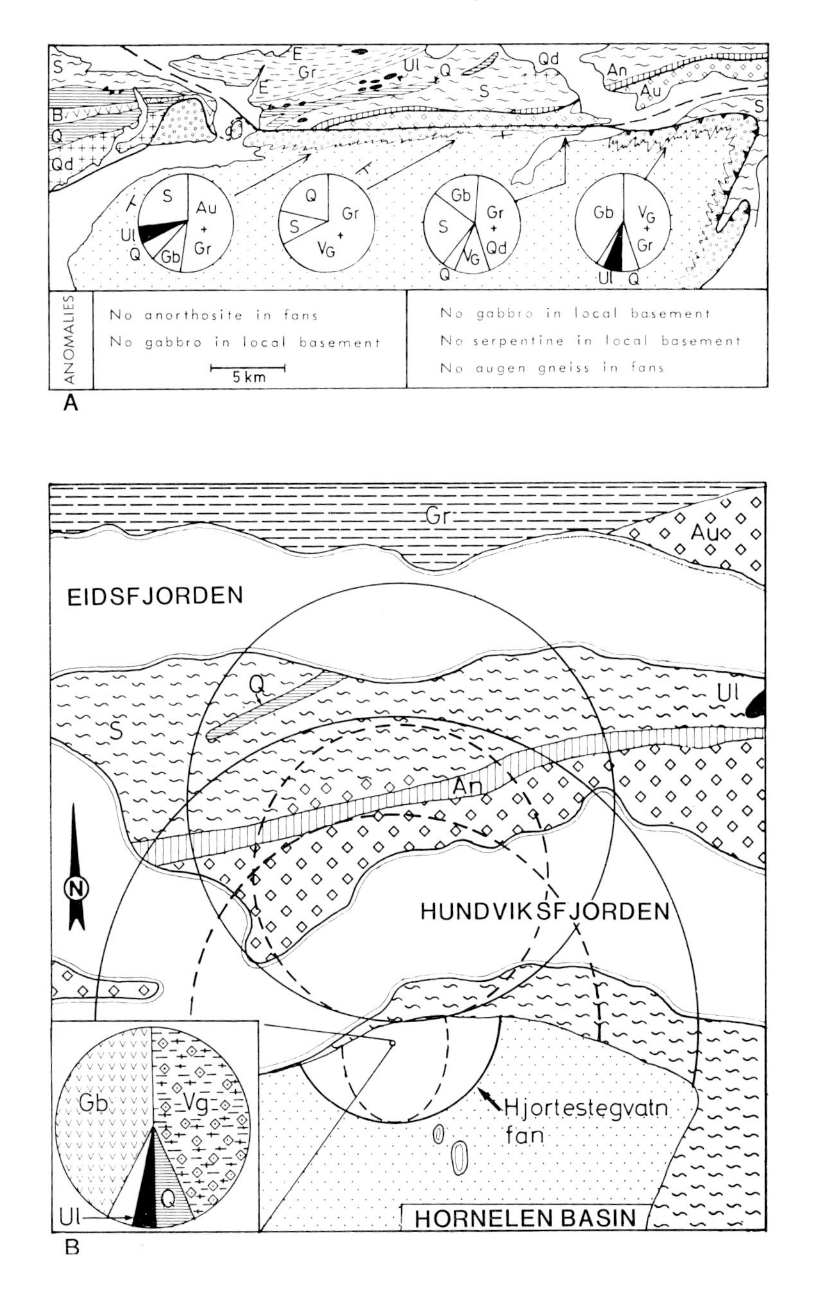

Fig. 11. (A) General clast composition anomalies along the northern margin of Hornelen Basin. (B) Hjortestegvatn Fan and its probable drainage basin area (see text for calculation). Both circular and semicircular areas are shown. Broken and unbroken lines refer to the minimum and maximum values in Table 1. Rock types are: Greenstone (B); Granodioritic (Gr), Augen (Au) and various banded (Vg) gneisses; Serpentinite (Ul); Quartzite (Q); Quartzdiorite (Qd); Anorthosite (An); Schist and metasediment (S); Gabbro (Gb); Eclogite (E).

Table 1. Hypothetical models for the relationships between Hjortestegvatn fan area and drainage basin area

Fan shape	Shortest half axis (km)	Longest half axis (km)	Fan area (km^2)	Drainage basin area (km^2)	Radius of drainage basin semi-circ. (km)	Radius of drainage basin circular (km)
1·5 km; 1·5 km; Semi-circular	1·5	1·5	3·5	33·1	4·6	3·3
1·25 km; 1·5 km; Semi-elliptic	1·25	1·5	2·9	26·9	4·1	2·9
0·75 km; 1·5 km; Semi-elliptic	0·75	1·5	1·8	15·8	3·2	2·2

the hinterland now immediately adjacent to the fan so that the encompassed geology can be compared with fan pebble types. Taking into account errors due to the exact shape of the fan being unknown and the possibility of the original basin margin having been as much as 1 km farther north than the present margin, this method gives a useful semi-quantitative demonstration of likely lateral movement. Of the fan pebble types both gabbro clasts (40%) and ultrabasic clasts (4%) cannot be found in the plotted drainage area whereas both augengneiss and charnockitic rocks presently exposed in the drainage area are not found as pebbles in the fan body (Fig. 11B). Pebble types in the fans along the southern margin, on the other hand, match reasonably well with the geology of the adjacent basement (Gloppen, 1978). This is consistent with the model of this edge being normal faulted (Fig. 14), and additionally suggests that the amount of post-Devonian thrusting of the basin has been very limited.

Skewed fan bodies

This criterion is based on the possibility of recognizing in the ancient record fan bodies which have been skewed laterally from their original position due to deposition having taken place preferentially on one flank after the fan had been moved laterally with respect to its source area. Such a situation has been documented on recent fans along the eastern edge of Death Valley (Fig. 12a) by Hooke (1972) where mapping of active washes on most of the fans shows that deposition is presently concentrated on their southern flanks (Fig. 12b). This has been attributed to right slip along the bounding fault during fan growth (Hooke, 1972). It is provisionally suggested here that a similar situation might well be recorded in many of the northern margin fans of Hornelen Basin, e.g. by the uppermost 60 m of sediment in Hjortestegvatn Fan sequence (Fig. 13). This part of the fan body is characterized by an overall upwards fining, despite being dominated by smaller scale upward coarsening sequences. The latter represent progradation of fan lobes, but the former probably reflects a gradual decrease of fan size or, as preferred here, a lateral shifting of the fan body so that the sampled profile includes a progressively more distal part of the fan radius through time. Even if the upwards fining records a true reduction of fan size with time, this also may have been conditioned by tectonic disturbance of the drainage area. The relationship between a fan and its drainage area is known to be a close one (Bull, 1964; Denny, 1965; Hooke, 1968), so that even small changes in the latter have important consequences for the former.

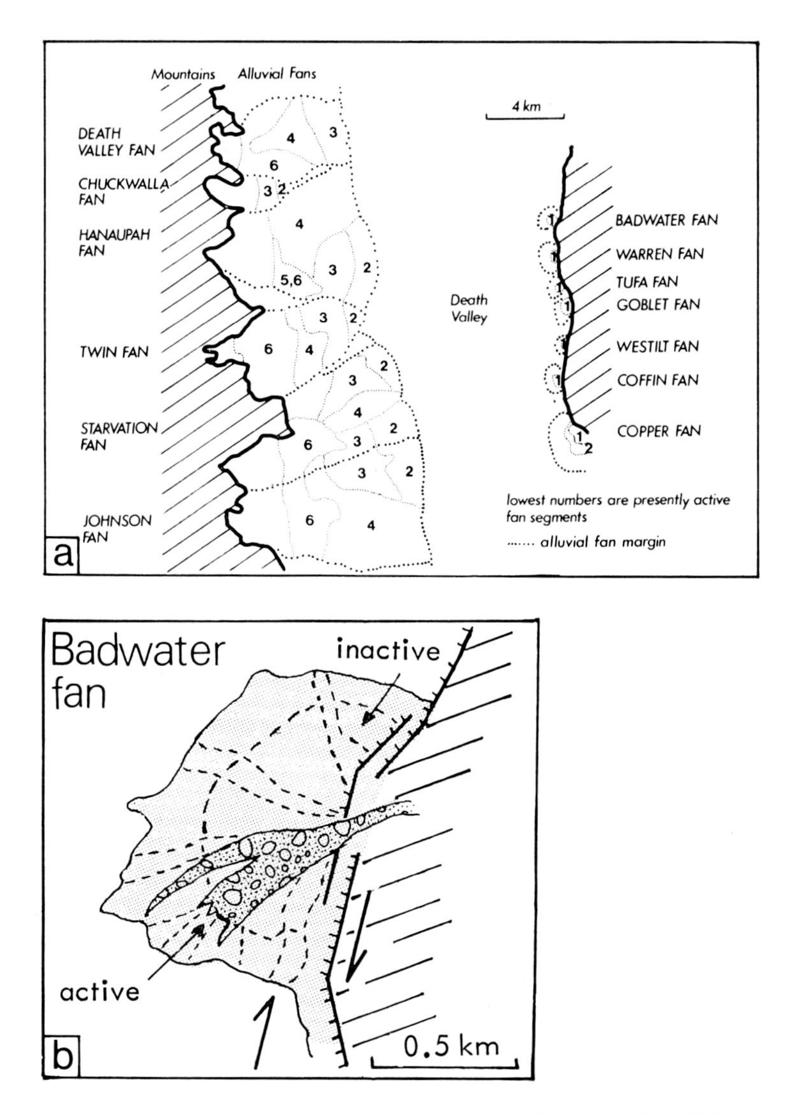

Fig. 12. Alluvial fans and their tectonic setting, Death Valley. (a) Large and small fans on west and east sides respectively, due largely to an eastwards tectonic tilt of basin floor. (b) Preferred deposition on the southern portions of fans, probably due to dextral slip along the bounding fault zone (after Hooke, 1972; Heward, 1978).

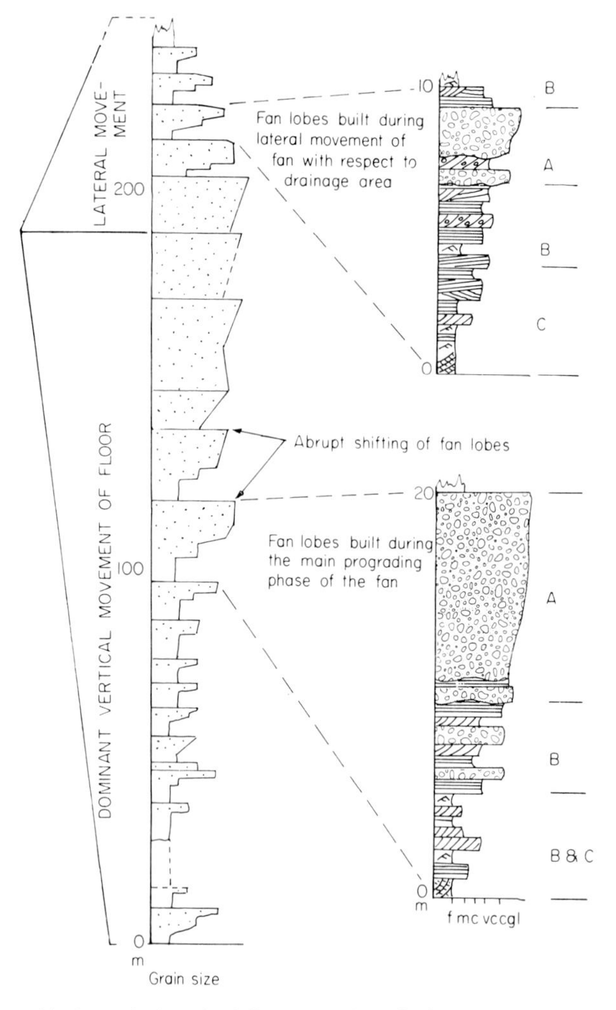

Fig. 13. Vertical profile through the middle/lower reaches of Hjortestegvatn Fan (located on Fig. 8). Note the overall upward coarsening-fining organization of the fan body despite the dominance of smaller upward coarsening sequences, the rigorous interfingering of floodbasin/lacustrine deposits with the fan lobes, and the tectonic interpretation.

CONCLUSIONS

Basin floor movements and sedimentation response

It is generally accepted, indeed axiomatic, that thick clastic sedimentary successions imply accommodating vertical movements of the floor in the receiving basin. On a scale at the other extreme it is also acceptable that individual seismic shocks, for example from a fault bounding a rapidly subsiding basin, may be reflected in the surface sediment by the presence of widespread soft sediment deformation structures.

Less commonly documented, however, are cases where tectonic events are recorded

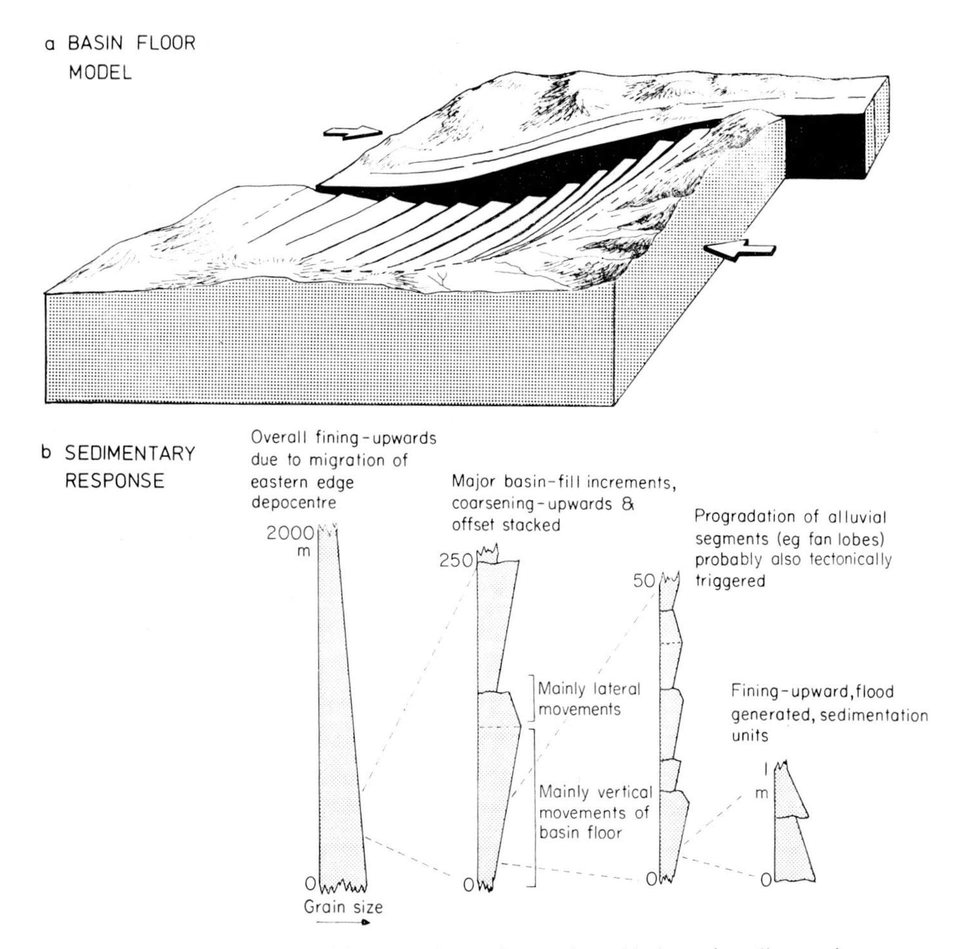

Fig. 14. A tentative tectonic model for Hornelen Basin together with the main sedimentation response on various scales.

by sedimentary sequences on a scale of metres to tens of metres, probably because such sequences are so commonly also explicable in terms of simpler causes, for example autocyclic shifting or climate changes. We know that in seismically active areas individual large earthquakes commonly produce vertical fault scarp displacements of several metres, such as in north-central Nevada where faults have a recurrence interval of 8000–10 000 years (Wallace, 1977b). Although the largest normal fault displacement recorded in the United States is about 13 m (Bonilla, 1970) it is commonly believed that clustering of events takes place in both time and space (Wallace, 1977b), so that with flurries of events scarp heights of more than 40 m can be produced (Wallace, 1977a). The relief created in this manner will presumably increase rates of erosion and sediment yield in the uplands and will produce a corresponding sedimentation response out across the fault-generated front.

It is argued here for Hornelen Basin that vertical movements of the basin floor are reflected in sedimentary sequences, most often upwards coarsening, on a scale of 5–20 m. Repetition of such sequences may well be an immediate reflection of lobe shifting on fans, of lake level changes in lacustrine sequences or of abandonment of alluvial tracts in fluvial sequences; but because they tend to be so widespread and on a similar scale within differing, laterally equivalent facies it is argued that they are ultimately controlled by periodic downward movements of the basin floor. It should be added that the thickness of an individual sequence says less about individual or cumulative fault displacement than about the thickness of sediment deposited prior to renewed base-level change.

There is no evidence of systematic lateral displacement of successive cyclothems on this scale (though this is often difficult to demonstrate because of the present tectonic tilt of the strata) so there is no suggestion of an additional horizontal component of movement of the basin floor at this level.

Lateral movement of the basin, interpreted to have taken place by several hundred phases of strike-slip movement along the present northern margin of Hornelen Basin (Fig. 14), is first recognized at a 100–200 m level in the sedimentary succession. Upward coarsening sequences on this scale mappably overlap each other, and imply an onlap of basement eastwards. Crude calculations based on areas such as shown in Fig. 10 imply an overlap of some 250 m per cyclothem, suggesting a horizontal component of movement up to two or three times the vertical one during episodes of subsidence and lateral migration of the depocentre. The geometric relationship of successive basin-fill increments to each other imply that the horizontal movements are largely recorded in the intervals *between* cyclothems. It has already been argued, however, that the asymmetry of some of the large-scale conglomeratic cyclothems along the northern edge of the basin may well reflect horizontal movements (separating fans from their drainage areas) already in the upper portion of such sequences.

At this stage it may be worth referring tentatively to the size of strike-slip displacement documented from known strike-slip faults on land, without wishing to suggest that the Devonian faulting and Californian Holocene faulting are closely analogous. Displacements occurring during individual earthquakes in historic times along segments of the San Andreas Fault are commonly of the order of several metres. Average displacements for the 1857 and 1906 earthquakes in central and southern California are 3·0 and 4·7 m respectively with maximum readings up to 9·5 m (Sieh, 1978).

Over a longer time interval, it has been suggested (Sieh, 1978) that there has been total right lateral displacement of some 40 m along a south-central reach of the San Andreas fault during nine large earthquake events from the sixth century A.D. until 1857. The point to be made here is that over a relatively short period of time a group of events in an active area can produce an amount of lateral movement of the order of size deduced (from the discrete amounts of mappable overlap of sequences) in Hornelen Basin. Less certain, however (and it cannot be demonstrated one way or the other yet) is the longer term pattern of events in strike-slip zones. For example is the pattern of alternating strike-slip and dip-slip dominance deduced for Hornelen Basin a realistic one? The clear evidence of Pleistocene or Pliocene rapid subsidence along segments of the palaeo-San Andreas zone in southern California (e.g. Crowell, 1974) may be significant here.

In Hornelen Basin, then, an episodic horizontal movement, during a more continuous background of vertical movement appears to have caused the accumulation of a stratigraphic thickness of some 25 km and a depocentre migration eastwards up to 50 km from its original positions. The lateral migration resulted in an overall upwards fining in the succession on the largest scale (Fig. 14). Depending on the east–west extent of 100–200 m increments of sedimentation, the true maximum vertical thickness of the succession at any point is likely to be less than 8 km. A tectonic model for the basin together with the main aspects of sedimentation response as seen now in the basin-fill succession are outlined in Fig. 14.

Rift *v*. strike-slip basin development

As already noted, there are difficulties involved in distinguishing sedimentary successions deposited under the control of purely vertical tectonic movements from those where there was also a significant strike-slip component of movement during deposition. The relative lack of published descriptions of ancient strike-slip basins as opposed to ancient rift basins is symptomatic of this. This is hardly surprising since most of the criteria listed and discussed above are also applicable, at least to a degree, to rift basins. The probable exceptions, and therefore most important in the identification of ancient strike-slip basins, are fan/drainage area mis-matches and evidence of depocentre migration with time. At the same time it should be said that the former can be ambiguous and that the latter can be achieved without great horizontal fault movements. Despite this, it is suggested that strike-slip movement between two blocks is a most simple and elegant way of causing basin migration.

While higher rates of sedimentation are predictable in strike-slip basins as compared to rift basins (see Miall, 1978) due to the lateral movement of both drainage and sedimentation areas, the reason for the frequent great depths (relative to their small size) of the former is less obvious. It may be that in the case of pure extension and rifting there will be an optimum depth of basin floor sinking, determined largely by isostatic balance, whereas with strike-slip faulting, where subsidence is often caused by one block being *forced* downwards by another, this optimum depth can be exceeded (J. Sales, personal communication, 1979).

The possible role of basinwide, upward coarsening sequences in strike-slip basin successions is of interest. Such sequences simply reflect, for example, progradation of alluvium into lacustrine/floodbasin tracts and require only abundant sediment availability and appropriate vertical movements of base level. For this reason they are also common in rift basins, perhaps particularly so in small, non-marine basins where there is rapid infilling and few 'basinal' processes to modify the alluvial sequences. On a small, but basinwide scale they may be closely related, as direct regressional/progradational response, to a series of basin floor movements. On a much larger scale and more crudely developed, they may form as a less direct response to tectonics, for example, due to progressively greater amounts of alluvial plain reworking as a consequence of decreasing basin floor subsidence through time (Allen, 1978). The striking feature in Hornelen Basin is not that upward coarsening sequences are present but that they are so well developed and so frequent throughout the 25 km pile. It is suggested that the periodic horizontal movements of the Hornelen depocentre greatly enhanced their development. The sourcewards (eastwards) retreat of the alluvial systems caused by dropping of the basin floor was periodically exaggerated by a wholesale eastwards shift of the depocentre. This resulted in lateral offsetting of successive 100–200 m increments and is now highlighted at this level in the succession by coarse-grained sediment at the top of one sequence being in contact with finer (more distal) than usual sediment in the immediately overlying sequence. To this extent, and because there were several hundred episodes of lateral movement during basin development, it can be said that the abundance of upward coarsening sequences in Hornelen Basin is a reflection of its strike-slip origin.

It is concluded that the characteristics of Hornelen Basin are more satisfactorily explained in terms of extension plus strike-slip movement rather than by extension alone. The present eastwards extension of the line of proposed strike-slip faulting has not previously been mapped as such, although Prost (1975) has recorded this zone as one of fundamental faulting. In addition, examination of satellite photographs along the zone suggests that there may be a series of *en echelon* faults or fractures present (I. Ramberg, personal communication, 1979), along which the lateral movement may have been taken up.

ACKNOWLEDGMENTS

This work grew out of the West Norway Devonian Project (financed largely by the Royal Norwegian Council for Scientific and Industrial Research) but was written up while the senior author was on research leave at the University of California at Santa Barbara. This was made possible by support from NATO (Grant 1724).

We are grateful to Vidar Larsen and to Sjur Aasheim for permission to use their unpublished data in Figs 6 and 11a respectively. We also thank Ellen Irgens, Jan Lien and John Gjelberg for drafting some of the figures. Various versions of the manuscript have benefited from discussion with and critical comments by Knut Bjørlykke, John Crowell, Peter Friend, Tor Nilsen, Eigill Nysæther, Ivar Ramberg, Harold Reading, John Sales and Art Sylvester.

REFERENCES

ALLEN, J.R.L. (1978) Studies in fluviatile sedimentation: an exploratory quantitative model for the architecture of avulsion-controlled alluvial suites. *Sediment. Geol.* **21,** 129–147.

BLUCK, B.J. (1978) Sedimentation in a late orogenic basin: the Old Red Sandstone of the Midland Valley of Scotland. In: *Crustal Evolution in Northwest Britain and Adjacent Regions* (Ed. by D. R. Bowes & B. E. Leake). *Geol. J. Spec. Issue*, **10,** 249–278.

BONILLA, M.G. (1970) Surface faulting and related effects. In: *Earthquake Engineering* (Ed. by R. L. Weigel), pp. 47–74. Prentice-Hall, New York.

BRYHNI, I. (1964) Relasjonen mellom senkaledonsk tektonikk og sedimentasjon ved Hornelens og Håsteinens devon. *Norg. geol. Unders.* **223,** 10–25.

BRYHNI, I. & SKJERLIE, F.J. (1975) Syndepositional tectonism in the Kvamshesten district (Old Red Sandstone), western Norway. *Geol. Mag.* **112,** 593–600.

BULL, W.B. (1964) Alluvial fans and near-surface subsidence in W. Fresno, California. *Prof. Pap. U.S. geol. Surv.* **437A.** 71 pp.

CROWELL, J.C. (1952) Probable large lateral displacement on San Gabriel Fault, Southern California. *Bull. Am. Ass. Petrol. Geol.* **36,** 2023–2035.

CROWELL, J.C. (1974) Origin of late Cenozoic basins in southern California. In: *Tectonics and sedimentation.* (Ed. by W. R. Dickinson). *Spec. Publ. Soc. econ. Paleont. Miner., Tulsa*, **22,** 109–204.

CROWELL, J.C. (1975) The San Gabriel Fault and Ridge Basin, southern California. In: *San Andreas Fault in Southern California* (Ed. by J. C. Crowell). *Spec. Rep. Calif. Div. Mines Geol.* **118** 208–219.

DENNY, C.S. (1965) Alluvial fans in the Death Valley region, California and Nevada. *Prof. Pap. U.S. geol. Surv.* **466.** 62 pp.

GLOPPEN, T.G. (1978) *Hornelen Basin (Devonian), western Norway: a study of various fan bodies and their deposits*. Unpublished Cand. real thesis, University of Bergen. 186 pp.

HARLAND, W.B. (1971) Tectonic transpression in Caledonian Spitzbergen. *Geol. Mag.* **108,** 27–42.

HARLAND, W.B. (1973) Tectonic evolution of the Barents Shelf and related plates. In: *Arctic Geology* (Ed. by M. G. Pitcher). *Mem. Am. Ass. Petrol. Geol.* **19,** 509–608.

HAWORTH, R.T. (1974) The development of Atlantic Canada as a result of continental collision: evidence from offshore gravity data. In: *Canada's Continental Margins and Offshore Petroleum Exploration* (Ed. by C. J. Yorath, E. R. Parker and D. J. Glass). *Mem. Can. Soc. Petrol. Geol.* **4,** 59–77.

HEWARD, A.P. (1978) Alluvial fan sequence and megasequence models: with examples from Westphalian D–Stephanian B coalfields, northern Spain. In: *Fluvial Sedimentology* (Ed. by A. D. Miall). *Mem. Can. Soc. Petrol. Geol.* **5,** 669–702.

HOOKE, R. LE B. (1968) Steady state relationships on arid-region alluvial fans in closed basins. *Am. J. Sci.* **266,** 609–629.

HOOKE, R. LE B. (1972) Geomorphic evidence for late-Wisconsin and Holocene tectonic deformation, Death Valley, California. *Bull. geol. Soc. Am.* **83,** 2073–2098.

KILDAL, E.S. (1970) Geologisk kart over Norge, berggrunnskart, Måløy, 1:250,000, norsk utgave. *Norg. geol. Unders.*

LARSEN, V. & STEEL, R.J. (1978) The sedimentary history of a debris flow dominated, Devonian alluvial fan: study of textural inversion. *Sedimentology*, **25,** 37–59.

LINK, M.H. & OSBORNE, R.H. (1978) Lacustrine facies in the Pliocene Ridge Basin Group, Ridge Basin, California. In: *Modern and Ancient Lake Sediments* (Ed. by A. Matter and M. E. Tucker) *Spec. Pub. int. Ass. Sediment.* **2,** 169–187.

MIALL, A.D. (1978) Tectonic setting and syndepositional deformation of molasse and other nonmarine-paralic sedimentary basins. *Can. J. Earth Sci.* **15,** 1613–1632.

MITCHELL, A.H.G. & READING, H.G. (1978) Sedimentation and tectonics. In: *Sedimentary Environments and Facies* (Ed. by H. G. Reading), pp. 439–476. Blackwell Scientific Publications, Oxford.

MORRIS, W.A. (1976) Transcurrent motions determined palaeomagnetically in the Northern Appalachians and Caledonides and the Acadian orogeny. *Can. J. Earth Sci.* **13,** 1236–1243.

NILSEN, T.H. (1968) The relationship of sedimentation to tectonics in the Solund area of southwestern Norway. *Norg. geol. Unders.* **259.** 108 pp.

NILSEN, T.H. (1973) Devonian (Old Red Sandstone) sedimentation and tectonics of Norway. In: *Arctic Geology* (Ed. by M. G. Pitcher). *Mem. Am. Ass. Petrol. Geol.* **19,** 471–481.

PHILLIPS, W.E.A., STILLMAN, C.J. & MURPHY, T. (1976) A Caledonian plate tectonic model. *J. geol. Soc. Lond.* **132,** 579–609.

PROST, A. (1975) Linéament au faille transformante: un modèle particulier dans le développement orogénique calédonien scandinave au niveau du 62°N 3. *Réunion Annuelle Sciences de la Terre.*

RAMBERG, I.B., GABRIELSEN, R.H., LARSEN, B.T. & SOLLI, A. (1977) Analysis of fracture patterns in southern Norway. *Geol. Mijnb.* **56,** 295–311.

READING, H.G. (1980) Characteristics and recognition of strike-slip systems. In: *Sedimentation in Oblique-Slip Mobile Zones* (Ed. by P. F. Ballance and H. G. Reading). *Spec. Publ. int. Ass. Sediment.* **4,** 7–26.

SIEH, K.E. (1978) Prehistoric large earthquakes produced by slip on the San Andreas Fault at Pallet Creek, California. *J. geophys. Res.* **83,** 3907–3939.

STEEL, R.J. (1976) Devonian basins of western Norway; sedimentary response to tectonism and varying tectonic context. *Tectonophysics*, **36,** 207–224.

STEEL, R.J. & AASHEIM, S. (1978) Alluvial sand deposition in a rapidly subsiding basin (Devonian, Norway). In: *Fluvial Sedimentology*. (Ed. by A. D. Miall). *Mem. Can. Soc. Petrol. Geol.* **5,** 385–413.

STEEL, R.J., MÆHLE, S., NILSEN, H., RØE, S.L. & SPINNANGR, Å. (1977) Coarsening-upward cycles in the alluvium of Hornelen Basin (Devonian), Norway: sedimentary response to tectonic events. *Bull. geol. Soc. Am.* **88,** 1124–1134.

WALLACE, R.E. (1977a) Profiles and ages of young fault scarps, north-central Nevada. *Bull. geol. Soc. Am.* **88,** 1267–1281.

WALLACE, R.E. (1977b) Time-history analysis of fault scarps and fault traces: a longer view of seismicity. *6th World Conf. on Earthquake Engineering, New Delhi, India.* **2,** 409–412.

ZIEGLER, P.A. (1978) North-Western Europe: tectonics and basin development. *Geol. Mijnb.* **57,** 589–626.

Spec. Publ. int. Ass. Sediment. (1980) **4,** 105–125

Deposits associated with a Hercynian to late Hercynian continental strike-slip system, Cantabrian Mountains, Northern Spain

ALAN P. HEWARD *and* HAROLD G. READING

Department of Geological Sciences, South Road, Durham DH1 3LE and Department of Geology and Mineralogy, Parks Road, Oxford OX1 3PR

ABSTRACT

Very thick Upper Carboniferous deposits in the Cantabrian Mountains, northern Spain overlie an almost continuous, widespread and thin Palaeozoic succession of shallow-marine sandstones, shales and limestones. The >15 km composite Upper Carboniferous succession is extremely variable in nature and extent and is punctuated by unconformities and disconformities due to localized phases of deformation. It can be divided into four facies associations: shallow-marine carbonate association, marine basinal association, marginal-marine clastic association, and non-marine clastic association. Periods of basin initiation during Namurian C to Westphalian A, Westphalian A to B, Westphalian D to Cantabrian and Stephanian A to C are superimposed on a trend in which Carboniferous basins develop progressively south-westwards and become increasingly continental. Accumulation took place on shelf areas and in small rapidly subsiding basins which are considered to have been associated with a Hercynian to late-Hercynian continental strike-slip system on the southern margin of the Biscay–North Pyrenean fault. Evidence for vertical movements is abundant. Evidence for lateral movement is less direct but is inferred from the dislocation of some alluvial fans and fan deltas from their source regions, the very rapid lateral and vertical facies changes, the almost universal contemporaneity of deposition with deformation and erosion close by, the types of deformation and the lack of metamorphism and igneous activity which characterize this Hercynian orogenic belt. The nature of deformation, lack of volcanicity, and uplift reflected in the increasing continentality of younger Upper Carboniferous deposits suggest dominantly transpressive strike-slip movement.

INTRODUCTION

The Cantabrian Mountains of northern Spain extend through the provinces of Oviedo (Asturias), León, Palencia and Santander, their present relief reflecting Tertiary uplift (Fig. 1a). Palaeozoic and late Precambrian sediments form the mountain range, within an envelope of Mesozoic and Tertiary rocks.

In northern Spain three palaeogeographical divisions of the Palaeozoic have been

0141-3600/80/0904-0105$02.00

recognized (Fig. 1b; Wagner, 1970; Julivert, 1971a; Wagner & Martínez-García, 1974). In the Asturian–Leonese and Galician–Castilian zones, thick and rather complete Palaeozoic basinal successions accumulated.

In contrast, the succession of the Cantabrian zone can be divided into two unequal parts, a relatively thin sequence of late Precambrian to Lower Carboniferous and a very thick sequence of Upper Carboniferous sediments which accumulated under active tectonic conditions. Although it is the Upper Carboniferous deposits with which this paper is concerned, the underlying succession is important because it forms the immediate basement on which the Upper Carboniferous was deposited; it is often deformed along with and influences the deformation of the Upper

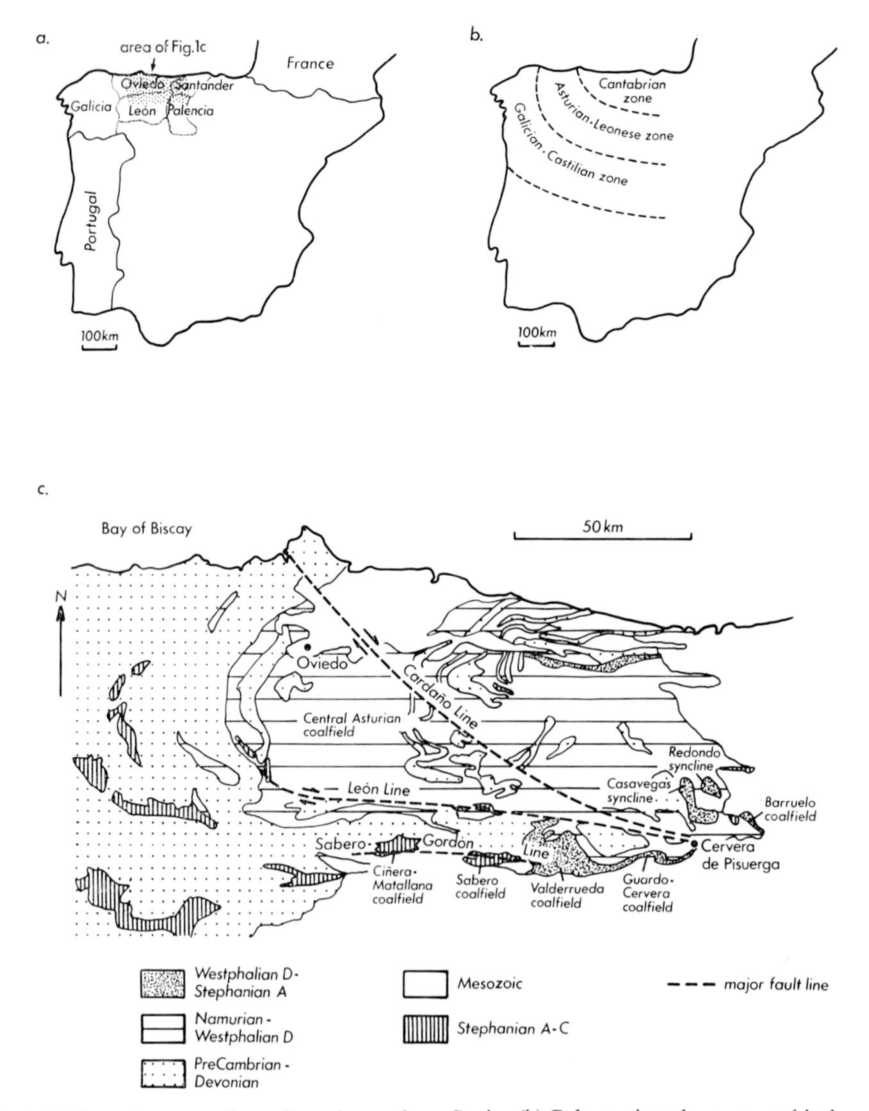

Fig. 1. (a) Location map of provinces in northern Spain. (b) Palaeozoic palaeogeographical zones, after Wagner & Martínez-García (1974). (c) General geological map of Cantabrian Mountains with localities mentioned in text, after Wagner (1970).

Carboniferous; its erosion provided most of the clastic sediments for the Upper Carboniferous and the Hercynian orogeny developed differently according to the underlying palaeogeography.

Prior to the Upper Carboniferous, the Cantabrian zone was a platform on which a relatively thin and somewhat incomplete succession formed. There is no evidence for a metamorphic basement; weakly or non-metamorphosed Precambrian shales and turbidites are the oldest rocks exposed. The overlying Cambrian to Lower Carboniferous is ca 2700 m thick, and consists of regionally widespread shallow-marine sandstones (quartzites), shales and limestones. Middle Ordovician–Lower Silurian (Llandovery) deposits are generally absent. Whilst some Devonian and Lower Carboniferous deposits are widespread, localized uplift, unconformities and disconformities occur, and the first evidence is seen for the delimitation of sedimentary facies along major structural lines (van Adrichem Boogaert, 1967; Evers, 1967; Boschma & van Staalduinen, 1968; Wagner, Winkler Prins & Riding, 1971).

Upper Carboniferous geology is in marked contrast to that of the earlier Palaeozoic. The Galician–Castilian zone underwent deformation, metamorphism and granitic intrusion (Matte, 1968; Bard *et al.*, 1973; Wagner & Martínez-García, 1974; Ries, 1978, 1979). These effects diminish eastward and the Cantabrian zone was the site of periodic deformation, contemporaneous with the accumulation of >15 km of sediment. Marked variations in thickness; rapid changes in depositional facies; thick conglomerates; and fold phases, unconformities and disconformities of variable extent typify the Upper Carboniferous. Sedimentation occurred on shelves and in rapidly subsiding basins (Reading, 1970, 1975; van de Graaff, 1971a, b; Maas, 1974; Young, 1976; N. A. Rupke, 1977; Heward, 1978a, b; Bowman, 1979). Metamorphism is generally absent or low-grade, and intrusive and volcanic rocks are volumetrically insignificant.

Upper Carboniferous sediments are overlain unconformably by relatively undeformed Permian and Triassic alluvial deposits. Throughout most of the rest of the Mesozoic the Cantabrian Mountain region seems to have been neither a source area, nor a site of deposition, with the exception of continental and marine Cretaceous deposits at its margins. Uplift of the present Cantabrian Mountains during and since the Eocene has provided a major source for Tertiary and recent alluvial sediments shed southwards over the Meseta (de Sitter, 1965; de Jong, 1971).

UPPER CARBONIFEROUS DEPOSITS

The great thickness, extreme variability of sedimentary facies and the considerable tectonic deformation of the Cantabrian Mountains have led to great difficulties in unravelling the Upper Carboniferous stratigraphy. However the excellence of some fossil groups, such as fusulinids and plant macro-floras, and the sheer effort of stratigraphers have enabled broad correlations to be made and a new stage to be erected, the Cantabrian Stage, at the base of the Stephanian Series, based on a stratotype in coal basins of northern Palencia (Table 1; Fig. 1; Wagner *et al.*, 1977).

Occurrence

Some Upper Carboniferous deposits, such as the Namurian A–C Caliza de Montaña, extend regionally; others are limited to small (20×5 km–50×20 km) elongate fault or fold basins. These basins are oriented E–W in the eastern Cantabrians and skirt the Asturian arc in the west (Figs 1c and 2). It is generally difficult and often impossible to determine which of these are structural remnants of larger basins and which represent original depositional basins. However, abundance of basin margin and slope deposits (alluvial fan, submarine fan, slope apron deposits), strongly contrasting lithological successions, and inability to correlate directly adjacent time-equivalent successions do suggest that some outcrops approximately correspond to original basins.

Thickness

The composite Upper Carboniferous succession exceeds 15 km although probably no more than 2–5 km of sediment accumulated in any single depositional phase. Approximate sedimentation rates range from about 0·005–0·035 m/1000 years for slowly deposited Namurian A–B low energy limestones of the Barcaliente Formation (lower formation of Caliza de Montaña, Wagner *et al.*, 1971), to 0·3–0·35 m/1000 years for the shelf/basinal carbonates and clastics of the Namurian C–Westphalian A San Emiliano Formation (Bowman, 1979), and 0·21–0·36 m/1000 years for Stephanian A-B alluvial fan and lacustrine deposits (Heward, 1978a).

Deposits vary markedly in extent and thickness. While the Barcaliente Formation limestone extends through most of the Cantabrian region and only varies from 50 to 350 m, the overlying higher energy limestones of the Valdeteja Formation (upper Caliza de Montaña) are less extensive and vary rapidly in thickness; at one place they thin from 700 m to nil over a distance of only 2 km (Wagner *et al.*, 1971).

It is, however, more common to see abrupt thickness variations between contrasting depositional facies and only detailed stratigraphic correlations can detect these. Wagner *et al.* (1977) describe a 2500 m coal-bearing clastic succession in the Westphalian D-lower Cantabrian which accumulated across a fault from a 90-m thick platform limestone sequence containing two disconformities. The variability in thickness and facies of Namurian C–Westphalian A or Westphalian A-D deposits over the Cantabrian Mountain region provide further examples with looser stratigraphic control (Fig. 2c and d).

Thick, repetitive regressive and transgressive marginal marine sequences also occur. Deltaic coarsening-upward sequences are 45–50 m thick in northern Palencia (lower Westphalian D, van de Graaff, 1971a), up to 78 m in the Guardo–Cervera coalfield (Fig. 1c, upper Westphalian D–Cantabrian) and commonly in excess of 100 m in the Casavegas syncline (upper Westphalian D–Cantabrian; Young, 1976). In the Casavegas syncline regressive barrier island/beach sequences are up to 60 m thick, transgressive shoreline sequences 30 m, and transgressive delta destructive deposits 3·5–13 m. These thicknesses are greater than the average for deltaic sequences of 40 m, and for barrier island sequences of 15 m (Klein, 1974). In addition, the preservation of transgressive shoreline deposits from shoreface erosion (Swift, 1968; Kraft, 1971) suggests high rates of subsidence.

Rapid changes in depositional facies

A very wide range of clastic and carbonate facies occurs within the Upper Carboniferous deposits. Some examples of rapid facies changes have already been described and others are apparent in the summary palaeogeographical maps (Fig. 2). Individual facies are commonly limited to separate depositional basins or shelf areas and abrupt lateral changes occur in the vicinity of fault lines. Even the widespread Namurian A-B Barcaliente Formation limestone, in its lower part, passes abruptly into a clastic and carbonate turbidite succession in the vicinity of the Sabero–Gordón line (Fig. 2b; Wagner *et al.*, 1971; van Staalduinen, 1973). Abrupt facies changes also occur in vertical successions. In the Redondo syncline (Fig. 1c), for instance, shales and carbonate debris flows are overlain by a 2000 m succession of turbidites intercalated with coal-bearing fan-delta deposits (Young, 1976).

Conglomerates

In contrast to the relatively fine-grained Precambrian–Lower Carboniferous succession, conglomerates are abundant within Upper Carboniferous deposits. They vary from minor interbeds in other depositional facies to conglomeratic formations (e.g. Westphalian B Curavacas Conglomerate, which is 500–1500 m thick and extends 20 km along strike). With the exception of some carbonate debris flows and olistostromes, conglomerates are generally clast-supported and consist of well rounded clasts of predominantly Palaeozoic, including older Upper Carboniferous, formations. Clast roundness may be the result of a multicycle history, but is more likely to be a function of the Carboniferous tropical climate and high clast concentrations within sediment gravity flows. No multicycle forms of conglomerate clasts (Tanner, 1976) have been observed. Thick conglomerates occur within alluvial fan (Heward, 1978a, b), submarine fan (N. A. Rupke, 1977) and slope deposits, while carbonate debris flows and olistostromes characterize most marine basinal successions. The common occurrence of conglomerates suggests continuous proximity to uplifted source areas. The thicknesses of alluvial fan conglomerates, repeated stacking of alluvial fan coarsening and fining-upward megasequences, and abrupt vertical facies changes from fan conglomerates to lacustrine and marine shales provide strong evidence for repeated vertical fault movements (Heward, 1978a, b).

Facies associations and palaeogeography

Upper Carboniferous deposits can be divided into four broad environmental facies associations: *shallow-marine carbonate association, marine basinal association, marginal-marine clastic association,* and *non-marine clastic association.* The summary palaeogeographical maps (Fig. 2) are convenient for summarizing broad thickness variations, facies changes and palaeogeography. No attempt has been made to restore those Upper Carboniferous deposits affected by nappe movement and thrusting to their original positions. Estimates of the extent of horizontal displacement are a few kilometres–few tens of kilometres.

The *Shallow-marine carbonate association* includes thin bedded micritic limestones, thicker bedded bioclastic limestones, small carbonate build-ups and interbedded thin clastics and coals. Abundant algae and other fossils suggest shallow-water, often low-energy conditions. These deposits are either regionally extensive, or limited to a specific basin or area. They are relatively thick (50–2000 m), can vary rapidly in thickness, but are commonly thin when compared with laterally equivalent clastic successions. Condensed deposits, whilst more typical of the Lower Carboniferous, also occur (Ricacabiello Formation; Sjerp, 1966, Fig. 2c), and disconformities are fairly common. The association decreases in importance in younger Carboniferous

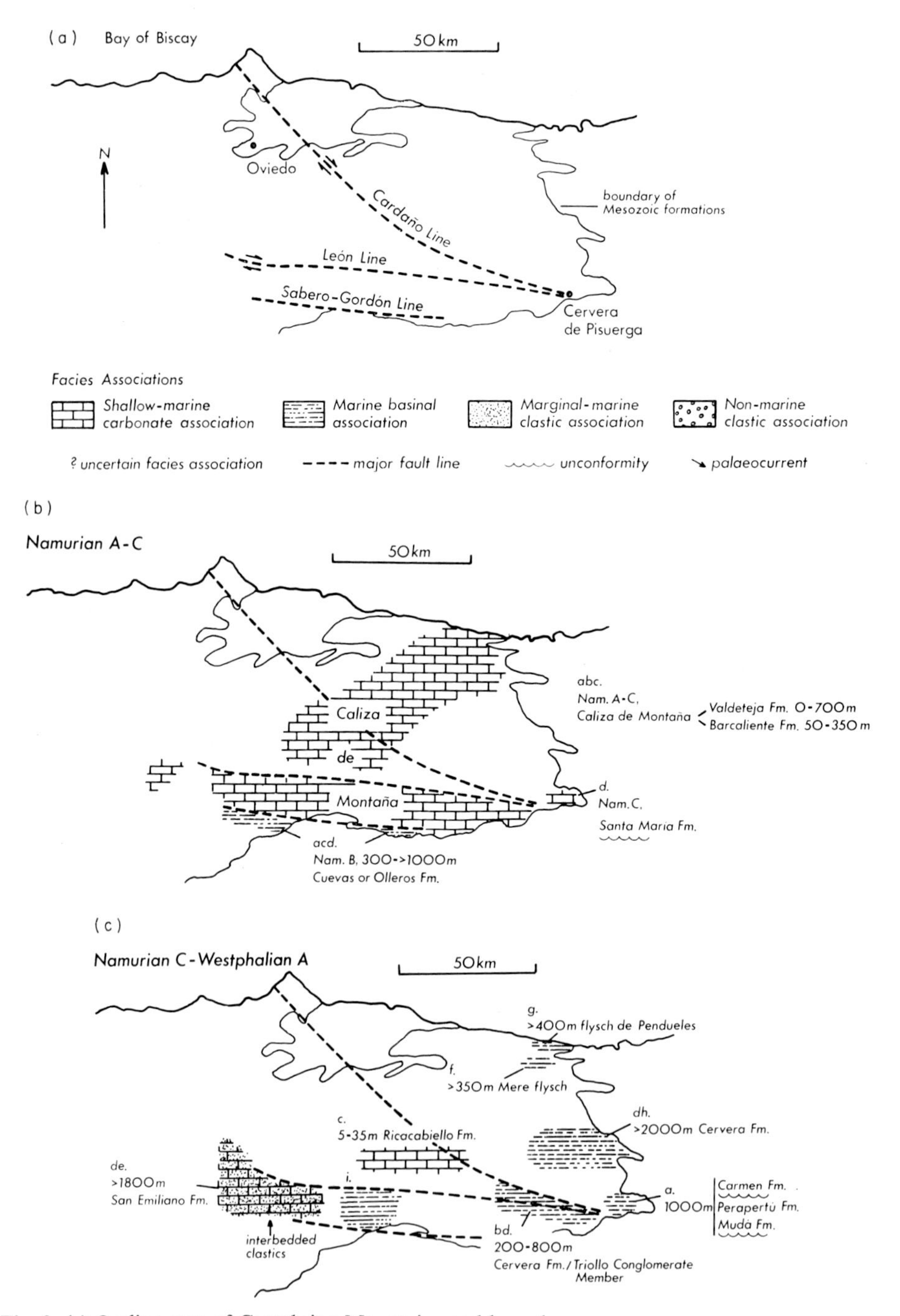

Fig. 2. (a) Outline map of Cantabrian Mountains and legend.

(b) Summary palaeogeographical map Namurian A-C. (Based on: a. Boschma & van Staalduinen, 1968; b. Martínez-García, 1971; c. Wagner & Fernández-García, 1971; d. Wagner *et al.*, 1971).

(c) Summary palaeogeographical map Namurian C-Westphalian A. (Based on: a. Wagner & Wagner-Gentis, 1963; b. J. Rupke, 1965; van Veen, 1965; c. Sjerp, 1966; d. Boschma & van Staalduinen, 1968; e. Moore *et al.*, 1971; Bowman, 1979; f. Martínez-García, 1971; g. Martínez-Garcia, Corrales & Carballeira, 1971; h. Maas, 1974; i. Bowman, personal communication).

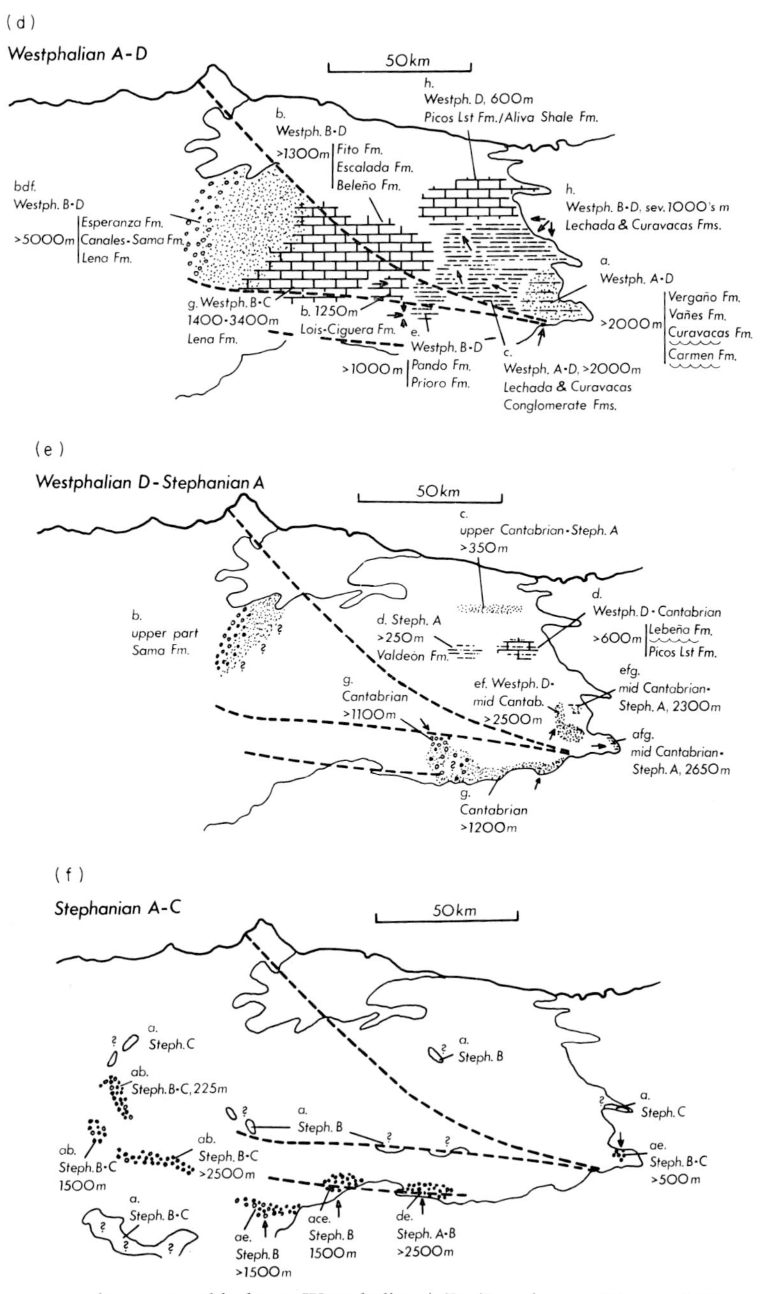

(d) Summary palaeogeographical map Westphalian A-D. (Based on: a. Wagner & Wagner-Gentis, 1963; van de Graaff, 1971a; b. van Ginkel, 1965; Sjerp, 1966; de Meijer, 1971; c. Savage, 1967; Lobato Astorga, 1977; d. García-Loygorri *et al.*, 1971; Pello & Corrales, 1971; e. van Loon, 1972; f. Bless & Winkler Prins, 1973; g. van Staalduinen, 1973; h. Maas, 1974; N. A. Rupke, 1977.

(e) Summary palaeographical map Westphalian D-Stephanian A. (Based on: a. Wagner & Winkler Prins, 1970; b. García-Loygorri *et al.*, 1971; c. Martínez-García & Wagner, 1971; d. Maas, 1974; e. Young 1976; f. Wagner *et al.*, 1977; g. Heward, unpublished).

(f) Summary palaeogeographical map Stephanian A-C. (Based on: a. Wagner, 1970; b. Corrales, 1971; c. Wagner, 1971a; d. Knight, 1975; e. Heward, 1978a, b and unpublished).

deposits. Examples are the Barcaliente and Valdeteja Formations (= Caliza de Montaña), San Emiliano, Lena, Lois–Ciguera, Caliza Massiva, Escalada, Picos de Europa, Pando, Piedras Luengas Limestone and Intermediate Limestone Formations (van Ginkel, 1965; van de Graaff, 1971b; de Meijer, 1971; Wagner *et al.*, 1971, 1977; Maas, 1974; Bowman, 1979).

The *Marine basinal association* is restricted to small basins. The deposits are extremely variable and consist of clastic and carbonate turbidites, coarse grained submarine fan deposits, slope apron conglomerates, carbonate debris flows, olistostromes and shales. Successions may be several thousands of metres thick with much penecontemporaneous deformation, some of which results from gravity sliding (Maas, 1974; Savage, 1967, 1979). Example are the Cuevas, Olleros, Cervera, Carmen, Parapertú, Flysch de Pendueles, Lechada, Curavacas, Vañes, Prioro and Brañosera Formations (Wagner & Wagner-Gentis, 1963; van de Graaff, 1971a; Wagner *et al.*, 1971, 1977; van Loon, 1972; van Staalduinen, 1973; Maas, 1974).

The *Marginal-marine clastic association* includes deltaic, fan-delta, and shoreline deposits, coals and thin interbedded carbonates. They are generally restricted to small basins but can be more widespread. Deposits several thousands of metres thick are often characterized by repeated thick regressive sequences and the preservation of thick transgressive sequences. Examples are the deposits of Central Asturian, Barruelo, Pisuerga, Guardo-Cervera, and Valderrueda coalfields (Reading, 1970; García-Loygorri *et al.*, 1971; Young, 1976; Wagner *et al.*, 1977).

The *Non-marine clastic association* mainly includes both alluvial fan conglomerates and sandstones and lacustrine shales and coals. It is restricted to small basins where accumulations 500–4000 m thick occur. Where studied in detail, alluvial fan deposits contain evidence of repeated fault movements. Examples are confined to Westphalian D and Stephanian deposits and include those in the north Valderrueda, Peña Cildá, Sabero, Ciñera–Matallana and La Magdalena coal basins, western margin of Central

Table 1. Upper Carboniferous stratigraphic subdivisions, approximate ages, and possible periods of basin initiation and fold phases for the Cantabrian Mountain region

		Series	Stages	m.y.	Periods of Basin initiation	Fold Phases
						Saalian or Uralian
				270–280	↑	
UPPER CARBONIFEROUS	Pennsylvanian	STEPHANIAN	Stephanian C Stephanian B Stephanian A Cantabrian		↓	Asturian
				290–295	↑	
		WESTPHALIAN	Westphalian D Westphalian C Westphalian B Westphalian A		↓ ↕	Leonian
				310–315	↑	
		NAMURIAN	Namurian C Namurian B Namurian A		↓	Sudetic or Palentian
				325		

Asturian coalfield, and possibly other Stephanian A-c coal basins (Wagner, 1970, 1971a; Corrales, 1971; Knight, 1975; Heward, 1978a, b).

It is apparent from the palaeogeographical maps (Fig. 2) that there were periods of basin initiation during the Namurian C–Westphalian A, Westphalian A-B, Westphalian D–Cantabrian and Stephanian A-C (Table 1). There is a general trend in which Carboniferous deposits become younger and increasingly continental towards the southwest. Shallow-marine carbonate and marine basinal deposits are dominant during Namurian and lower Westphalian to be superseded by marginal-marine and non-marine clastic deposits during Westphalian D and Stephanian. In addition to a dominant and persistent supply from south of the present Cantabrian Mountains, local clastic sediment sources operated at different times.

Fold phases, unconformities and disconformities

Stratigraphic subdivision of the Upper Carboniferous has emphasized up to four major phases of deformation and unconformity (Table 1) (de Sitter, 1962, 1965; de Sitter & Boschma, 1966; Boschma & van Staalduinen, 1968; Wagner, 1966, 1970; Wagner & Martínez-García, 1974).

These phases are spectacularly developed locally, but their timing is often imprecise. Thick conglomerates which overlie unconformities have been interpreted as post-orogenic deposits and correlated with other conglomerates not necessarily overlying unconformities, so that they appear to have a great regional extent. In detail these 'major' phases of deformation and unconformity are variable in extent, timing and significance and there are a large number of other fold phases, unconformities and disconformities. The presence of thick conglomerates merely indicates proximity to uplifted source areas.

In northern Palencia (Fig. 1a), for instance, the type region for the *Palentian* phase, there are Namurian B-C, Namurian C and Westphalian A unconformities prior to the emphasized Westphalian B (Palentian) phase of folding and erosion (Wagner & Wagner-Gentis, 1963; Wagner, 1971b; Wagner *et al.*, 1971). The Palentian unconformity is overlain by the thick Los Cintos Conglomerate. Attempts have been made to correlate the Los Cintos conglomerate with the Curavacas Conglomerate which also overlies an unconformity, and thus ascribe to it a regional significance (de Sitter & Boschma, 1966; Boschma & van Staalduinen, 1968; Wagner, 1970; Wagner & Martínez-García, 1974). There were and remain doubts whether the Westphalian B Los Cintos Conglomerate is time-equivalent to the Curavacas Conglomerate, or whether the enormous accumulation of the latter is Westphalian A and perhaps follows a Westphalian A phase of deformation (Wagner, 1970; Wagner & Martínez-García, 1974). To the west of Palencia, in the province of León, a disconformity occurs locally between basal Westphalian A and Westphalian B deposits (Moore *et al.*, 1971) and, south of the León Line, thrust and nappe movements occurred some time between Namurian and Westphalian D (J. Rupke, 1965; de Sitter, 1965; Julivert, 1971a); in Asturias there is no evidence of Westphalian B deformation (Julivert, 1971a); in southern Santander Maas (1974) describes a number of unconformities and disconformities; and in northern Santander there is a major Namurian–Westphalian D disconformity. Thus the Westphalian B 'Palentian phase' can only be demonstrated in northern Palencia. Elsewhere, although unconformities have been equated with the Palentian phase, there is no firm evidence that they occurred during the Westphalian B.

The effects of the upper Westphalian D *Leonian* phase are only clearly seen in northeastern León, where Westphalian B to lower Westphalian D deposits were deformed and eroded prior to accumulation of Upper Westphalian D conglomerates (Wagner, 1970; Wagner & Martínez-García, 1974). Elsewhere, the upper Westphalian D phase has been equated with refolding, and with a number of disconformable palaeo-karst surfaces in limestones in northern Palencia (Wagner *et al.*, 1977); however folding here during Westphalian D time cannot be directly demonstrated.

The *Asturian* phase includes a number of spectacular unconformities below conglomeratic successions whose bases range in age from Stephanian A-C. In northern Palencia there is evidence of the timing of deformation, where Cantabrian to Stephanian A sediments of the Barruelo coalfield were isoclinally folded prior to the accumulation of late Stephanian B deposits (Wagner & Winkler Prins, 1970). However, there is little difference in the age of floras from the pre-deformation succession of the Barruelo coalfield and from the post-deformation Stephanian succession in the Sabero coalfield, 70 km to the west (Fig. 1c; Knight, 1975). Elsewhere the timing of deformation is uncertain but the onset of Stephanian sedimentation occurs progressively later in a westwards direction from Sabero (Wagner, 1970). Thus it seems improbable that there was a synchronous 'Asturian phase' of deformation but rather that deformation affected one area whilst sedimentation occurred in another.

Little is known of the *Saalian* or *Uralian* phase(s) of deformation except that Stephanian A-C deposits were folded prior to the Permian (de Sitter, 1965; Wagner, 1970).

Thus throughout the Upper Carboniferous, whilst it is probable that some periods were characterized by more intense deformation than others, the overriding impression is that deformation and erosion occurred contemporaneously with sedimentation close by (Maas, 1974; Reading, 1975; Savage, 1979).

Structural style

The Cantabrian Mountains are characterized by E–W trending structures which become orientated to a N–S direction towards the west as the Asturian arc is approached (Fig. 1b and c). Deformation is intense, with extensive thrusting and overturning, but is essentially superficial with metamorphism absent and cleavage only locally developed. The presence of highly competent horizons such as the Cambro-Ordovician quartzites and Upper Carboniferous conglomerates and limestones interbedded with incompetent horizons such as the Cambrian and Lower Carboniferous nodular limestones and shales, known as 'griotte', leads to widespread décollement and disharmonic folding.

A diversity of structural style includes thrusts and nappes, possibly gravity-induced, folds and refolded folds, resulting from both vertical movements and from lateral compression, and gravity-induced flap and cascade folds. Structures rarely conform to the patterns of a classical compressive orogen and vertical movements appear at least as important as laterally compressive ones. The sporadic development of cleavage emphasises the local nature of stress fields responsible for much of the deformation (Savage, 1979).

A large number of nappe and thrust slices have been described (de Sitter, 1962, 1965; J. Rupke, 1965; Sjerp, 1966; Evers, 1967; van den Bosch, 1969; Julivert, 1971a, b; Wagner, 1971b; van Staalduinen, 1973; Maas, 1974; Savage, 1979). Nappe movement and thrusting occurred from the south, west and north at a number of periods during the Upper Carboniferous (Fig. 3a). Estimates of the extent of horizontal movement are of a few kilometres–few tens of kilometres (Julivert, 1971a; Wagner, 1971b; Maas, 1974). The converging centripetal pattern of nappe movement is atypical of classical orogenic belts and gravity has been considered the only possible driving force (Savage, 1979), although Julivert (1971a) favoured basement control. Nappes and thrust sheets have been refolded along E–W and, less commonly, N–S axes.

Most Upper Carboniferous deposits occur in E–W oriented synclines or in synclines which skirt the Asturian arc. The predominant E–W orientation of folding persists throughout the Upper Carboniferous (de Sitter, 1962, 1965; Wagner & Martínez-García, 1974; Savage, 1979). Savage (1967) and Maas (1974) consider that the synclines they analysed formed by vertical movement of fault defined basement blocks rather than by N–S compression, although Maas also documents compressive folding. In the well known (because of mining) southern Cantabrian Mountain coalfield synclines, the northern limbs are relatively complete and the southern limbs are thrust out indicating some S–N compression (Figs 1c, 3b; Wagner & Winkler Prins, 1970; Wagner, 1971a; Knight, 1975). In the less well known sedimentary basins of the northern Cantabrian Mountains the northern limbs have been thrust out (Martínez-García & Wagner, 1971).

Gravity-induced flap and cascade folds and collapse faults resulting from instability of uplifted weakly consolidated sediments are common features of thick Upper Carboniferous marine basinal successions and also some of the Picos de Europa nappes (Fig. 3a, c; Savage, 1967, 1979; Maas, 1974). In the latter case, tectonic transport accompanying gravity folding is from S to N, contrasting to the direction of nappe movement. Flap and cascade folds have horizontal axial planes and orientations which reflect underlying topography. Maas (1974) in describing thick basinal deposits suggested a continuum from synsedimentary slumps and olistostromes, to later flap and cascade folding and collapse faulting.

The major structural lines or zones of Figs 1c and 2 (Cardaño, León, Sabero–Gordón Lines and similar smaller features) are points of contention between those who believe them to be late Hercynian fractures which have subsequently been reactivated (e.g. Marcos, 1968; Wagner, 1970; Julivert, 1971b; Moore *et al.*, 1971; Wagner & Martínez-García, 1974), and de Sitter and co-workers who suggest that they controlled mid-Devonian-Upper Carboniferous sediment distributions and also influenced Upper Carboniferous deformation (e.g. de Sitter, 1962, 1965; de Sitter & Boschma, 1966; Savage, 1967, 1979; Boschma & van Staalduinen, 1968; Kullman & Schönenberg, 1978). Assessment of the significance of these features appears warranted.

Metamorphism and igneous activity

Metamorphism in the Upper Carboniferous and underlying Palaeozoic succession is generally absent or of low-grade greenschist facies (van Veen, 1965; Lobato Astorga, 1977; N. A. Rupke, 1977). Upper Carboniferous igneous rocks are rare and deeply weathered at surface exposures. Most occurrences are close to major fault lines.

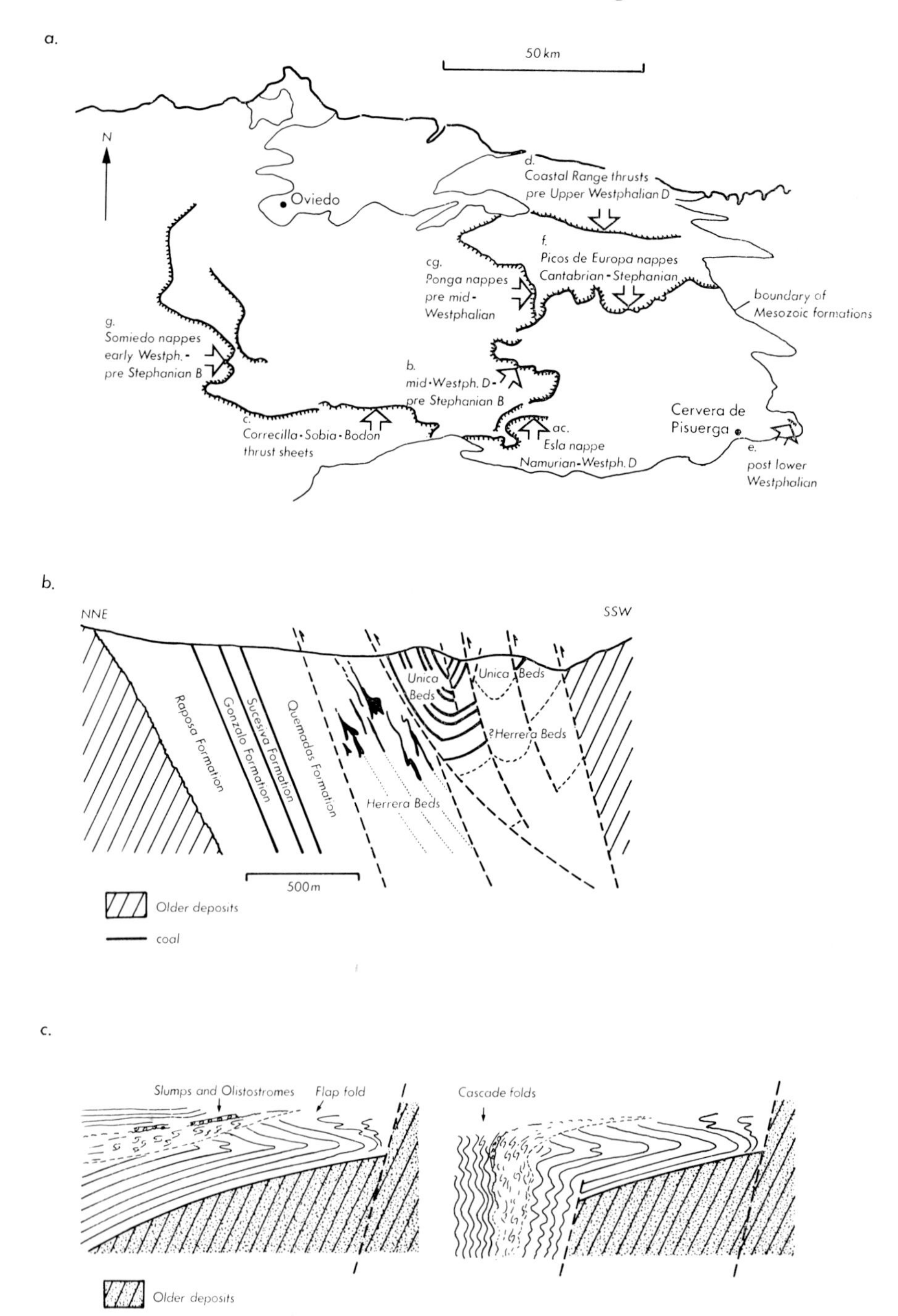

Fig. 3. Characteristics of Upper Carboniferous deformation. (a) Nappes and thrust sheets, after Julivert (1971a). (Direction of movement and timing based on a. J. Rupke, 1965; b. Sjerp, 1966; c. Julivert, 1971a; d. Martínez-García & Wagner, 1971; e. Wagner, 1971b; f. Maas, 1974; g. Savage, 1979.) (b) NNE-SSW section through the Sabero coalfield syncline illustrating the relatively complete northern limb and thrust out southern limb, after Knight (1975). (c) Flap and cascade folds in unstable weakly consolidated sediments initiated due to basement faulting. Based on examples from S. Santander, after Maas (1974).

Diorites, granodiorites, dolerites and quartz porphyry intrusives have been recorded, as have lavas and tuffs (including tonsteins/cinerites). In certain cases intrusions can be demonstrated to have occurred in sedimentary successions prior to deformation (van Veen, 1965; Wagner, 1971a; Knight, 1975; Lobato Astorga, 1977; Savage, 1979).

UPPER CARBONIFEROUS DEPOSITS AS AN EXPRESSION OF STRIKE-SLIP MOVEMENT

Palaeo-tectonic setting

A large number of tectonic and plate-tectonic reconstructions have been made for the Hercynian orogeny (e.g. Laurent, 1972; Bard *et al.*, 1973; Dewey & Burke, 1973; Johnson, 1973; Riding, 1974; Dvorak *et al.*, 1977). These reconstructions usually involve the subduction of substantial amounts of oceanic crust. More recently, Arthaud and Matte (1975, 1977) and Le Pichon, Sibuet & Francheteau (1977) have discussed tectonic and plate-tectonic patterns for the late Hercynian. Arthaud & Matte suggest that following the formation of a rigid continental block by continental plate collision during the Hercynian, the late Hercynian was a period of major right-lateral shear between an American–European plate and one that includes Africa and

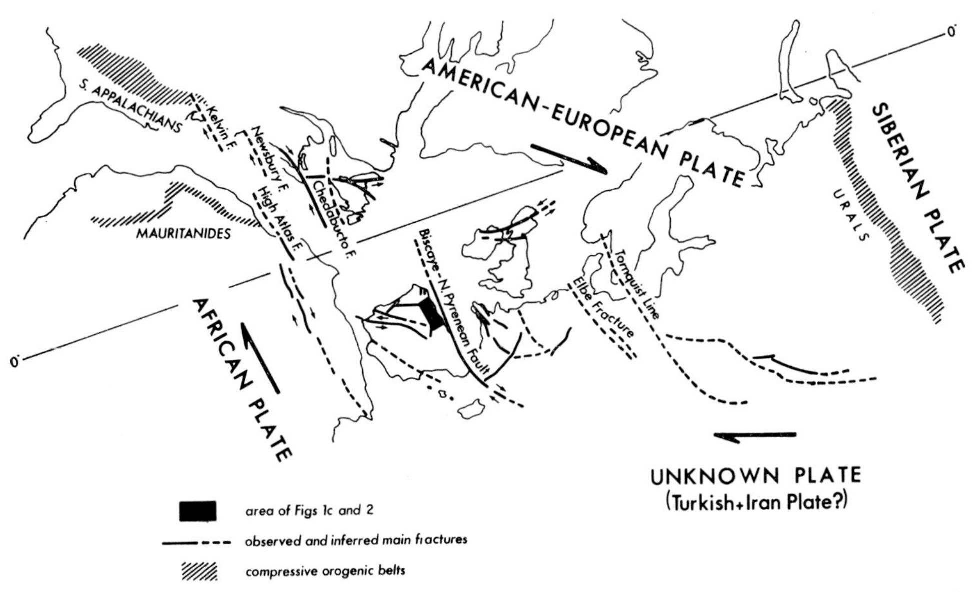

Fig. 4. Palaeo-tectonic map of the late Hercynian, after Arthaud & Matte (1977).

an unknown extension (Fig. 4). The Urals, Mauritanides and Southern Appalachians formed as compressive orogenic belts at the ends of this shear zone. The Biscay–North Pyrenean fault, running parallel to the north coast of Spain, is one of the first-order strike-slip systems within this shear zone with approximately 150 km of dextral offset. Between the first-order dextral strike-slip systems, there are regions with smaller conjugate wrench faults, thrusts, folds and sedimentary basins. It is within such a region that the Upper Carboniferous deposits of the Cantabrian Mountains accumulated and were deformed.

Evidence for vertical movements

Most strike-slip faults have a component of dip-slip movement and normal faults are abundant within strike-slip systems. It is this vertical movement that leads to the proximity of uplifted source areas and sedimentary basins (Kingma, 1958; Lensen, 1958; Clayton, 1966; Crowell, 1974a, b). Rapid uplift and subsidence result in thick sedimentary successions (e.g. >1000 m of Quaternary sediments, Hanmer Plains, Hope Fault, New Zealand, Freund, 1971; >12000 m of Miocene–early Pleistocene sediments, Ridge Basin, California, Link & Osborne, 1978) and very high sedimentation rates (0·4–1·5 m/1000 years, Schwab, 1976; Miall 1978). Uplift of weakly consolidated sediments can cause gravity deformation (Babcock, 1974).

Numerous features of the Upper Carboniferous deposits of the Cantabrian Mountains can be attributed to rapid vertical movements and to the proximity of source areas and depositional basins. Abrupt thickness and facies changes; thick, rapidly accumulated, basinal and shelf successions; repetitively thick regressive and transgressive sequences; the vertical stacking of deltaic environments in the Guardo–Cervera coalfield (Fig. 1c; Heward, unpublished, cf. Fisher & McGowen, 1969) and the vertical stacking of alluvial fan megasequences (Heward, 1978b; cf. Steel, 1976; Steel & Aasheim, 1978); the presence of localized unconformities and disconformities and common gravity induced deformation are consistent with rapid vertical movements. High sedimentation rates, time-equivalent sedimentary successions and adjacent unconformities, ubiquitous conglomerates and basin margin deposits, and numerous local source areas reflect the proximity of source areas and depositional basins.

Approximate sedimentation rates were up to 0·36 m/1000 years. These are high for many types of sedimentary basin but are low compared to recent strike-slip basins (Schwab, 1976; Miall, 1978). This may be due to uncertainty and approximations of dating, and the difficulty of determining the length of phases of non-deposition and erosion which separate phases of sedimentation. On the other hand such rates may simply reflect smaller lateral movement within these Hercynian strike-slip belts as compared with movements on the Tertiary San Andreas and Alpine faults.

Evidence for lateral movements

In Recent and Tertiary strike-slip fault systems, lateral movement can be demonstrated by the offsetting of geomorphological features, by the mismatch of present source areas and depositional basins, by the offsetting of originally continuous lithologies and depositional systems, by the recognition of conglomerates strewn along fault lines, by reconstruction of plate movement by use of magnetic anomalies and by the occurrence of synchronously forming zones of uplift and of subsidence along curving, offsetting and splaying fault systems.

In the Upper Carboniferous deposits of the Cantabrian Mountains the direct identification of lateral movements is hindered by the complexity of the geology, by rapid facies changes and by the regional extent and lack of distinctive lithologies within the underlying Palaeozoic source rocks. However, at our present state of knowledge, which still lacks much detailed sedimentology, two basins are known where it is unlikely that deposits were derived from the regions with which they are now juxtaposed.

The Tejerina syncline at the northern end of the Valderrueda coalfield (Fig. 1c) contains more than 1100 m of clast-supported conglomerates and coal-bearing sandstones and shales of lower Cantabrian age (Wagner, 1978 personal communication; Figs 2e, 5a). Cross-bedding and clast imbrication within these alluvial fan deposits indicate a source to the NW. The traceable downcurrent extent of conglomerate beds implies accumulation on alluvial fans having a radius of >7 km and an approximate fan area of 75 km². Conglomerate clasts suggest a source area consisting of Cambro-Ordovician quartzites, Devonian limestones and sandstones, Namurian Caliza de Montaña and a minor contribution of lower Westphalian limestones and granulestones (van Loon, 1972; Goester, 1973). One 60–120 m conglomerate extending downcurrent >7 km consists almost exclusively of Cambro-Ordovician quartzite clasts. No multicycle conglomerate clasts (Tanner, 1976) have been observed.

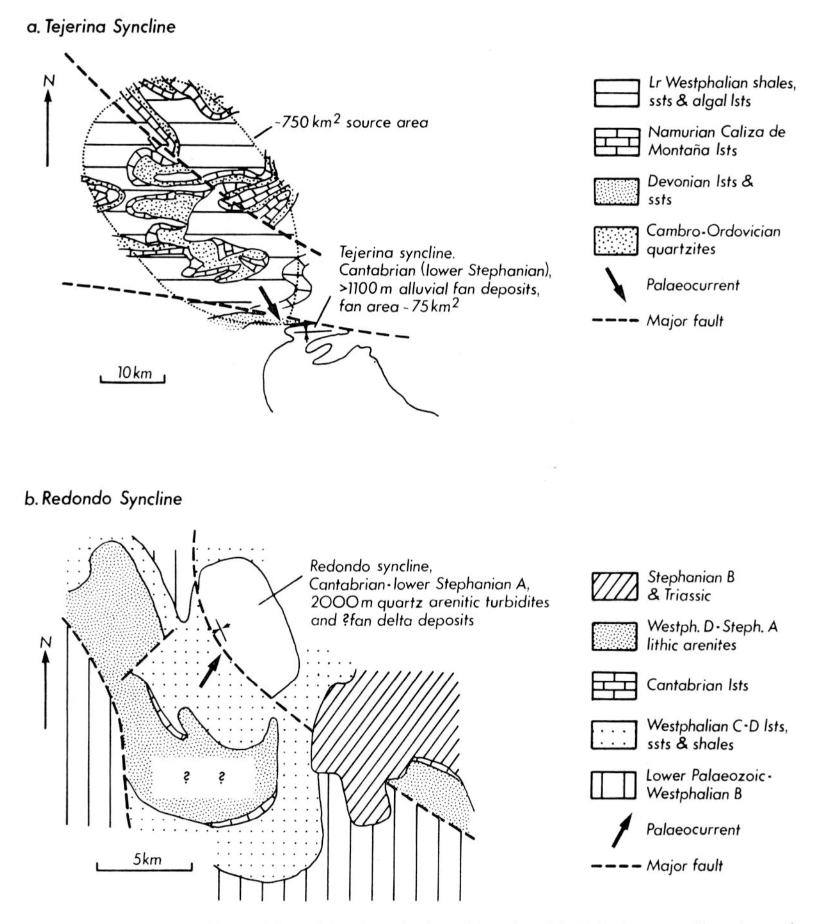

Fig. 5. Problem source area/depositional basin relationships for (a) Tejerina syncline (map based on Helmig, 1965; Sjerp, 1966; Wagner, 1970) and (b) Redondo syncline (map based on Wagner *et al.*, 1977). Evidence for lateral movements discussed in text.

Empirical alluvial fan area/drainage area relationships derived from modern semi-arid fans (Bull, 1977, Fig. 14) imply a source area of 40–750 km² for the Tejerina alluvial fan deposits. As alluvial fans derived from harder and less fractured lithologies are generally small in relation to drainage basin area (Bull, 1977; Hooke & Rohrer,

1977), the clast-supported quartzite and limestone conglomerates imply a source area nearer the larger than the smaller figure.

The ca. 750 km² region now to the NW of the Tejerina syncline consists predominantly of lower Westphalian shales, sandstones and algal limestones, with some Cambro-Ordovician quartzites and Namurian Caliza de Montaña, and very minor amounts of Devonian lithologies (Fig. 5a; Helmig, 1965; Sjerp, 1966; de Meijer, 1971). This region is unlikely to have been the source area for the Tejerina deposits (Savage, 1978, personal communication). We suggest their present juxtaposition results from lateral movement.

The Redondo syncline, northern Palencia, contains more than 2000 m of mid Cantabrian–lower Stephanian A (Wagner *et al.*, 1977) turbidites and intercalated fan-delta deposits (Figs 2e, 5b; Young, 1976). Sole marks and cross-bedding indicate sediment supply from the SW. The quartzose composition of the sandstones contrasts with contemporaneous lithic sandstones in adjacent basins derived from the regional source to the S and SW. Fan-deltas are generally small and occur close to sediment sources (McGowen, 1971; McGowen & Scott, 1974; Erxleben, 1975; Flores, 1975). Thus there should have been a nearby uplifted source area immediately to the SW of the Redondo syncline capable of supplying a thick succession of quartzose sandstones and shales. While it is possible that contemporaneous shelf sands might have provided a source, this is unlikely since the area now consists of Westphalian C-D limestones, sandstones and shales (van de Graaff, 1971a), lower Cantabrian platform limestones (Wagner *et al.*, 1977) and litharenetic fluvially-dominated deltaic deposits of approximately the same age as the quartzose sandstones of the Redondo syncline (Fig. 5b). It is therefore improbable that the presently adjacent area was the sediment source and again lateral movement is suggested.

Some of the previously described abrupt facies and thickness changes within the Upper Carboniferous deposits of the Cantabrian Mountains may also be the result of lateral movements tectonically juxtaposing contrasting deposits. Lateral movements along curving and splaying faults which lead to contemporaneous basin formation and source area deformation are probably reflected in the contemporaneity of some unconformities and disconformities with thick sedimentary successions close by. Periods of basin initiation and phases of more intense deformation may have resulted from phases of regional transtension and transpression as in Recent and Tertiary strike-slip systems (Harland, 1971; Lowell, 1972; Nilsen & Clarke, 1975; Blake *et al.*, 1978; Norris, Carter & Turnbull, 1978).

On a broader scale the major Hercynian features of the Cantabrian Mountains are atypical of compressive orogenic belts which result from subduction and continental collision, and support an origin through lateral movement. In particular there is a lack of calc-alkaline intrusive and extrusive igneous rocks and of regional metamorphism. Deformation consists of abundant vertical movement and compressive episodes, including several phases (Fig. 3a) of nappe movement and thrusting directed centripetally inwards.

Location of strike-slip faults

Whilst many features of the Upper Carboniferous geology of the Cantabrian Mountains are explicable in terms of deposition and deformation associated with strike-slip faults, the identification of the causative faults is a problem. It may be that the major structural lines such as the León Line (Fig. 1c) played some part, but the exact location of many of the faults will probably remain obscure owing to later fragmentation and deformation. Similar problems of locating causative faults are encountered by Nilsen (1978) in his description of Late Cretaceous and Early Tertiary deposits associated with the proto-San Andreas fault system in California.

CONCLUSIONS

Very thick and variable Upper Carboniferous deposits in the Cantabrian Mountains, northern Spain, probably accumulated on shelf areas and in small rapidly subsiding basins associated with a Hercynian to late Hercynian continental strike-slip system. The association of thick marine and continental sediments is typical of strike-slip systems at the margins of continental areas.

Tectonic and plate tectonic reconstructions have suggested that strike-slip, probably transpressive, movement, was the cause of intense fracturing in the crust of southern Europe in late-Hercynian times from about 290 to 250 m.y. BP (=Stephanian to lower Permian; Arthaud & Matte, 1977). If we are correct in attributing late Namurian, Westphalian and Stephanian sedimentation and orogenesis to strike-slip motion, then, at least in northern Spain, this strike-slip system was operating from early Upper Carboniferous times perhaps along lines which governed sedimentation as early as the Middle Devonian. Strike-slip movement of a similar, Devonian to Upper Carboniferous age range has been reported from eastern Canada (Belt, 1969; Webb, 1969).

The extent of lateral movement inferred for these Carboniferous strike-slip systems is a few tens of km to 100–200 km (Belt, 1969; Webb, 1969; Arthaud & Matte, 1975, 1977). This is much less than that reported from well known Recent and Tertiary fault systems. For example displacement along the San Andreas fault system has been 450 km in the last 15 m.y. (Blake *et al.*, 1978) and along the Alpine fault system ca 480 km since the Eocene (Norris *et al.*, 1978). This difference may account for the thinner basinal successions and slower apparent sedimentation rates in the Carboniferous. In addition the relatively uniform underlying platform Palaeozoic succession in the Cantabrian Mountains probably had less influence on the pattern of faults and sedimentary basins than in California where fault patterns and basin development may be strongly influenced by the heterogeneous underlying geology (Blake *et al.*, 1978).

In the Cantabrian Mountains regional transtensile strike-slip movement may be indicated by the periods of sedimentary basin initiation during Namurian C–Westphalian A, Westphalian A-B, Westphalian D–Cantabrian and Stephanian A-C (Table 1). Regional transpressive strike-slip movement may be indicated by the intervening deformation (Table 1; Fig. 3a). These alternations may have been superimposed on a transpressive regime reflected in the trend towards continental conditions of deposition.

As stratigraphical correlation becomes more refined within the 45–55 m.y. of the Upper Carboniferous, it may be possible to define more closely the phases when either basin development was dominant or uplift was dominant and thus distinguish regional phases of transtension, of pure strike-slip and of transpression, and to separate their effects from those due to local complexities of the fault pattern.

ACKNOWLEDGMENTS

A. P. Heward gratefully acknowledges financial support for field work from a Shell postgraduate grant and subsequently from the Royal Society and the University of Durham Research Fund. H. G. Reading is grateful for a research grant from the U.K. Natural Environment Research Council. Mike Bowman, Bob Carter, John Knight, Dick Norris and Bob Wagner commented on drafts of this paper and suggested many improvements. The responsibility for errors and mistaken opinions however is ours.

REFERENCES

ADRICHEM BOOGAERT, H.A. VAN (1967) Devonian and Lower Carboniferous conodonts of the Cantabrian Mountains and their stratigraphic application. *Leidse geol. Meded.* **39,** 130–189.

ARTHAUD, F. & MATTE, PH. (1975) Les décrochements tardi-hercyniens du sud-ouest de l'Europe. Géométrie et essai de reconstitution des conditions de la déformation. *Tectonophysics*, **25,** 139–171.

ARTHAUD, F. & MATTE, PH. (1977) Late Palaeozoic strike-slip faulting in southern Europe and northern Africa: result of a right-lateral shear zone between the Appalachians and the Urals. *Bull. geol. Soc. Am.* **88,** 1305–1320.

BABCOCK, E.A. (1974) Geology of the northeast margin of the Salton Trough, Salton Sea, California. *Bull. geol. Soc. Am.* **85,** 321–332.

BARD, J.P., CAPDEVILA, R., MATTE, PH. & RIBEIRO, A. (1973) Geotectonic model for the Iberian Variscan orogen. *Nature (Phys. Sci.)*, **241,** 50–52.

BELT, E.S. (1969) Newfoundland Carboniferous stratigraphy and its relation to the Maritimes and Ireland. In: *North Atlantic-Geology and Continental Drift* (Ed. by M. Kay). *Mem. Am. Ass. Petrol. Geol.* **12,** 734–753.

BLAKE, M.C., JR, CAMPBELL, R.H., DIBBLEE, T.W., JR, HOWELL, D.G., NILSEN, T.H., NORMARK, W.R., VEDDER, J.C. & SILVER, E.A. (1978) Neogene basin formation in relation to plate-tectonic evolution of San Andreas fault system, California. *Bull. Am. Assoc. Petrol. Geol.* **62,** 344–372.

BLESS, M.J.M. & WINKLER PRINS, C.F. (1973) Palaeoecology of Upper Carboniferous strata in Asturias (N. Spain). *C.r. 7th Congr. Av. Strat. géol. Carbon.*, Krefeld 1971, **II,** 129–137.

BOSCH, W.J. VAN DEN (1969) Geology of the Luna–Sil region, Cantabrian Mountains (NW Spain). *Leidse geol. Meded.* **44,** 137–225.

BOSCHMA, D. & STAALDUINEN, C.J. VAN (1968) Mappable units of the Carboniferous in the southern Cantabrian Mountains. *Leidse geol. Meded.* **43,** 221–232.

BOWMAN, M.B.J. (1979) The depositional environments of a limestone unit from the San Emiliano Formation (Namurian/Westphalian), Cantabrian Mts., NW Spain. *Sediment. Geol.*

BULL, W.B. (1977) The alluvial fan environment. *Prog. phys. Geogr.* **1,** 222–270.

CLAYTON, L. (1966) Tectonic depressions along the Hope fault, a transcurrent fault in North Canterbury, New Zealand. *N.Z. J. Geol. Geophys.* **9,** 95–104.

CORRALES, I. (1971) La sedimentación durante el Estafaniense B-C en Cangas de Narcea, Rengos y Villablino (NW de España). *Trab. Geol. Oviedo*, **3,** 69–73.

CROWELL, J.C. (1974a) Sedimentation along the San Andreas fault, California. In: *Modern and Ancient Geosynclinal Sedimentation* (Ed. by R. H. Dott JR & R. H. Shaver). *Spec. Publ. Soc. econ. Paleont. Miner., Tulsa*, **19,** 292–303.

CROWELL, J.C. (1974b) Origin of late Cenozoic basins in Southern California. In: *Tectonics and Sedimentation* (Ed. by W. R. Dickinson). *Spec. Publ. Soc. econ. Paleont. Miner, Tulsa*, **22,** 190–204.

DE JONG, J.D. (1971) Molasse and clastic-wedge sediments of the southern Cantabrian Mountains (NW Spain) as geomorphological and environmental indicators. *Geol. Mijnb.* **50,** 399–416.

DE MEIJER, J.J. (1971) Carbonate petrology of algal limestones (Lois-Ciguera Formation, Upper Carboniferous, León, Spain). *Leidse geol. Meded.* **47,** 1–97.

DE SITTER, L.U. (1962) The Hercynian orogenes in northern Spain. In: *Some Aspects of the Variscan Fold Belt* (Ed. by K. Coe), pp. 1–18. University of Manchester Press,

De Sitter, L.U. (1965) Hercynian and alpine orogenies in northern Spain, *Geol. Mijnb.* **44,** 373–383.

De Sitter, L.U. & Boschma, D. (1966) Explanation geological map of the Palaeozoic of the southern Cantabrian Mountains, 1:50,000. Sheet 1 Pisuerga. *Leidse geol. Meded.* **31,** 191–238.

Dewey, J.F. & Burke, K.C.A. (1973) Tibetan, Variscan and Precambrian basement reactivation: products of continental collision. *J. Geol.* **81,** 683–692.

Dvorak, J., Mirouse, R., Paproth, E., Pelhate, A., Ramsbottom, W.H.C. & Wagner, R.H. (1977) Relations entre la sédimentation Eodevono-Carbonifère et la tectonique Varisque en Europe Centrale et Occidentale. In: *La Châine Varisque d'Europe Moyenne et Occidentale, Coll. intern. CNRS, Rennes,* **243,** 241–273.

Erxleben, A.W. (1975) Deltaic and related carbonate systems in the Pennsylvanian Canyon Group of north-central Texas. In: *Deltas, Models for Exploration* (Ed. by M. L. Broussard), *Houston geol. Soc.* 399–425.

Evers, H.J. (1967) Geology of the Leonides between the Bernesga and Porma rivers, Cantabrian Mountains, NW Spain. *Leidse geol. Meded.* **41,** 83–151.

Fisher, W.L. & McGowen, J.H. (1969) Depositional systems in the Wilcox Group (Eocene) of Texas and their relationship to occurrence of oil and gas. *Bull. Am. Ass. Petrol. Geol.* **53,** 30–54.

Flores, R.M. (1975) Short-headed stream delta: model for Pennsylvanian Haymond Formation, west Texas. *Bull. Am. Ass. Petrol. Geol.* **59,** 2288–2301.

Freund, R. (1971) The Hope fault, a strike-slip fault in New Zealand. *Bull. N.Z. geol. Surv.* **86,** 49 pp.

Garcia-Loygorri, A., Ortuno, G., Caride de Linan, C., Gervilla, M., Greber, Ch. & Feys, R. (1971) El Carbonífero de la Cuenca Central Asturiana. *Trab. Geol. Oviedo,* **3,** 101–150.

Ginkel, A.C. van (1965) Spanish Carboniferous fusulinids and their significance for correlation purposes. *Leidse geol. Meded.* **34,** 172–225.

Goester, F. (1973) *Sedimentologie van Cea-Afzettingen: bij Tejerina (prov. León, Spanje).* Unpublished M.Sc. Thesis, University of Leiden.

Graaff, W.J.E. van de (1971a) Three Upper Carboniferous, limestone-rich, high-destructive delta systems, with submarine fan deposits, Cantabrian Mountains, Spain. *Leidse geol. Meded.* **46,** 157–235.

Graaff, W.J.E. van de (1971b) The Piedrasluengas Limestone, a possible model of limestone facies distribution in the Carboniferous of the Cantabrian Mountains. *Trab. Geol. Oviedo,* **3,** 151–159.

Harland, W.B. (1971) Tectonic transpression in Caledonian Spitzbergen. *Geol. Mag.* **108,** 27–42.

Helmig, H.M. (1965) The geology of the Valderrueda, Tejerina, Ocejo and Sabero coal basins (Cantabrian Mountains, Spain). *Leidse geol. Meded.* **32,** 75–149.

Heward, A.P. (1978a) Alluvial fan and lacustrine sediments from the Stephanian A and B (La Magdalena, Ciñera-Matallana and Sabero) coalfields, northern Spain. *Sedimentology,* **25,** 451–488.

Heward, A.P. (1978b) Alluvial fan sequence and megasequence models: with examples from Westphalian D-Stephanian B coalfields, northern Spain, In: *Fluvial Sedimentology* (Ed. by A. D. Miall), *Mem. Can. Soc. Petrol. Geol.* **5,** 669–702.

Hooke, R. Le B. & Rohrer, W.L. (1977) Relative erodibility of source-area rock types, as determined from second-order variations in alluvial fan size. *Bull. geol. Soc. Am.* **88,** 1177–1182.

Johnson, G.A.L. (1973) Closing of the Carboniferous sea in western Europe. In: *Implications of Continental Drift to the Earth Sciences* (Ed. by D. H. Tarling & S. K. Runcorn), pp. 843–850. Academic Press, London.

Julivert, M. (1971a) Décollement tectonics in the Hercynian Cordillera of northwest Spain. *Am. J. Sci.* **270,** 1–29.

Julivert, M. (1971b) L'évolution structurale de l'arc asturien. In: *Histoire Structurale du Golfe de Gascogne, Inst. Français Petr. Colloques et Séminaires,* **22–1,** 1–28.

Kingma, J.T. (1958) Possible origin of piercement structures, local unconformities and secondary basins in the Eastern Geosyncline, New Zealand. *N.Z. J. Geol. Geophys.* **1,** 269–274.

Klein, G. de V. (1974) Estimating water depths from analysis of barrier island and deltaic sedimentary sequences. *Geology,* **2,** 409–412.

Knight, J.A. (1975) *The systematics and stratigraphic aspects of the Stephanian flora of the Sabero coalfield.* Unpublished Ph.D. Thesis, University of Sheffield.

Kraft, J.C. (1971) Sedimentary facies patterns and geologic history of a Holocene marine transgression. *Bull. geol. Soc. Am.* **82,** 2131–2158.

KULLMANN, J. & SCHÖNENBERG, R. (1978) Facies differentiation caused by wrench deformation along a deep-seated fault system (León line, Cantabrian Mountains, North Spain). *Tectonophysics*, **48,** T15–T22.

LAURENT, R. (1972) The Hercynides of south Europe, a model. *Proc. 24th Int. geol. Congr. Montreal*, **3,** 363–370.

LENSEN, G.J. (1958) A method of graben and horst formation. *J. Geol.* **66,** 579–587.

LE PICHON, X., SIBUET, J.C. & FRANCHETEAU, J. (1977) The fit of the continents around the North Atlantic ocean. *Tectonophysics*, **38,** 169–209.

LINK, M.H. & OSBORNE, R.H. (1978) Lacustrine facies in the Pliocene Ridge Basin Group: Ridge Basin, California. In: *Modern and Ancient Lake Sediments* (Ed. by A. Matter & M. E. Tucker), *Spec. Publ. int. Ass. Sediment.* **2,** 169–187.

LOBATO ASTORGA, L. (1977) *Geología de los valles altos de los ríos Esla, Yuso, Carrión y Deva*. Unpublished Ph.D. Thesis, University of Oviedo.

LOON, A.J. VAN (1972) A prograding deltaic complex in the Upper Carboniferous of the Cantabrian Mountains (Spain): The Prioro-Tejerina basin. *Leidse geol. Meded.* **48,** 1–81.

LOWELL, J.D. (1972) Spitsbergen Tertiary orogenic belt and the Spitsbergen fracture zone. *Bull. geol. Soc. Am.* **83,** 3091–3102.

MAAS, K. (1974) The geology of Liebana, Cantabrian Mountains, Spain: deposition and deformation in a flysch area. *Leidse geol. Meded.* **49,** 379–465.

MARCOS, A. (1968) Nota sobre el significado de la 'León line'. *Brevoria geol. Astur.* **7,** 1–5.

MARTINEZ-GARCIA, E. (1971) The age of the Caliza de Montaña in the eastern Cantabrian Mountains. *Trab. Geol. Oviedo*, **3,** 267–276.

MARTINEZ-GARCIA, E., CORRALES, I. & CARBALLEIRA, J. (1971) El flysch carbonífero de Pendueles (Asturias). *Trab. Geol. Oviedo*, **3,** 277–283.

MARTINEZ-GARCIA, E. & WAGNER, R.H. (1971) Marine and continental deposits of Stephanian age in eastern Asturias (NW Spain). *Trab. Geol. Oviedo*, **3,** 285–305.

MATTE, PH. (1968) La structure de la virgation hercynienne de Galice (Espagne). *Trav. Lab. Géol. Fac. Sci. Univ. Grenoble*, **44,** 1–128.

MCGOWEN, J.H. (1971) Gum Hollow fan delta, Neuces Bay, Texas. *Rep. Invest. Bur. econ. Geol. Texas*, **69,** 91 pp.

MCGOWEN, J.H. & SCOTT, A.J. (1974) Fan-delta deposition: processes, facies and stratigraphic analogues. (*Abstr.*) *Ann. Mtg. Am. Ass. Petrol. Geol.* 60–61.

MIALL, A.D. (1978) Tectonic and syndepositional deformation of molasse and other non marine-paralic sedimentary basins. *Can. J. Earth Sci.* **15,** 1613–1632.

MOORE, L.R., NEVES, R., WAGNER, R.H. & WAGNER-GENTIS, C.H.T. (1971) The stratigraphy of Namurian and Westphalian rocks in the Villamanín area of northern León, NW Spain. *Trab. Geol. Oviedo*, **3,** 307–363.

NILSEN, T.H. (1978) Late Cretaceous geology of California and the problem of the proto-San Andreas fault. In: *Mesozoic Palaeogeography of the Western United States* (Ed. by D. G. Howell & K. A. McDougall), *Pacific Section Soc. econ. Paleont. Miner., Los Angeles*, 559–573.

NILSEN, T.H. & CLARKE, S.H., Jr. (1975) Sedimentation and tectonics in the early Tertiary continental borderland of Central California. *Prof. Pap. U.S. geol. Surv.* **925,** 56 pp.

NORRIS, R.J., CARTER, R.M. & TURNBULL, I.M. (1978) Cainozoic sedimentation in basins adjacent to a major continental transform boundary in southern New Zealand. *J. geol. Soc. Lond.* **135,** 191–205.

PELLO, J. & CORRALES, I. (1971) Characteristics of the sedimentation of early Westphalian D rocks near the north-western border of the Central Asturian coalfield (Cordillera Cantábrica). *Trab. Geol. Oviedo*, **4,** 365–372.

READING, H.G. (1970) Sedimentation in the Upper Carboniferous of the southern flanks of the central Cantabrian Mountains, *Proc. Geol. Ass.* **81,** 1–41.

READING, H.G. (1975) Strike-slip fault systems: an ancient example from the Cantabrians. *IXth Int. Congr. Sedim. Nice 1975*, Thème **4,** 287–292.

RIDING, R. (1974) Model of the Hercynian fold belt. *Earth planet. Sci. Lett.* **24,** 125–135.

RIES, A.C. (1978) The opening of the Bay of Biscay—a review. *Earth Sci. Rev.* **14,** 35–63.

RIES, A.C. (1979) Variscan metamorphism and K-Ar dates in the Variscan fold belt of S Brittany and NW Spain. *J. geol. Soc. Lond.* **136,** 89–103.

RUPKE, J. (1965) The Esla Nappe, Cantabrian Mountains (Spain). *Leidse geol. Meded.* **32,** 1–74.

RUPKE, N.A. (1977) Growth of an ancient deep-sea fan. *J. Geol.* **85,** 725–744.

SAVAGE, J.F. (1967) Tectonic analysis of Lechada and Curavacas synclines, Yuso basin, León, NW Spain. *Leidse geol. Meded.* **39,** 193–247.

SAVAGE, J.F. (1979) The Hercynian orogeny in the Cantabrian Mountains, N Spain. *Krystalinikum,* **14,** 91–108.

SCHWAB, F.L. (1976) Modern and ancient sedimentary basins: comparative accumulation rates. *Geology,* **4,** 723–727.

SJERP, N. (1966) The geology of the San Isidro-Porma area (Cantabrian Mountains, Spain). *Leidse geol. Meded.* **39,** 55–128.

STAALDUINEN, C.J. VAN (1973) Geology of the area between the Luna and Torío rivers, southern Cantabrian Mountains, NW Spain. *Leidse geol. Meded.* **49,** 167–205.

STEEL, R.J. (1976) Devonian basins of western Norway—sedimentary response to tectonism and to varying tectonic context. *Tectonophysics,* **36,** 207–224.

STEEL, R.J. & AASHEIM, S.M. (1978) Alluvial sand deposition in a rapidly subsiding basin (Devonian, Norway). In: *Fluvial Sedimentology* (Ed. by A. D. Miall). *Mem. Can. Soc. Petrol. Geol.* **5,** 385–412.

SWIFT, D.J.P. (1968) Coastal erosion and transgressive stratigraphy. *J. Geol.* **76,** 445–456.

TANNER, W.F. (1976) Tectonically significant pebble types: sheared, pocked and second-cycle examples. *Sediment. Geol.* **16,** 69–83.

VEEN, J. VAN (1965) The tectonic and stratigraphic history of the Cardaño area, Cantabrian Mountains, northwest Spain. *Leidse geol. Meded.* **35,** 45–104.

WAGNER, R.H. (1966) Palaeobotanical dating of Upper Carboniferous folding phases in NW Spain. *Mem. Inst. geol. min. España,* **66,** 1–169.

WAGNER, R.H. (1970) An outline of the Carboniferous stratigraphy of Northwest Spain. *Congr. Coll. Univ. Liège,* **55,** 429–463.

WAGNER, R.H. (1971a). The stratigraphy and structure of the Ciñera-Matallana coalfield (prov. León, N.W. Spain). *Trab. Geol. Oviedo,* **4,** 385–429.

WAGNER, R.H. (1971b) Carboniferous nappe structures in north-eastern Palencia (Spain). *Trab. Geol. Oviedo,* **4,** 431–459.

WAGNER, R.H. & WAGNER-GENTIS, C.H.T. (1963) Summary of the stratigraphy of Upper Palaeozoic rocks in NE Palencia, Spain. *Proc. Kan. Nederl. Akad. Westensch. Amsterdam,* (B) **66,** 149–163.

WAGNER, R.H. & WINKLER PRINS, C.F. (1970) The stratigraphic succession, flora and fauna of Cantabrian and Stephanian A rocks at Barruelo (prov. Palencia), NW Spain. *Congr. Coll. Univ. Liège,* **55,** 487–551.

WAGNER, R.H. & FERNANDEZ-GARCIA, L. (1971) The Lower Carboniferous and Namurian rocks north of La Robla (León). *Trab. Geol. Oviedo,* **4,** 507–531.

WAGNER, R.H. & MARTINEZ-GARCIA, E. (1974) The relationship between geosynclinal folding phases and foreland movements in North-west Spain. *Stud. Geol.* **vii,** 131–158.

WAGNER, R.H., WINKLER PRINS, C.F. & RIDING, R.R. (1971) Lithostratigraphic units of the lower part of the Carboniferous in northern León, Spain. *Trab. Geol. Oviedo,* **4,** 603–663.

WAGNER, R.H., PARK, R.K., WINKLER PRINS, C.F. & LYS, M. (1977) The post-Leonian basin in Palencia: a report on the stratotype of the Cantabrian Stage. In: *Symposium on Carboniferous Stratigraphy* (Ed. by V. M. Holub & R. H. Wagner), *Spec. Pub. geol. Surv. Prague,* 89–146.

WEBB, G.W. (1969) Paleozoic wrench faults in Canadian Appalachians. In: *North Atlantic—Geology and Continental Drift* (Ed. by M. Kay). *Mem. Am. Ass. Petrol. Geol.* **12,** 754–786.

YOUNG, R. (1976) *Sedimentological studies in the Upper Carboniferous of north-west Spain and Pembrokeshire.* Unpublished D. Phil. Thesis, University of Oxford.

Spec. Publ. int. Ass. Sediment. (1980) **4,** 127–145

Strike-slip related sedimentation in the Antalya Complex, SW Turkey

A. H. F. ROBERTSON *and* N. H. WOODCOCK

Grant Institute of Geology, West Mains Road, Edinburgh EH9 3JW and Department of Geology, Downing Street, Cambridge CB2 3EQ

ABSTRACT

A variety of mostly ophiolite-derived rudaceous rocks was deposited during a major phase of strike-slip faulting which affected the Antalya Complex, SW Turkey, in latest Cretaceous and early Palaeogene times. The area originated in the late Triassic as marginal igneous crust associated with the early stages of continental separation and ocean basin formation. Overlying and adjacent sedimentary rocks record passive margin conditions prior to onset of wrench tectonics in the latest Cretaceous.

The sediments range from well bedded ophiolite-derived arenites to volumetrically abundant rudites, including massive to poorly stratified disorganized conglomerates and breccias, which are mostly matrix supported. The rudites, which are often sheared, range from monomict (e.g. gabbro or limestone breccia) to polymict with clasts from an entire ophiolite suite and its former sedimentary cover.

The sediments record a history of locally alternating transpression and transtension in an overall strike-slip regime. Transpression is represented by pervasive shearing, cataclasis and low-grade metamorphism associated with emplacement of mafic and ultramafic sheets. In contrast, transtensile gaps were largely filled passively by upward diapiric protrusion of serpentinite liberated from the ultramafic rocks of the underlying ophiolite suite. The strike-slip belt represents part of a major transcurrent zone, probably a fossil transform, developed along a former passive continental margin. Its overall sedimentary and structural features are distinct from both intra-oceanic and intra-continental transform belts.

INTRODUCTION

The Antalya Complex records the initiation, construction and later tectonic disruption of part of the continental margin of a small Mesozoic–Cainozoic ocean basin (Woodcock & Robertson, 1978). Structural and sedimentological data and regional comparisons (Robertson & Woodcock, 1980) show that a large segment of this margin underwent major strike-slip faulting during late Cretaceous to early Tertiary times. Here we focus on ophiolite-derived clastic sediments and serpentinite

0141-3600/80/0904-0127$02.00

protrusions associated with the strike-slip faulting. These are located within a zone of mafic igneous rocks representing late Triassic oceanic crust formed during the early stages of continental rifting.

GEOLOGICAL SETTING

The Antalya Complex comprises sedimentary rocks, mafic and ultramafic igneous rocks and volumetrically minor metamorphic rocks, previously termed the Antalya

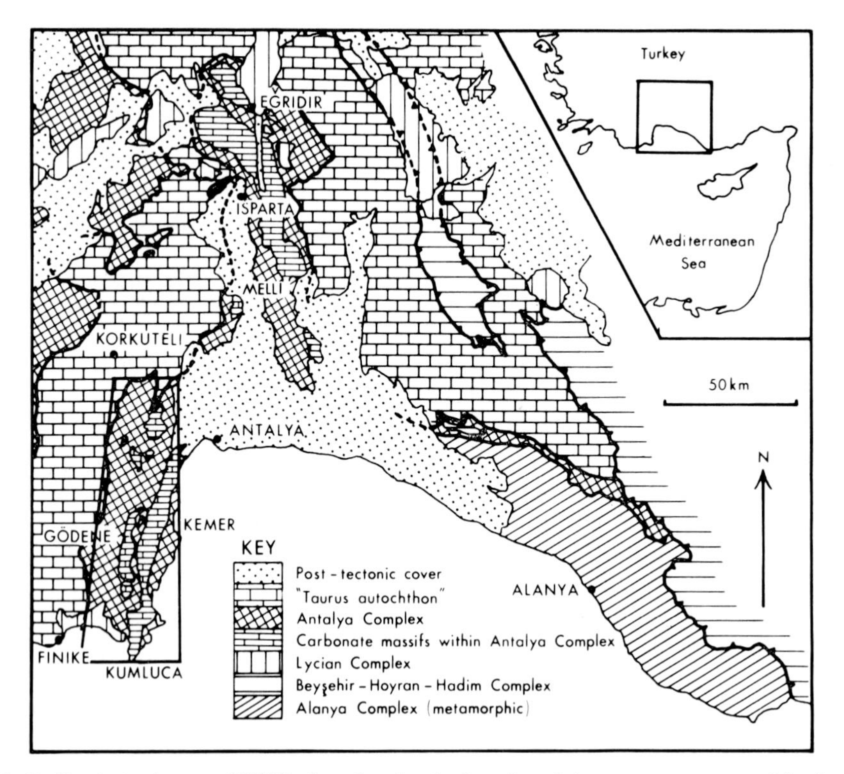

Fig. 1. Outline tectonic map of SW Turkey showing the location of the western segment of the Antalya Complex in relation to the East Mediterranean (inset). The eastern and western segments of the Antalya Complex are separated by the northward closing 'Isparta Angle'. West of Isparta, Antalya and Lycian rocks not fully differentiated.

Nappes (Fig. 1; Brunn *et al.*, 1971; Dumont *et al.*, 1972; Delaune-Mayere *et al.*, 1977). Whereas earlier studies have emphasized the dominance of low-angle thrust tectonics (Brunn *et al.*, 1970; Monod *et al.*, 1974; Monod, 1976; Ricou, Argyriadis & Lefèvre, 1974; Ricou, Argyriadis & Marcoux, 1975), recent work (Woodcock & Robertson, 1978, 1980) has demonstrated the importance of steep-dipping structures with a strike-slip component of movement. On the basis of detailed mapping, we have subdivided the southern part of the western Antalya Complex outcrop into five N–S trending zones (Woodcock & Robertson, 1980). Regional reconnaissance allows these zones to be extended northwards (Fig. 2) to encompass the whole of the western Antalya Complex outcrop. The zones are distinguished

Table 1. Characteristics of the tectonic zones of the western segment of the Antalya Complex

Zone	Bey Dağlari	Kumluca	Gödene	Kemer	Tekirova
Structural style	Upright megascopic folds, mesoscopic folds near Kumluca Zone contact, complex deformation near major N–S vertical faults	Cover stacked westward by N–S striking, E dipping imbricate thrusts, mesoscopic folding	Subvertical lozenges of basement + cover (mesoscopically folded) separated by major vertical strike-slip faults (many with protruded serpentinite)	Thick vertical or E dipping sheets, separated by major N–S high-angle faults	Detailed structure uncertain
Environmental interpretation	Mesozoic–Palaeogene carbonate platform on continental crust; Miocene 'flysch'	Mesozoic continental margin sediments originally on 'transitional' continental–oceanic basement	Late Triassic early rifting volcanics with overlying distal margin sediments or off-margin 'reefs'	Mesozoic carbonate margin sediments on continental crust	Late Cretaceous oceanic crust
'Cover' sequence	Very thick Mesozoic–Palaeogene dominantly neritic carbonates and overlying Miocene terrigenous clastics	Thin L. Triassic to U. Cretaceous clastic and hemipelagic sediments	Mainly U. Triassic to ?Cretaceous pelagic and hemipelagic sediments + some clastics and carbonate build-ups	Thick Triassic to Cretaceous, dominantly clastic carbonates continuous with 'basement'	Minor clastics of uncertain age
'Basement' sequence	Not exposed; Palaeozoic sediments occur further N	Not present? cover allochthonous	U. Triassic mafic extrusives and intercalated pelagic sediments	Partial Ordovician to Permian sedimentary sequence continuous upward with cover	U. Cretaceous mafic/ultramafic intrusive sequence

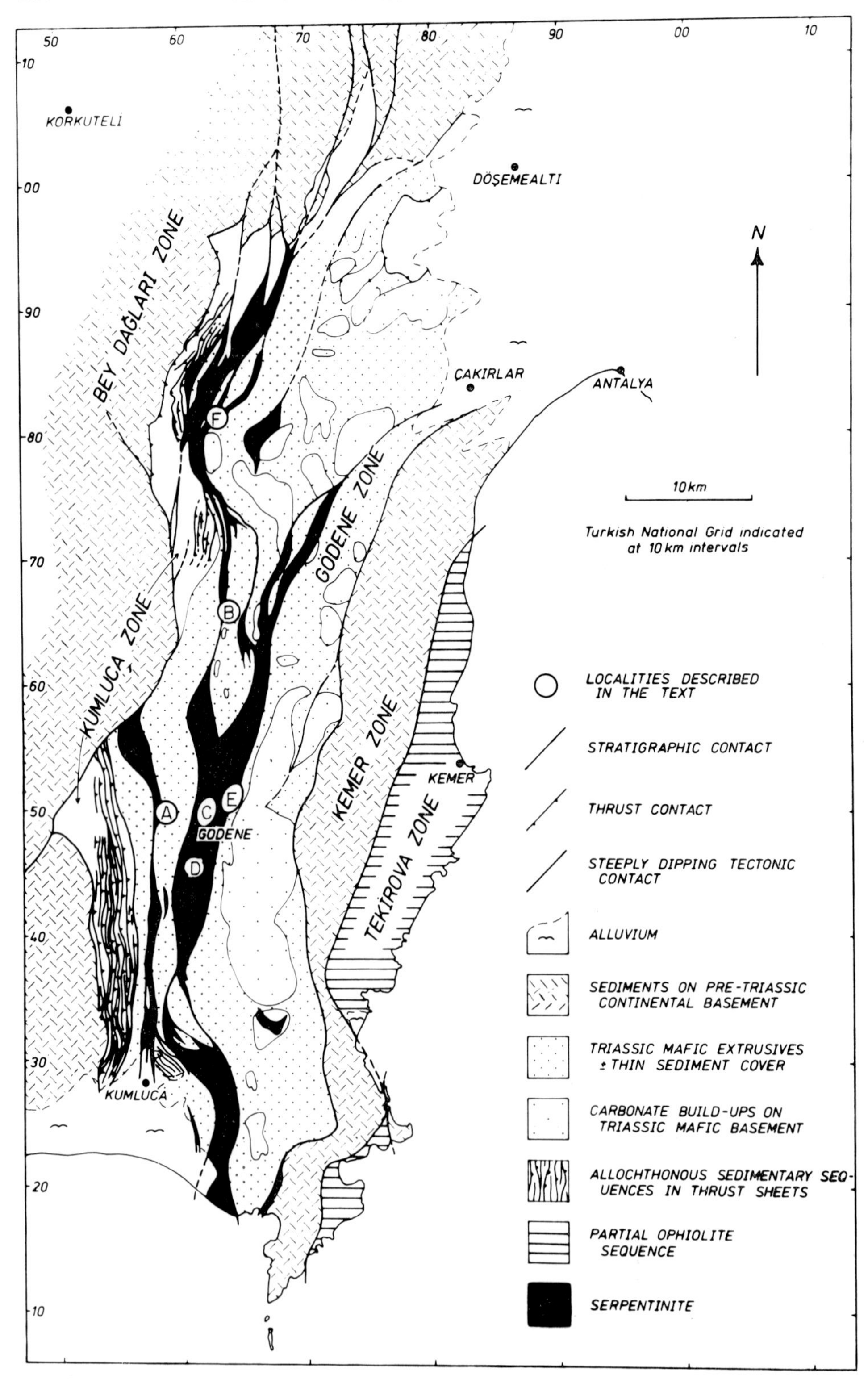

Fig. 2. Simplified geological map of the western segment of the Antalya Complex including the strike-slip belt, with locations of ophiolite-derived clastic sediments indicated

on the basis of basement type, sedimentary and igneous sequences and tectonic style (Table 1). All interzone contacts are tectonic except locally where stratigraphical continuity can be traced between parts of the Kumluca and the Gödene Zone, and possibly between the Gödene and the Kemer Zone.

As presently exposed, the Antalya Complex represents a tectonically dismembered segment of a Mesozoic passive continental margin (Woodcock & Robertson, 1978; Robertson & Woodcock, 1980). From west to east the tectonic zones record a transition from a Mesozoic carbonate platform probably floored by continental crust (Bey Dağlari Zone), across Mesozoic continental margin sediments (Kumluca Zone) into oceanic crust formed during the initial stages of continental rifting (Gödene Zone) (Table 1). The zones further east (Fig. 2) are seen as being tectonically displaced with respect to the western zones: they are composed of carbonate platform and basement lithologies (Kemer Zone) and portions of late Cretaceous (Thuizat & Montigny, in press) oceanic crust (Juteau *et al.*, 1977) (Tekirova Zone).

Most of the strike-slip related sedimentary rocks described here are tectonically intercalated with or overlie the Mesozoic rocks of the Gödene Zone. Lithologically these rocks are mostly mafic extrusive rocks, gabbro and serpentinized ultramafic rocks which are interpreted as a tectonically dismembered ophiolite suite, although a well developed sheeted dyke complex is probably not present (cf. Penrose Conference definition; Anon, 1972). The extrusive rocks comprise sequences up to 1·4 km thick of mafic pillow lavas and lava breccias (Juteau, 1975), intercalated with radiolarian cherts, late Triassic *Halobia*-bearing calcilutites, turbiditic sandstones and detached blocks of reef limestone. The extrusives are overlain by a variety of pelagic and hemi-pelagic siliceous and calcareous sedimentary rocks, including quartzose sandstones and siltstones, radiolarian mudstones and cherts, calcilutites, marls and chalks. Located within, or adjacent to, the main outcrop areas of the late Triassic ophiolite-derived rocks are major masses of shallow water and hemipelagic limestone up to several thousand metres thick, mostly ranging in age from late Triassic to late Cretaceous. Several of these massifs, including the Bakırlı Dağ (Delaune-Mayere *et al.*, 1977) discussed below, pass down with normal contact into pre-rift Permian sedimentary rocks (Kalafatçioğlu, 1973 and unpublished data). Consequently the now tectonically dissected carbonate massifs are interpreted as remnants of substantial carbonate build-ups on slivers of continental crust, isolated within the zone of late Triassic ophiolitic rocks. This evidence, combined with local westward lateral sedimentary transitions to the Kumluca Zone continental margin sediments and the alkaline geochemical trends in the extrusive sequences as a whole (Juteau, 1975), confirm that the Gödene Zone was located adjacent to a major continental landmass (Bey Dağları Zone) in the Mesozoic. In this respect the Gödene Zone differs from the late Cretaceous Tekirova Zone and the Troodos Complex (Cyprus), which are seen as representing younger, more axial, oceanic crust (Robertson & Woodcock, 1980).

STRUCTURAL EVIDENCE FOR STRIKE-SLIP TECTONICS

Structural style of serpentinites

In the southern part of the Antalya Complex type area the Gödene Zone is characterized by subvertical N–S screens of sheared serpentinite which cut the mafic lava sequences (Fig. 2; Woodcock & Robertson, in preparation). Structural features of the serpentinite screens collectively implying strike-slip displacements along them include (a) their braided, anastomosing pattern, (b) the subvertical attitude of their contacts and internal fabric, (c) their continuity with major high-angle faults, (d) the highly sheared nature of the serpentinite, (e) the presence of oblique, steeply plunging, or truncated structures in the lozenges between serpentinite screens. Several other features further imply that the serpentinites were emplaced upwards into the strike-slip fault zones as low-temperature diapiric intrusions ('protrusions'): (a) presence of tectonic inclusions of brecciated gabbro, harzburgite, and diabase (b) the absence of any along-strike high-level source for the serpentinite, (c) the absence of high-temperature contact aureoles.

Relationships with adjacent areas

Although strike-slip faulting apparently does not affect the southern segment of the Kumluca Zone (Fig. 2), the northern continuation of some Gödene Zone faults, which extend up to 80 km, can be traced through the Kumluca Zone and for 30 km into the adjacent Bey Dağları Zone carbonate platform. To the north they disappear under early Miocene flysch which underlies the Lycian Nappes (Fig. 1, Poisson, 1978). To the east, the Gödene Zone is in high-angle tectonic contact with the Kemer Zone, which comprises a series of mostly steeply E-dipping, N–S striking *en echelon* tectonic slabs up to 20 km long and 4 km wide. Sequences include Mesozoic shallow water and hemipelagic limestones resting on Palaeozoic basement lithologies (Dumont *et al.*, 1972; Brunn *et al.*, 1971; Baykal & Kalafatçioğlu, 1973; Kalafatçioğlu, 1973; Delaune-Mayere *et al.*, 1977). At least in the south, the Gödene Zone–Kemer Zone contact is marked by tectonic slivers of metamorphic low-grade pelites and psammites (Juteau, 1975) interpreted by us as metamorphosed continental margin lithologies. These, and the marked stratigraphic contrast across the contact, imply substantial displacement along strike-slip faults. Likewise the high angle tectonic contact between the Kemer Zone and the late Cretaceous Tekirova Zone may involve substantial lateral displacement.

Nature and timing of offset

Pervasive strike-slip faulting which affected at least the western part of the Antalya Complex is best documented in the mafic and ultramafic rocks of the Gödene Zone. Although conclusive evidence of the relative offset direction is lacking, regional comparisons (Robertson & Woodcock, 1980) suggest that during the late Cretaceous the Antalya margin accommodated a N–S sinistral transform fault associated with the northward subduction of oceanic crust, continuous with the Troodos Massif, beneath southern Turkey. In the south of the area the strike-slip faults more or less parallel the former Mesozoic passive margin, but in the north some fault strands diverge along the axis of the Isparta Angle (Fig. 1), while minor ones cut through the margin into the carbonate platform.

The timing of strike-slip faulting is constrained by continuity of pelagic sedimentation within the Gödene Zone until mid-late Cretaceous and by the presence of later Palaeogene and early Neogene (Akbulut, 1977) undeformed ophiolite-derived rudites (Asku Çay Formation) in the axis of the Isparta Angle. Comparable strike-slip faulting associated with serpentinite protrusion which occurs in geologically similar parts of SW Cyprus took place wholly within the late Cretaceous (Swarbrick, 1980). The bulk of the strike-slip faulting in the Gödene Zone was probably associated with deformation and emplacement of the Antalya Complex which took place in latest Cretaceous and early Palaeogene times.

STRIKE-SLIP RELATED SEDIMENTS

Clastic sediments related to strike-slip faulting have been encountered in five widely spaced areas in the Gödene Zone, marked A–F in Fig. 2: (A–B) minor wedges of oligomict ophiolite-derived breccias situated relatively close to the Kumluca Zone contact, (C) substantial deposits of unsheared ophiolite-derived rudites overlying sheared Gödene Zone rocks, (D) breccias associated with major masses of gabbro, (E) oligomict ophiolite-derived rudites locally overlying Mesozoic massive limestones, (F) sheared and unsheared oligomict rudite sand arenites tectonically intercalated with serpentinite and gabbro. Below, the term 'ophiolite-derived' is abbreviated to 'ophiolitic'.

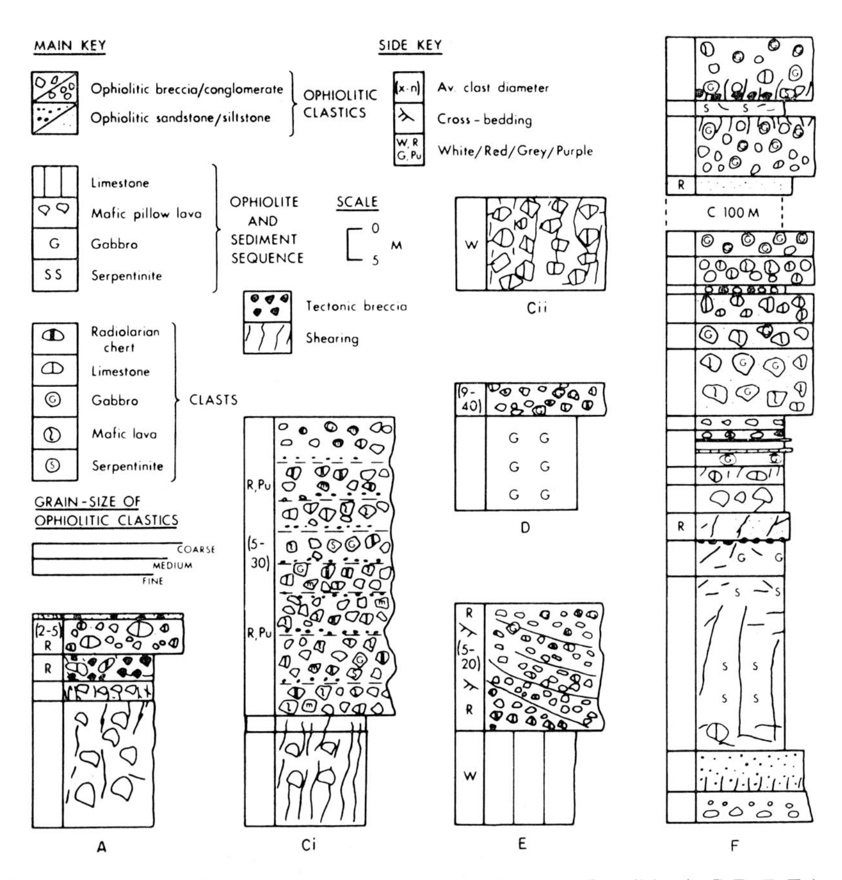

Fig. 3. Sedimentary logs of the ophiolite-derived clastic sediments at Localities A, C, D, E, F (see Figs 2, 5): Locality A, minor breccia wedges. Locality C(i), Çınarcık breccias, unsheared ophiolite-derived rudites. Locality C(ii), sheared limestone breccias. Locality D, plutonic rock breccias overlying gabbro, Kirkadinomek Hill. Locality E, Çınarcık breccias overlying Tekedağ limestones. Locality F, sheared ophioliotic clastics intercalated with serpentinite.

Localities A, B; minor breccia wedges

Small wedges of stratified ophiolitic breccias occur at several localities near the Gödene Zone–Kumluca Zone contact. In one case (Locality A, Figs 2 and 3) typical pillowed lavas with intercalated calcareous and siliceous pelagic sediments become increasingly sheared and brecciated upsection, then pass into oligomict tectonic breccia composed of angular clasts of crushed radiolarite and pelagic calcilutite in a sheared friable lava matrix. Overlying compositionally similar breccias show sedimentary fabrics. Clasts up to 0·25 m in diameter are supported in a matrix of silt and sand-sized derived radiolarian chert. The breccias show a tendency to fine upwards; clasts are uniformly angular and poorly sorted. Some individual beds are virtually monomict.

Numerous wedges of coarser-grained breccias also crop out towards the Kumluca Zone contact in the NW of the area (Locality B, Fig. 2). There, clasts up to 1·5 m in diameter of mafic lavas with subordinate diabase, gabbro, serpentinite and pelagic sedimentary rocks occur in a matrix of structureless volcaniclastic silt. Most of these breccias are strongly sheared, but a crude sedimentary fabric is still visible in most cases. Clast type is normally closely related to adjacent rocks (cf. below).

Locality C; ophiolitic rudites resting on sheared ophiolite

Much more extensive developments of mostly ophiolitic clastics were identified by Yilmaz (1978) near Çınarcık, N of Gödene (Fig. 2). Mapping of the type area (Fig. 6) reveals up to 70 m thick sequences of poorly consolidated undeformed ophiolitic rudites which are crudely stratified and only gently to moderately inclined. The sediments range from massive and crudely bedded matrix-supported rudites (units 1–2 m thick), to laminated ophiolitic arenites showing traces of grading. Clasts in the

Fig. 4. Positive print of a thin section of the Çınarcık breccias, Locality C. Note the highly angular, poorly sorted ophiolitic clasts including fresh and altered mafic lava, altered gabbro, radiolarian siltstone and sheared chert. See text for additional explanation.

rudites reach 0·4 m in diameter and locally show weakly developed imbrication (Fig. 7). They comprise lithologies from the whole igneous suite and its former sedimentary cover, including serpentinite, variably serpentinized ultramafics, gabbro, diabase, mafic lava, radiolarian chert, pelagic limestone and rare quartzose sandstone (Fig. 7).

The predominantly ophiolitic rudites show interstratification of horizons grading from ophiolitic rudites with only scattered limestone clasts (Fig. 7) to almost pure limestone breccias (Fig. 3). These breccias, which are generally more lithified than their ophiolitic counterparts, are composed of clasts of calcarenite and calcilutite

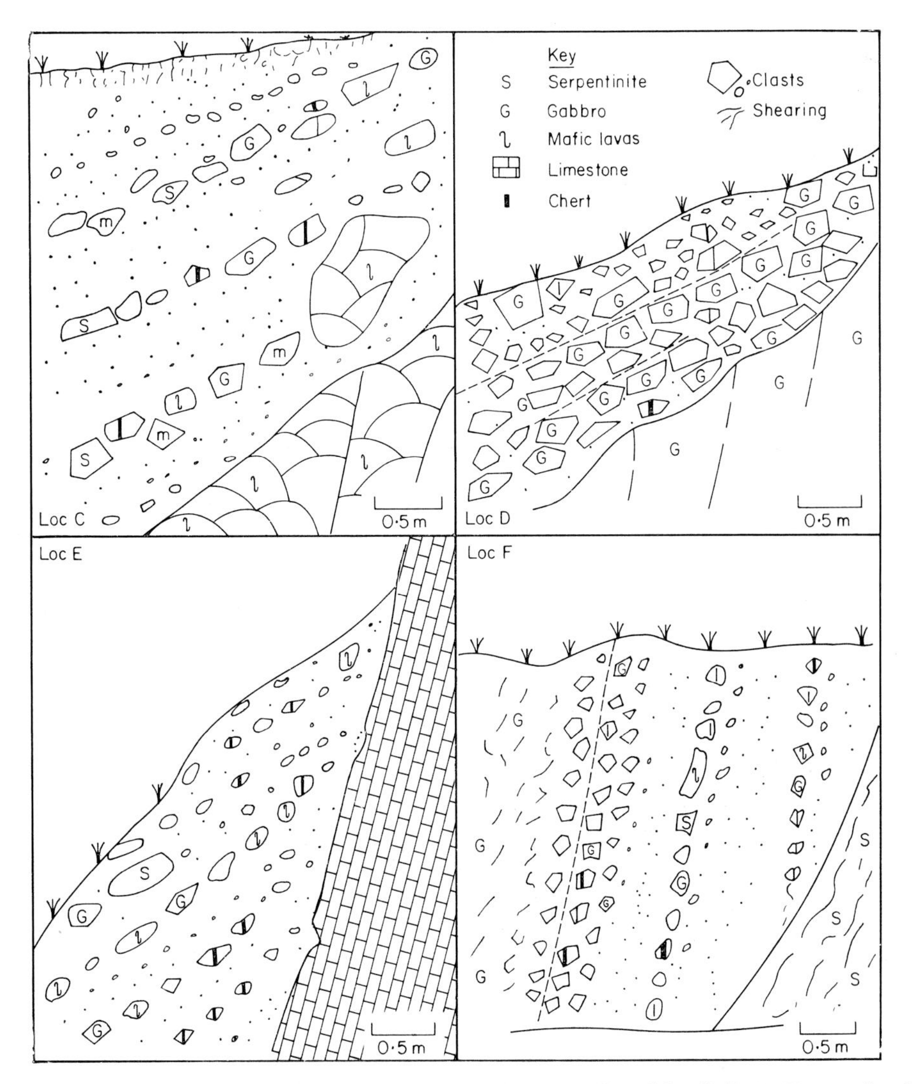

Fig. 5. Field sketches of ophiolite-derived clastic sediments at Localities C, D, E, F: Locality C, Oligomict ophiolite-derived rudites overlying pillow lavas at the type area near Çınarcık. Locality D, Gabbroic breccias overlying gabbro intruded by diabase dykes, Kirkadinomek Hill. Locality E, Steeply dipping oligomict rudites locally unconformably overlying the Mesozoic Tekedağ limestones, Locality F, Sheared oligomict rudites intercalated with serpentinite.

lithologically comparable to the Tekedağ Massif, now located several kilometres to the east beyond a major N–S trending sheet of serpentinite (Fig. 6).

In thin section (Fig. 4) the finer ophiolitic rudites consist of clasts of all the ophiolitic and sedimentary rocks as well as grains of polycrystalline quartz presumably derived from turbiditic sandstones intercalated within the mafic lavas. Scattered grains of sheared radiolarite, siltstone, lava and gabbro are abundant. Ferromagnesian minerals often show a network of fine cracks. Deformed grains are extensively replaced by microspar-sized calcite. Carbonate grains show extensive pressure solution. In contrast, many of the unstrained grains of basalt and gabbro are surprisingly fresh and unrecrystallized. Sheared rocks are represented by grains of recrystallized limestone and cleaved chloritic siltstone.

At Çınarcık (Locality C) the oligomict rudites are seen to rest unconformably on steeply inclined sheared and brecciated calcilutites and mafic pillow lavas typical of the Gödene Zone (Fig. 5a). Close to the contact the calcilutites exhibit a marked cleavage oblique to bedding, transitional over several metres into zones of tectonic breccia with sheared angular clasts up to 0·05 m in diameter. Detached blocks of the sheared and brecciated rocks are seen as rafts in the basal ophiolitic rudites, which are themselves undeformed. The sheared and brecciated limestones are recognizable as late Triassic *Halobia* limestones originally deposited within the lavas. In the surrounding area the mafic lavas and associated sedimentary rocks are again intensely sheared and extensively brecciated. To the east the exposures are obliquely truncated by a major N–S striking sheet of vertically dipping serpentinite which separates Locality C from the Tekedağ limestone massif and associated ophiolitic rudites (Locality E).

Locality D; plutonic rock breccias

Distinctive breccias composed predominantly of gabbroic and ultramafic rocks overlie pillow lavas and serpentized harzburgite in several areas (Figs 2, 5). Near Kirkadinomek Tepe, first described by Yilmaz (1978), medium to mafic gabbros cut by occasional diabase dykes are then depositionally overlain with normal contact by virtually matrix-free clast-supported breccias. These are composed mostly of gabbro but with subordinate volumes of other mafic igneous and sedimentary rocks (Figs 3, 5). The relationship of these breccias to the Çınarcık rudites is unknown.

Locality E; ophiolitic rudites on limestone

Ophiolitic rudites are again found to the NE along the contact of the major Tekedağ carbonate massif (Fig. 2). As depicted in Figs 3 and 5, massive subvertical recrystallized Mesozoic limestones (?Jurassic) are unconformably overlain by steeply dipping crudely cross-bedded ophiolitic breccias. At least locally, derivation was from the NE. Radiolarite is particularly abundant in the basal horizons but surprisingly, in contrast to Locality C, clasts from the stratigraphically underlying limestones are not present.

Locality F; ophiolitic rudites and serpentinites

A variety of ophiolitic clastics are found intercalated with serpentinite and gabbro within a ca 8×3 km, N–S trending zone located north of the major Bakırlı Dağ massif (Figs 2, 8). This massif, like the Tekedağ to the south, originated as part of a major carbonate build-up within the Gödene Zone and rests in part on pre-rift Permian

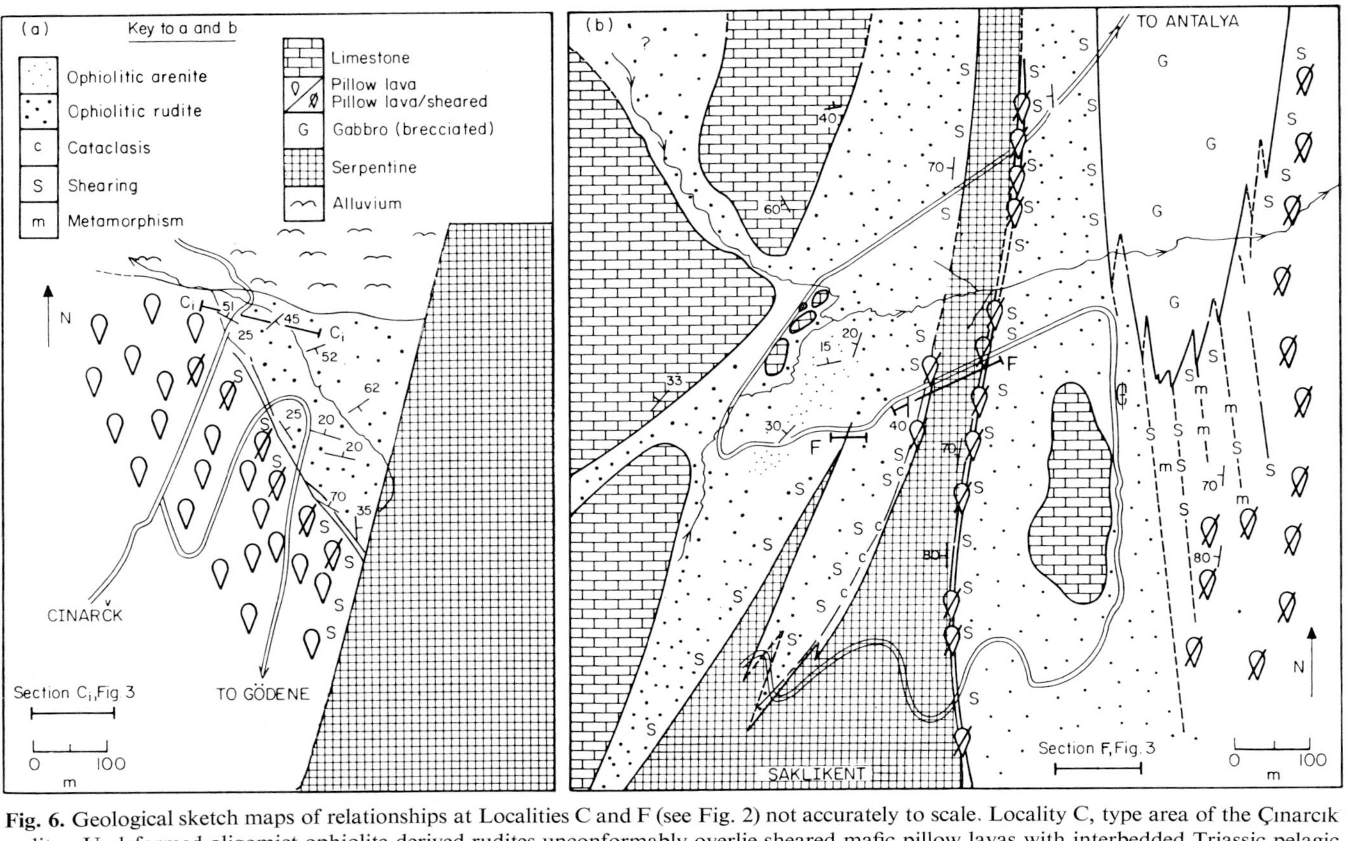

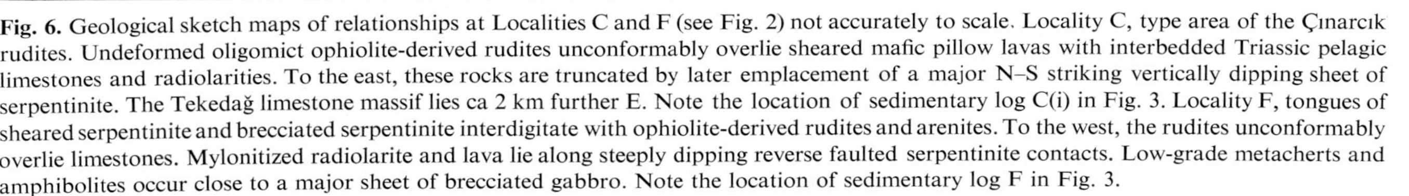
Fig. 6. Geological sketch maps of relationships at Localities C and F (see Fig. 2) not accurately to scale. Locality C, type area of the Çınarcık rudites. Undeformed oligomict ophiolite-derived rudites unconformably overlie sheared mafic pillow lavas with interbedded Triassic pelagic limestones and radiolarities. To the east, these rocks are truncated by later emplacement of a major N–S striking vertically dipping sheet of serpentinite. The Tekedağ limestone massif lies ca 2 km further E. Note the location of sedimentary log C(i) in Fig. 3. Locality F, tongues of sheared serpentinite and brecciated serpentinite interdigitate with ophiolite-derived rudites and arenites. To the west, the rudites unconformably overlie limestones. Mylonitized radiolarite and lava lie along steeply dipping reverse faulted serpentinite contacts. Low-grade metacherts and amphibolites occur close to a major sheet of brecciated gabbro. Note the location of sedimentary log F in Fig. 3.

lithologies (Delaune-Mayere *et al.*, 1977). Other smaller carbonate bodies (Fig. 6) mostly located west of the ophiolitic clastic outcrops are probably tectonically fragmented and subsided remnants of another large, once coherent, carbonate complex. Ophiolitic rudites are particularly well exposed along the road from Saklikent to Antalya (Fig. 6).

The ophiolitic rudites interfinger with N–S striking sheets of serpentinized ultramafic rocks which dip at 70°–90° to the W (Fig. 6). Internally the serpentinites are massive or blocky; towards the margins a marked pseudoconglomeratic texture is seen (Fig. 7). Intensity of shearing increases towards the contacts until in the marginal several metres the original igneous textures are completely destroyed. Where relatively competent lithologies, for example radiolarite, are present along the contact, bedding planes have been rotated parallel to the serpentinite fabric. Serpentinite has been locally intruded along and across bedding planes (Figs 7, 8), then sheared, flattened (up to 10:1) and deformed along with the chert beds into angular folds. Extensive cataclasites are generally found along contacts where ultramafic sheets have been

Fig. 7. Field photographs of Localities C and F (Fig. 2). (a) Crudely stratified matrix-supported ophiolite-derived rudite. Note scattered limestone clasts (white) and virtual absence of sorting. (b) Almost massive ophiolite-derived rudite rich in limestone clasts (white). Note angular to sub-rounded nature of most clasts. (c) Sheared serpentinite close to the contact with ophiolite-derived rudites. Note the high angle deformation fabric and pseudoconglomeratic texture. (d) Flattened and subsequently kinked radiolarian cherts within 1 m of a major sheet of ultramafic rocks. The pale grey zones (lower middle) are serpentinite injected along bedding planes, then sheared.

Fig. 8. Field photographs of Locality F. (a) Bakırlı Dağ carbonate massif (in distance), part of a former major carbonate complex constructed suprajacent to the Triassic Gödene Zone mafic lavas. Ophiolitic rudites crop out in the wooded area in the middle distance. (b) Stratified ophiolite-derived fine rudites unconformably overlying a tectonized subsided block of massive Mesozoic limestone composed mostly of reef talus. (c) Almost massive, matrix supported ophiolite-derived rudites steeply dipping in conformity with adjacent serpentinite sheet. (d) Unusually fine-grained, well stratified, undeformed ophiolite-derived arenites deposited in the axis of a small basin (see Fig. 9). A lateral passage to sheared ophiolite-derived rudites is observed over ca 10–30 m.

emplaced along dip-slip reverse faults. Some degree of thermal metamorphism is indicated by local occurrence of low-grade amphibolite and recrystallized radiolarian chert coloured brilliant purple rather than its usual red. These low-grade metamorphic rocks are well developed immediately north of a major mass of tectonically brecciated layered gabbro (Fig. 6).

The serpentinites are typically sheathed by a narrow zone of sheared lava (Fig. 6) and then pass laterally into ophiolitic rudites. Close to this contact, which is tectonic, the rudites consist of matrix-supported disorganized conglomerates and breccias with a crude superimposed tectonic fabric. These conglomerates vary from stratified to massive and from virtually monomict (e.g. gabbro or mafic lava) to oligomict with all the ophiolitic rocks present as clasts (Fig. 8). One of these oligomict horizons, for example, comprises subrounded clasts of lava and gabbro in a soft brown unconsolidated silty ophiolitic matrix. Gabbro blocks reach diameters of 1 m (Fig. 3) but typically the ophiolitic clasts range up to 0·2 m in diameter. Much larger isolated blocks of pelagic calcilutite are taken to be tectonically incorporated sedimentary intercalations in the original late Triassic lava sequence.

Traced westward away from the serpentinite contact the ophiolitic sediments are more gently inclined, more organized and generally finer grained (Figs 6, 8). Sequences up to 7 m thick of almost flat-lying well bedded ophiolitic arenites with stringers of pebbles are seen to alternate with grey-green fine ophiolitic arenites showing traces of grading. The grey-green colour of these sediments contrasts with the red-purple of the Çınarcık ophiolitic rudites in the south (Localities C, E). Further west the ophiolitic arenites pass into rudites with increased numbers of limestone clasts and locally overlie unconformably a major tectonically fragmented Mesozoic limestone mass (Fig. 6).

To the south of the mapped area (Fig. 6) the interdigitating strands of serpentinite merge to form a single steeply dipping arcuate sheet which can be traced over 20 km to the south (Fig. 2). Northwards the serpentinite and gabbro sheets extend over 5 km, together with extensive ophiolitic rudites.

INTERPRETATION

The braided subvertical serpentinite screens are interpreted as low-temperature protrusions of serpentinite driven upwards by the volume increase and density decrease accompanying serpentinization of ultramafic rocks presumed to originate below the Gödene Zone lava pile. Serpentinite protrusion was probably favoured by a component of extension perpendicular to a strike-slip fault strand: the transtensile regime of Harland (1971). Local segments of transtension and transpression must occur during displacement along any non-linear strike-slip fault, and regional transtension or transpression will result from gross non-parallelism of the fault zone with its slip vector. The deep-rooted Gödene Zone faults probably allowed ingress of water to the lower levels of the marginal oceanic crust, and the resulting serpentinite was then protruded upwards, particularly to fill transtensional gaps.

Features of the ophiolitic clastic sediments consistent with an origin related to strike-slip faulting include (a) extremely localized distribution in elongate zones parallel to the tectonic grain of the area, (b) occurrence at a variety of structural levels, precluding simple deposition as an unconformable blanket over Mesozoic rocks, (c) extreme heterogeneity of clast types, sometimes including lithologies not now seen in immediately adjacent areas, (d) rapid deposition from a series of tectonically active zones, (e) a history of progressive but localized post-depositional deformation, (f) association with cataclastic and low-grade metamorphic rocks not known elsewhere in the Gödene Zone. Consequently we envisage the various ophiolitic clastics as being the result of major tectonic movements characteristic of strike-slip faulting.

Specific tectonic histories can be inferred from individual sediment occurrences. For example, several of the volumetrically minor ophiolitic breccias (Locality A) can be interpreted as the fill of small V-shaped depressions which formed in the pillow lava surface, perhaps along transtensional fault strands. The absence of contained serpentinite in this area implies that the basins were superficial and short-lived. Any further deformation in the area was taken up along separate strands. In contrast the other rudites, including both the sheared and unsheared ophiolitic breccias, limestone breccias and plutonic rock breccias, are volumetrically substantial and involve all parts of the ophiolite suite and its sedimentary cover.

The Çınarcık breccias (Localities C, E, Fig. 2) accumulated rapidly along ephemeral fault scarps. Non-marine deposition of the sediments is indicated by the uniformly oxidized nature of the sediments, the absence of fossils and of interbedded carbonates. Deposition as a series of water-laden slurry flows is suggested by the sedimentary structures, especially the disorganized matrix-supported ophiolitic rudites with local grading and clast imbrication, and also the initially highly variable angles of sediment repose.

In the type area a steeply dipping 'parent' face of limestones and mafic lavas can be identified, but other clasts must have been derived from lithologies not now exposed in immediately adjacent areas. To the east the ophiolitic rudites are now truncated by a major vertical N–S trending ophiolite sheet, which must have been emplaced later in the tectonic history of the area.

Ophiolitic rudites without limestone clasts which locally overlie the Tekedağ

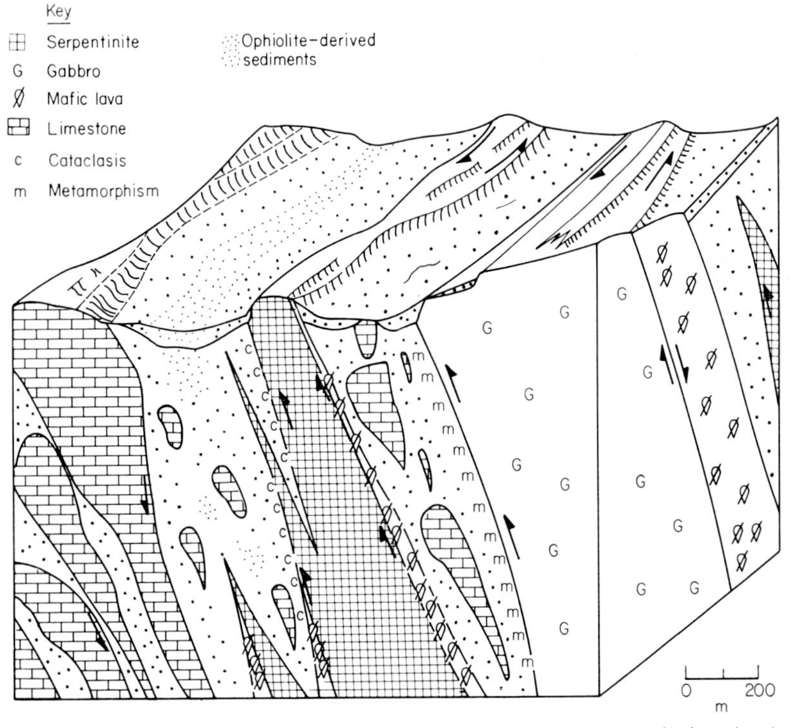

Fig. 9. Block diagram to illustrate the tectonic setting of ophiolite-derived rudites inferred from field relations at Locality F. Interdigitating sheets of ultramafic rocks and gabbro were emplaced by strike-slip faulting with a local reverse dip-slip component. Limestone masses formerly suprajacent to the ophiolitic rocks were dissected then subsided. Mylonite developed along reverse dip-slip ultramafic rock contacts. Low-grade metacherts and amphibolites formed along gabbro sheet margins. Coarse rudites were shed from uplifted ophiolitic rocks and limestones into small elongate basins where finer grained locally undeformed ophiolite-derived arenites accumulated. In contrast, transtensile areas were marked by extensive vertical serpentinite protrusion with minimal sediment deposition.

margin attest to the strong relative subsidence of the limestones which formerly lay suprajacent to the ophiolitic rocks. The extent of tectonic movements is confirmed by the pervasively sheared and brecciated nature of the Gödene Zone rocks in the vicinity of the ophiolitic clastics. In places plutonic rocks, for example gabbro, were exposed and eroded, probably subaerially, to produce scree deposits which were originally steeply inclined.

A comparable mode of origin is envisaged for the ophiolitic clastics at Locality F, where interdigitation with serpentinite and gabbro breccia is seen. The approximate geometry of an original elongate, ca 1–2 km wide, N–S trending sedimentary basin can be inferred (Fig. 9). The basin margins were bounded by steep scarps of serpentinite, lava and brecciated gabbro in the east and by tectonically dissected and subsided limestone blocks in the west. Intense shearing associated with the uplift of ultramafic rocks gave rise to cataclasis along contacts. Possibly complete serpentinization did not take place until late in the tectonic history. The low-grade metacherts and amphibolites are best developed close to the contact with a major mass of tectonically brecciated and sheared gabbro. Heat was probably produced by friction during emplacement of the relatively competent gabbros. From the uplifted scarps coarse poorly stratified rudites were shed, mostly by mass flow mechanisms. Variable degrees of mixing of clast type took place to form a range of rudites from virtually monomict to polymict. In more axial areas of the basin, almost flat-lying ophiolitic arenites and rudites were laid down. The green colour and sedimentary structures suggest subaqueous deposition, possibly under the influence of ephemeral streams in an otherwise arid area. Continued faulting deformed the more proximal rudites, especially close to the serpentinites, but locally the more axial arenites escaped deformation (Fig. 9).

Despite these interpretations, the tectonic settings of individual deposits tend to record only the site of relative uplift and subsidence, not the underlying tectonic mechanism nor the deformation history along the belt. For instance, superficially similar sedimentary fills might be produced in either a transtensile basin bounded by normal faults or in a transpressive basin bounded by reverse faults. The vast majority of the serpentinite screens, those which are vertical with no marginal metamorphic effects, probably represent passive protrusion of serpentinite along transtensile strands. The unsheared rudites in the SE (Çınarcık breccias, Locality C) and some of the minor sediment occurrences (Locality A) may have developed in transtensile basins. In contrast at Locality F the ultramafic sheets were emplaced with a marked reverse dip-slip component with strong localized compression leading to genesis of mylonites and low-grade metamorphic rocks. Consequently the resulting sediments were probably laid down in a basin intimately related to transpressive fault strands.

Even such considerations may only be of local significance in a setting where any part of the strike-slip belt may suffer a number of transpressive and transtensile episodes during its total history (Fig. 10). The ophiolitic rudites represent only a minute part of the entire complex. In most areas serpentinites were apparently rapidly protruded up available gaps, preventing the development of deep sedimentary basins. Most of the major ophiolitic rudite occurrences lie close to the major carbonate masses, interpreted as tectonically dissected portions of formerly extensive carbonate complexes constructed on slivers of continental crust within the ophiolite zone. Much of the strong uplift associated with strong local compression, as at Locality F, may be a response to impingement by strike-slip strands on continental slivers and their thick carbonate cover.

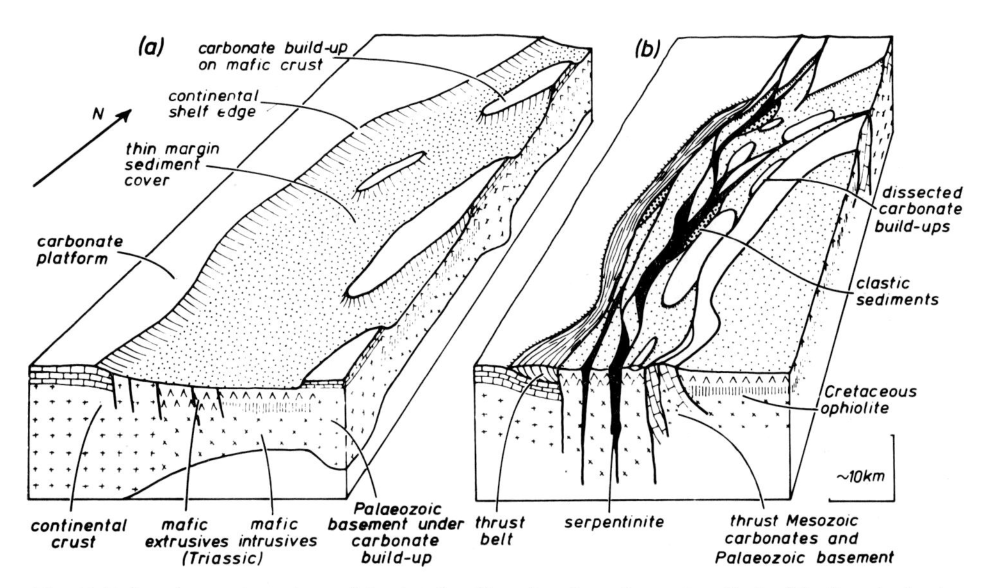

Fig. 10. Inferred tectonic settings of the Antalya Complex along the western limb of the Isparta Angle. (a) During the Mesozoic passive margin phase which preceded wrench faulting, (b) in the latest Cretaceous or early Tertiary towards the end of the phase of strike-slip faulting. Note the locally developed zones of transpression and transtension, and the tectonic dissection of the offshore carbonate massifs.

GENERAL COMPARISONS

The combination of serpentinite protrusion and ophiolitic clastic sediment deposition relate to an important phase of strike-slip tectonics in the Antalya Complex, probably active during late Cretaceous and early Palaeogene times. The strike-slip faults essentially paralleled and probably exploited structural weaknesses in a former Mesozoic passive continental margin (Fig. 10). Comparable linear serpentinite bodies in geologically analogous rocks of the Mamonia Complex, SW Cyprus (Robertson & Woodcock, 1979) have recently been reinterpreted (Swarbrick, 1980) in terms of Cretaceous strike-slip tectonics. On a larger scale, the strike-slip belt we document may lie along an important Mesozoic–Cainozoic plate boundary. Biju-Duval, Dercourt & Le Pichon (1977) postulate that the boundary between the 'Apulian microplate' to the west and the 'Anatolian microplate' to the east may run N–S through the Isparta Angle (Fig. 1). Consistent with this, Mesozoic rocks including ophiolites along the western limb of the Isparta Angle (Melli area) show pervasive vertical tectonic fabrics imposed in either late Cretaceous or early Tertiary time. The strike-slip faulting in the Gödene Zone and the possible strike-slip displacements involved in emplacement of the Kemer and Tekirova Zones can all be attributed to a major inter-plate transform fault (Fig. 10). More geological data are needed to test this hypothesis.

Comparison of the Gödene Zone strike-slip zone with possible ancient and modern analogues reveals some features in common with both intra-oceanic and intra-continental strike-slip zones, but there are also important differences. Serpentinite protrusions in intra-oceanic transform zones (e.g. Thompson & Melson, 1972; Bonatti & Honnorez, 1976) are known to fill transtensile gaps in the way that we envisage for the Gödene Zone. However, along intra-oceanic transform faults much of the topographic relief is produced by differences in age and thus thermal maturity across the fracture zone, whereas the Gödene Zone marginal oceanic rocks had reached thermal equilibrium long before onset of wrench tectonics, hence the absence of contact aureoles. By contrast, intracontinental transform faults commonly intersect relatively strong, low-ductility, salic rocks at depth: upward infill of transtensile zones by protrusion is less characteristic, and thick, lithologically heterogeneous sedimentary infill can form in pull-apart basins (e.g. Crowell, 1974; Steel, 1976; Norris, Carter & Turnbull, 1978). An interesting feature of the Gödene Zone strike-slip belt is the presence of both oceanic and continental characteristics, consistent with its proposed continental margin setting (Fig. 10) involving a number of offshore carbonate massifs founded on small blocks of continental basement. Compared with intra-continental transforms the serpentinite protrusions through the marginal oceanic crust restrict the depth and consequent thickness and extent of sediments, yet, particularly where the strike-slip strands impinge on small continental fragments, there is more evidence of transpression and varied sediment infill than in the oceanic environment (Fig. 10). A generally analogous setting may have existed during the Jurassic in the foothills metamorphic belt of the Sierra Nevada, California (Saleeby 1977, 1979). However, ancient examples of strike-slip dominated continental margins are only sparsely recorded in the literature.

ACKNOWLEDGMENTS

This paper stems from a project funded by the British Natural Environment Research Council and the Turkish Geological Survey (M.T.A.). We particularly thank E. Demirtaşli and R. Kengil for logistical support in the field. P. Yilmaz first drew our attention to the Çınarcık breccias. J. Waldron commented on the manuscript.

REFERENCES

Akbulut, A. (1977) Étude géologique d'une partie du Taurus occidental au sud d'Egridir (Turquie). *Thèse 3ème cycle, Université Paris-Sud, Orsay.*

Anon (1972) Penrose Field Conference: Ophiolites. *Geotimes*, **17**, 24–25.

Baykal, F. & Kalafatçioğlu, A. (1973) New geological observations in the area west of Antalya Bay. *Bull. Miner. Res. Explor. Inst. Turkey*, **80**, 33–42.

Biju-Duval, B., Dercourt, J. & Le Pichon, X. (1977) From the Tethys Ocean to the Mediterranean Sea: A plate tectonic model of the evolution of the western Alpine system. In: *Structural history of the Mediterranean basins* (Ed. by B. Biju-Duval and L. Montadert), pp. 143–164. Éditions Technip, Paris.

Bonatti, E. & Honnorez, J. (1976) Sections of the earth's crust in the equatorial Atlantic. *J. geophys. Res.* **81**, 4104–4116.

Brunn, J.H., De Graciansky, P.Ch., Gutnic, M., Juteau, T., Lefèvre, R., Marcoux, J., Monod, O. & Poisson, A. (1970) Structures majeures et corrélations stratigraphiques dans les Taurides occidentales. *Bull. Soc. géol. Fr.*, **12**, 515–556.

Brunn, J.H., Dumont, J.F., De Graciansky, P.Ch., Gutnic, M., Juteau, T., Marcoux, J., Monod, O. & Poisson, A. (1971) Outline of the geology of the western Taurides. In: *Geology and history of Turkey*. (Ed. by A. S. Campbell), pp. 225–255. Petrol. Exploration Soc. Libya, Tripoli.

Crowell, J.C. (1974) Sedimentation along the San Andreas Fault. In: *Modern and ancient geosynclinal sedimentation* (Ed. by R. H. Dott Jr. & R. H. Shaver), *Spec. Publ. Soc. econ. Paleont. Miner. Tulsa*, **19**, 292–303.

Delaune-Mayere, M., Marcoux, J., Parrot, J.-F. & Poisson, A. (1977) Modèle d'évolution Mésozoique de la paléomarge Tethysienne au niveau des nappes radiolaritiques et ophiolitiques du Taurus Lycien, d'Antalya et du Baër-Bassit. In: *Structural history of the Mediterranean basins.* (Ed. by B. Biju-Duval & L. Montadert), pp. 79–94. Éditions Technip, Paris.

Dumont, J.F., Gutnic, M., Marcoux, J., Monod, O. & Poisson, A. (1972) Le Trias des Taurides occidentales (Turquie). Définition du bassin pamphylien: Un nouveau domaine à ophiolithes à la marge externe de la chaîne taurique. *Z. dt. geol. Ges.* **123,** 385–409.

Harland, W.B. (1971) Tectonic transpression in Caledonian Spitsbergen. *Geol. Mag.* **108,** 27–42.

Juteau, T. (1975) Les ophiolites des nappes d'Antalya (Taurides occidentales, Turquie). *Mém. Sciences Terre,* 32.

Juteau, T., Nicolas, A., Dubessy, J., Fruchard, J.C. & Bouchez, J.L. (1977) Structural relationships in the Antalya ophiolite complex, Turkey: Possible model for an oceanic ridge. *Bull. geol. Soc. Am.* **88,** 1740–1748.

Kalafatçioğlu, A. (1973) Geology of the western part of Antalya Bay. *Bull. Miner. Res. Explor. Inst. Turkey,* **81,** 31–84.

Monod, O., Marcoux, J., Poisson, A. & Dumont, J-F. (1974) Le domaine d'Antalya, témoin de la fracturation de la plateforme africaine au cours du Trias. *Bull. Soc. géol. Fr.* **16,** 116–127.

Monod, O. (1976) La 'courbure d'Isparta': une mosaïque de blocs autochthones surmontés de nappes composites à la jonction de l'arc hellénique et de l'arc taurique. *Bull. Soc. géol. Fr.* **18,** 521–532.

Norris, R.J., Carter, R.M. & Turnbull, I.M. (1978) Cainozoic sedimentation in basins adjacent to a major continental transform boundary in southern New Zealand. *J. geol. Soc. Lond.* **135,** 191–205.

Poisson, A. (1978) Recherches géologiques dans las Taurides occidentales (Turquie). *Thèse de Docteur des Sciences, Université de Paris-Sud.*

Ricou, L.-E., Argyriadis, I. & Lefèvre, R. (1974) Proposition d'une origine interne pour les nappes d'Antalya et le massif d'Alanya (Taurides occidentales, Turquie). *Bull. Soc. géol. Fr.* **16,** 107–111.

Ricou, L.-E., Argyriadis, I. & Marcoux, J. (1975) L'axe calcaire du Taurus, un alignement de fenêtres arabo-africaines sous des nappes radiolaritiques, ophiolitiques et métamorphiques. *Bull. Soc. géol. Fr.* **17,** 1024–1044.

Robertson, A.H.F. & Woodcock, N.H. (1979) The Mamonia Complex, southwest Cyprus; the evolution and emplacement of a Mesozoic continental margin. *Bull. geol. Soc. Am.* **90,** 651–665.

Robertson, A.H.F. & Woodcock, N.H. (1980) Tectonic setting of the Troodos massif in the east Mediterranean. *Proc. Int. Ophiolite Sym. Cyprus,* 1979.

Saleeby, J.B. (1977) Fracture zone tectonics, continental margin fragmentation and emplacement of the Kings-Kaweah ophiolite belt, southwest Sierra Nevada. In: *North American ophiolite volume* (Ed. by R. G. Coleman & W. P. Irwin), *Bull. Oregon Dept. Geol. Min. Indust.* **95,** 141–160.

Saleeby, J.B. (1979) Kaweah serpentinite melange, southwest Sierra Nevada foothills, California. *Bull. geol. Soc. Am.* **90,** 29–46.

Steel, R.J. (1976) Devonian basins of western Norway—sedimentary response to tectonism and to varying tectonic context. *Tectonophysics,* **36,** 207–224.

Swarbrick, R.E. (1980) The Mamonia Complex of SW Cyprus: A Mesozoic continental margin and its relationship with the Troodos Complex. *Proc. Int. Ophiolite Sym. Cyprus,* 1979.

Thompson, G. & Melson, W.G. (1972) The petrology of the oceanic crust across fracture zones in the Atlantic ocean: evidence of a new kind of sea-floor spreading. *J. Geol.* **80,** 529–538.

Thuizat, R. & Montigny, R. (1980) K-Ar geochronology of three Turkish ophiolites. *Proc. Int. Ophiolite Sym. Cyprus,* 1979.

Woodcock, N.H. & Robertson, A.H.F. (1978) Imbricate thrust belt tectonics and sedimentation as a guide to emplacement of part of the Antalya Complex, SW Turkey. *Proc. 6th Coll. Aegean Geol. Izmir,* 1977. Abstract only.

Woodcock, N.H. & Robertson, A.H.F. (1980) Wrench-related thrusting along a Mesozoic-Cainozoic continental margin: Antalya Complex, SW Turkey. In: *Thrust and Nappe Tectonics* (Ed. by K. R. McClay & N. J. Price). *Spec. Publ. geol. Soc. Lond.* **9.**

Yilmaz, P.O. (1978) Alakir Çay unit of the Antalya Complex (Turkey): an example of ocean floor obduction. Unpublished M.Sc. Thesis, Bryn Mawr College, U.S.A., 91 pp.

Spec. Publ. int. Ass. Sediment. (1980) **4**, 147–170

New Zealand and oblique-slip margins: tectonic development up to and during the Cainozoic

K. B. SPÖRLI

Geology Department, University of Auckland, New Zealand

ABSTRACT

The tectonic development of New Zealand is reviewed with the aim of assessing the role of oblique-slip movements.

Early and mid-Palaeozoic rock units of the 'foreland province' have been affected by the late Devonian to Carboniferous Tuhua Orogeny. Structures across the regional trends may record the earliest oblique-slip regime. Oblique structures were also developed during deposition and deformation of the Carboniferous to late Jurassic Rangitata Sequence. Up to the early Cretaceous Rangitata Orogeny the history of New Zealand was that of the Pacific margin of Gondwanaland. After this orogeny New Zealand separated from Gondwanaland. The basins formed in the process of separation can be grouped as follows: (i) Cretaceous, early rifting basins (e.g. New Caledonia Basin, Bounty Trough). (ii) Late Cretaceous–early Tertiary basins (e.g. Tasman Sea Basin). (iii) Eocene–Oligocene basins (e.g. South Fiji Basin). (iv) Basins which have opened during the last 10 m.y. (e.g. Lau–Havre Trough). Emplacement of the Northland Allochthon took place in the late Oligocene, when the pole of relative plate movement between the India and Pacific plates was close to New Zealand. Cainozoic calcalkaline arc volcanism was first established in the Miocene of northern New Zealand. The subduction was possibly associated with transform movement along a predecessor of the Alpine Fault. The presently active regime of compressional oblique slip on the Alpine Fault was initiated 10–15 m.y. ago.

Cainozoic block fault systems are dominated by oblique rhombic patterns in broad zones on both sides of the Alpine Fault. A rectangular pattern prevails in northernmost New Zealand. Along the East Coast of the North Island, décollement combined with dextral oblique-slip has led to strong imbrication and *en echelon* folding. Two major tectonic features of the North Island are the Hawke's Bay and Wanganui monoclines.

INTRODUCTION

New Zealand in its present configuration is a microcontinent which straddles the active boundary between the Indian and Pacific plates (Le Pichon, 1968). Its onland geology provides a unique opportunity to study the processes at such a boundary.

0141-3600/80/0904-0147$02.00

Furthermore, plate boundary processes have played an important role during its entire geological history, which can be traced back to the late Precambrian. The present paper reviews knowledge about the development of the New Zealand continent, with special emphasis on the relative movements of adjacent plates in the New Zealand region and the effect of these movements on the pattern of Cainozoic faults and sedimentary basins.

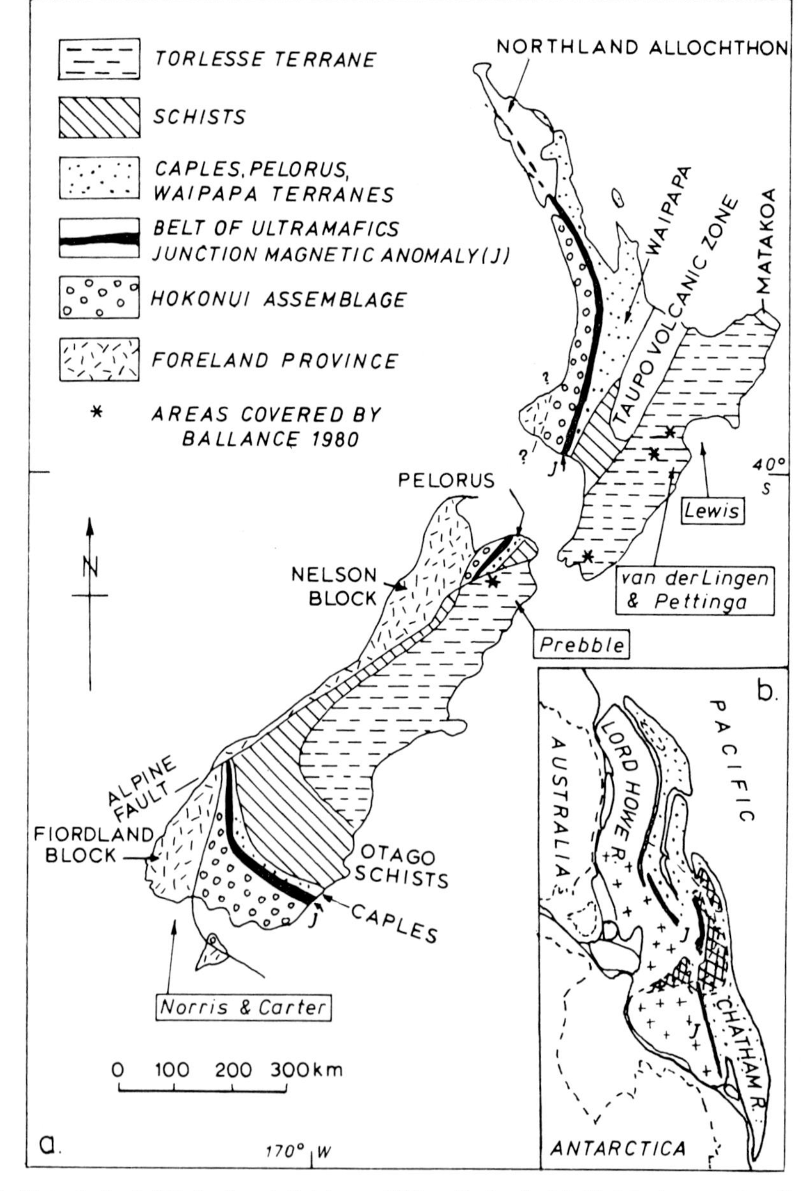

Fig. 1. Main tectonic-lithologic subdivisions of New Zealand. (a) Mainly basement units. Areas treated in other papers in this volume are indicated by arrows and names in boxes. (b) Margin of Gondwanaland after the end of the Rangitata Orogeny. Reconstruction after Griffith (1974). Crosses: foreland province. Dots: Rangitata sequence.

The Alpine Fault is a major tectonic feature of New Zealand (Wellman, 1955; Suggate, 1963). It has a dextral displacement of at least 480 km. The displacement may be as large as 1300 km if the sigmoidal bending of New Zealand (Fig. 1) is also taken into account (Wellman, 1974). The age of the Alpine Fault is still under dispute. Some authors (Ballance, 1976; Carter & Norris, 1976) set its initiation in the Tertiary; others (Suggate, 1972; Grindley, 1974) prefer a Cretaceous age. However, the sigmoidal bending of New Zealand, which indicates a regime of predominantly ductile shear, most likely occurred in early Cretaceous time (Rangitata Orogeny, see below) when plutonic processes (Landis & Coombs, 1967; Aronson, 1968) may have 'softened up' the continental margin. It is possible that the Alpine Fault may have originated by reactivation of a pre-existing Palaeozoic fault, of the type described by Harrington, Burns & Thompson (1973) and Cooper (1975).

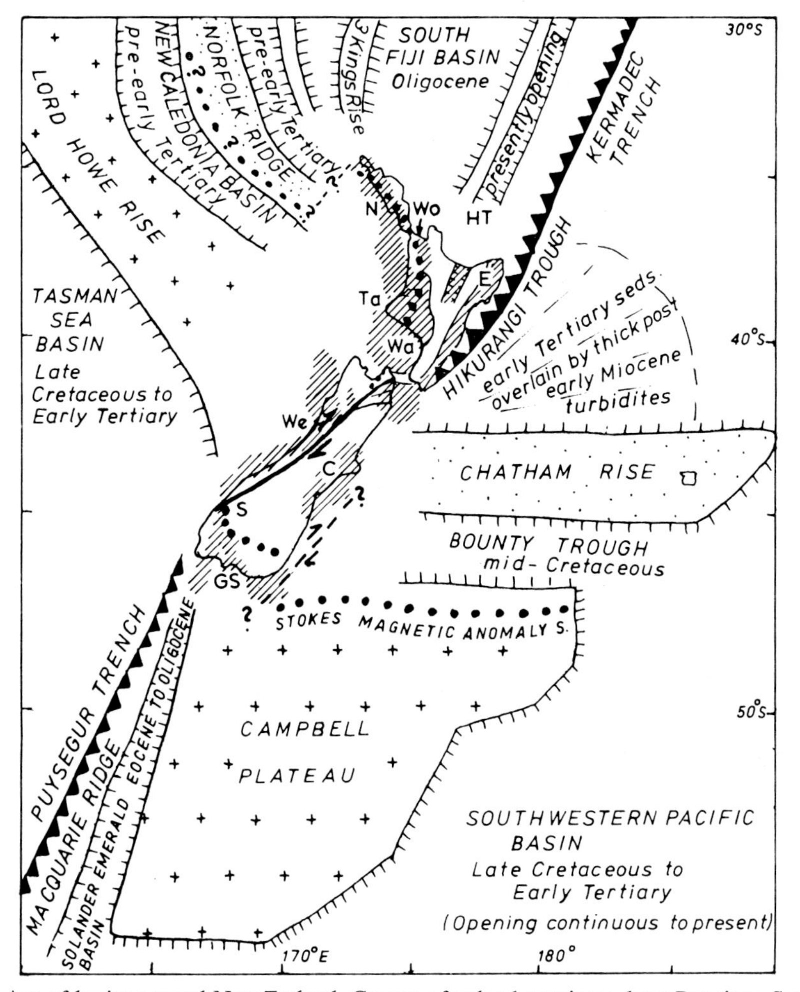

Fig. 2. Age of basins around New Zealand. Crosses : foreland province; dots: Rangitata Sequence, offshore. Note: Stokes Magnetic Anomaly System includes Junction Magnetic Anomaly shown in Fig. 1. Shaded: major post-Rangitata sedimentary basins (after Katz, 1976). N: Northland Basin. Ta: Taranaki Basin. Wa: Wanganui Basin. E: East Coast Basin. We: West Coast Basin, C: Canterbury Basin. GS: Great South Basin. S: Solander Basin. Wo: Waikato Basin.

A second, large dextral fault may lie just east of the South Island along the western margin of the Campbell Plateau (Cullen, 1970; Christoffel, 1978; Fig. 2, this paper). It causes an offset of the Stokes Magnetic Anomaly (Wellman, 1973; Hatherton, 1975).

Three major orogenies have been recognized in New Zealand: (i) the Devonian to Carboniferous Tuhua Orogeny, (ii) the early Cretaceous Rangitata Orogeny and (iii) the now active Kaikoura orogeny, which commenced in mid Tertiary time.

Up to the end of the Rangitata Orogeny the geological history of New Zealand was that of the margin of Gondwanaland (Craddock, 1975; Fig. 1, this paper). After the Rangitata Orogeny New Zealand separated from Gondwanaland and eventually became part of the boundary between the Indian and Pacific Plates.

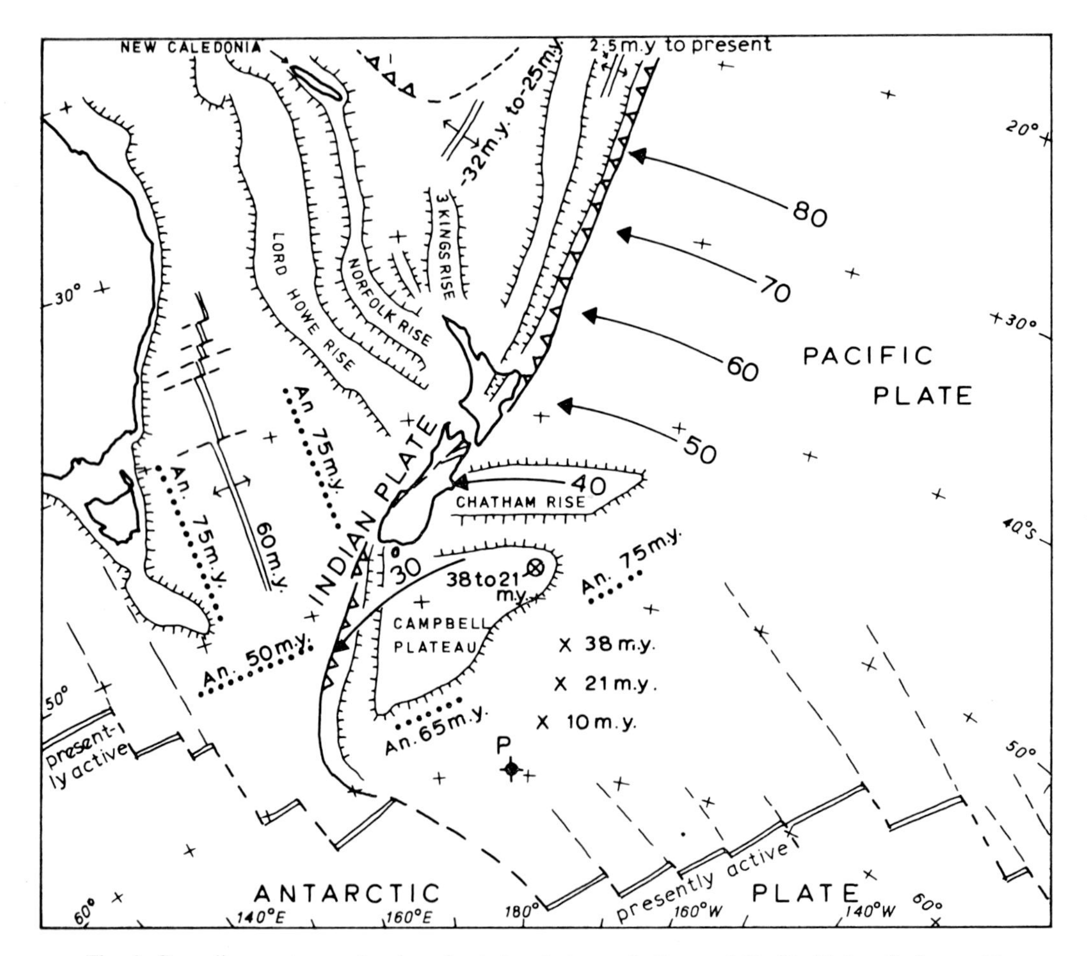

Fig. 3. Spreading centres and poles of rotation between Indian and Pacific Plates. Orthographic projection about a pole 60°S lat. and 180° long. P: present instantaneous pole; crosses: finite poles (ages indicated); circled cross: stage pole, with age span; curved arrows: direction of present convergence, with rate in mm/year, after Walcott (1978a). Double lines: spreading centres. Lines with teeth: trenches. Broken lines: faults. Dotted lines: magnetic anomalies (An: with age). All ages are listed in m.y. BP.

TUHUA OROGEN

Early and mid Palaeozoic rock units are part of the Tuhua Sequence (Carter *et al.*, 1974) and are confined to the Foreland Province (Aronson, 1968, = Western Province of Landis & Coombs, 1967) which on land is offset to form the Fiordland and Nelson blocks (Fig. 1). In the offshore, the Fiordland block extends out to the Campbell Plateau and the Nelson block out to the Lord Howe Rise (Figs 2, 3). This older part of New Zealand has affinities both with Australia and Antarctica (Craddock, 1975; Cooper, 1975).

In the Nelson block, Cooper (1975) has recognized three north–south trending belts: (i) a western, non-volcanic, greywacke-rich belt. (ii) a central belt of volcanic origin. (iii) an eastern, non-volcanic rich belt. Tectonic activity occurred in the Late Cambrian (Haupiri disturbance), in the Late Ordovician and culminated in the Late Devonian to Carboniferous Tuhua Orogeny.

Early, east–west trending recumbent folds are clearly oblique to the major tectonic and lithological trends (Grindley, 1961a, b) and may have originated as *en echelon* folds along the Palaeozoic margin, thus perhaps providing the earliest indication of oblique-slip in the New Zealand geological record.

RANGITATA OROGEN

The Rangitata Sequence (Carter *et al.*, 1974) of Carboniferous to Early Cretaceous age adjoins the Pacific side of the Foreland Province. An inner belt of arc and fore arc sequences in the immediate vicinity of the foreland (Hokonui Assemblage) can be distinguished from an outer belt (Caples, Pelorus, Waipapa, Torlesse terrains of Coombs *et al.*, 1976; Spörli, 1978; Fig. 1) which in part consists of more distal deposits, and in part represents sediments accreted to the margin of Gondwanaland during subduction of the underlying oceanic crust (Coombs *et al.*, 1976; Spörli, 1978). The inner belt has a simple structure and is separated from the more complex outer belt by the mostly ultramafic Dun Mountain Ophiolite Belt (Coombs *et al.*, 1976) which is the cause of the Junction Magnetic Anomaly, a part of the Stokes Magnetic Anomaly System (Wellman, 1973; Hatherton, 1975). Spörli (1978) has postulated that the complex structure in the outer belt is due to oblique subduction.

CRETACEOUS–CAINOZOIC DEVELOPMENT OF NEW ZEALAND

At the end of the Rangitata Orogeny the continental crust of New Zealand had more or less attained its present extent. Pre-early Cretaceous rocks now form the basement for the Cretaceous–Cainozoic sedimentary sequence.

A ubiquitous Early Cretaceous gap in sedimentation indicates a period of strong post-metamorphic erosion, which was followed by renewed clastic sedimentation (Fleming, 1970; Grindley, 1974) accompanied by extensional block faulting (Bishop, 1974; Moore, 1978; Pilaar & Wakefield, 1978), intrusion of basic dyke swarms and extrusion of alkaline, basaltic to rhyolitic lavas (Wellman & Cooper, 1971; Grapes,

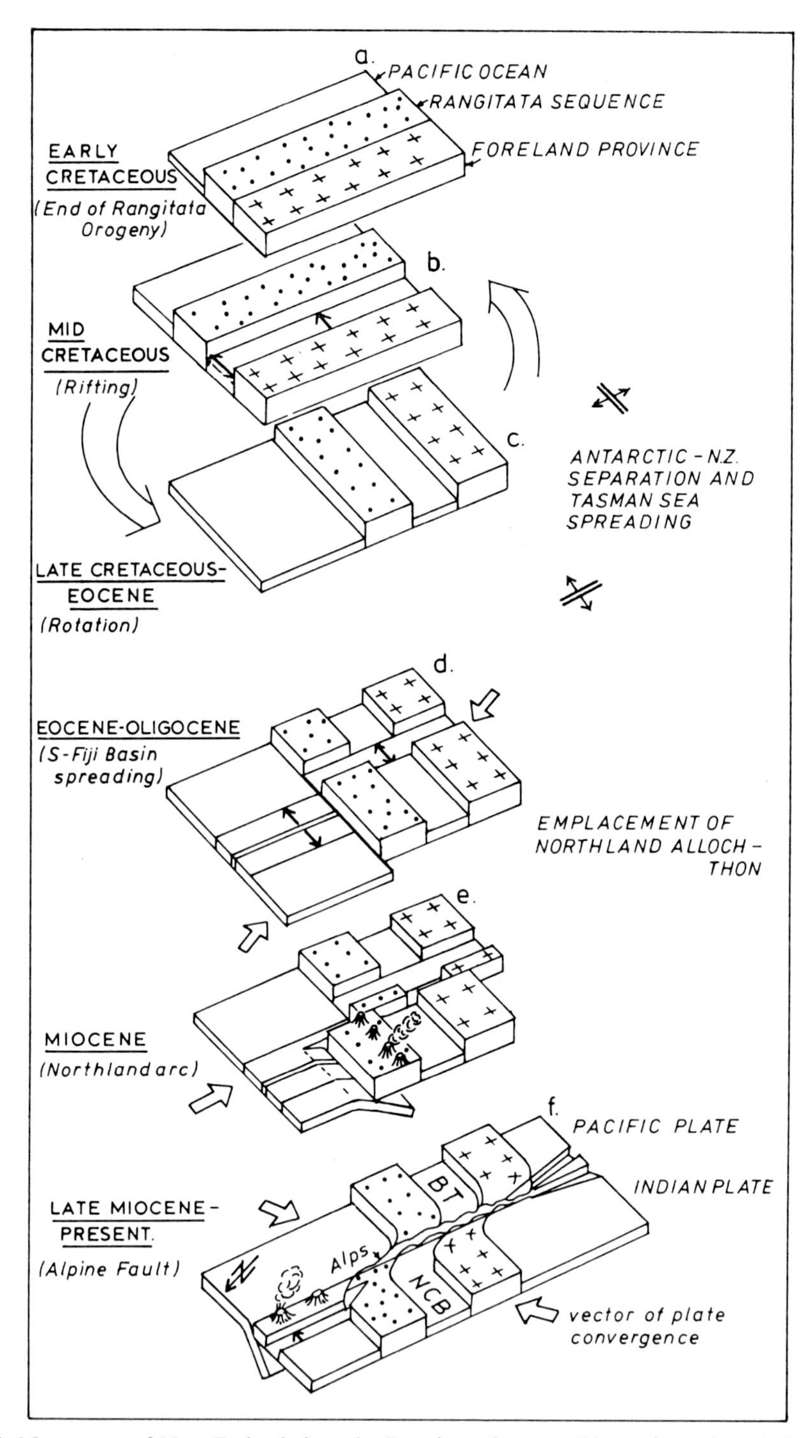

Fig. 4. Movements of New Zealand since the Rangitata Orogeny. Very schematic and simplified. Oblique aerial view to the present southeast. Overlaps between events neglected. Some oblique geometric relationships have been simplified to right angles for ease of illustration. Double lines and solid arrows show directions of spreading and rifting. Thinnest part of model is not necessarily all oceanic crust. If the sigmoidal bending of New Zealand is an early Cretaceous and not a Late Cainozoic event (see text), this complication has to be incorporated in all the stages of development shown in the figure.

1975, Hunt & Nathan, 1976). These events indicate the initial rifting of New Zealand away from Gondwanaland.

Movement of New Zealand away from Gondwanaland can best be understood by considering the ages of basins adjoining the continent (Fig. 2). While the time spans of formation of the various basins overlap considerably, four age groups can be distinguished on the basis of the main phases of basin-spreading or rifting:

1. *Cretaceous basins:* On the basis of subbottom reflectors and deep sea drilling information, Davey (1977) considers the New Caledonia Basin and the Bounty Trough to have initially opened in the Cretaceous. These basins probably formed a continuous system during the early rifting of the Gondwanaland margin.

2. *Late Cretaceous–Early Tertiary basins:* This group consists of the Tasman Sea Basin west of New Zealand and the Southwestern Pacific Basin south of New Zealand. Both basins contain sets of magnetic lineations and a spreading ridge (Hayes & Ringis, 1973; Molnar *et al.*, 1975; Weissel, Hayes & Herron, 1977; Fig. 3, this paper). The Tasman Sea system is now inactive, but spreading is still taking place along the Pacific–Antarctic Ridge south of New Zealand. The simultaneous opening of the Tasman Sea and the westward propagation of the Pacific–Antarctic Ridge caused the New Zealand continent to begin a sinistral rigid body rotation (Cullen, 1970; Fig. 4b, this paper). This rotation continued beyond the early Tertiary (Packham & Andrews, 1975).

3. *Eocene–Oligocene Basins:* In Eocene–Oligocene time major spreading took place in the South Fiji Basin (Packham & Terrill, 1975). In the northern part of the basin, a ridge–ridge–ridge triple junction was active during the Oligocene (Watts, Weissel & Davey, 1977). Oligocene spreading also has been recognized adjacent to northern New Zealand. (A. Malahoff, personal communication). The Solander–Emerald Trough to the south of the South Island (Fig. 2) also had its major phase of rifting in the Eocene–Oligocene (Davey, 1977). The Oligocene spreading systems are practically at right angles to the Cretaceous rifts. They may have been continuous through the New Zealand continental mass in the manner shown in Fig. 4d and connected with the actively westward propagating Pacific–Antarctic Ridge system.

4. *Presently opening basins:* Active back arc spreading is taking place at present in the Lau Basin (Sclater *et al.*, 1972; Weissel, 1977). This basin extends south into the Taupo Volcanic Zone (Cole, 1978) via the Havre Trough. Active faults in the Taupo Volcanic Zone indicate that it is an extensional, rift-like feature too (Grindley, 1960). Active extension is also presently taking place along the Pacific–Antarctic Ridge System south of New Zealand.

Plate reconstructions by Weissel, Hayes & Herron (1977) suggest that until latest Cretaceous time New Zealand lay along the boundary between a Lord Howe Plate and the Pacific Plate. Seemingly convergent movement across the boundary may be due to uncertainties in some of the rotations performed (Weissel *et al.*, 1977) rather than indicating the true situation, especially since there is no geological evidence for compressive deformation in New Zealand during the Late Cretaceous.

Oblique extension was predominant in New Zealand from the end of the Cretaceous to the end of the Oligocene (Weissel *et al.*, 1977). During this time interval spreading in the Tasman Sea ceased and the basin became part of the Indian Plate. No discontinuity in plate motion which would correspond to the Late Oligocene–Early Miocene emplacement of the Northland Allochthon (Ballance & Spörli, 1979) can yet be recognized. However, a new tectonic regime, dominated by dextral strike-slip along, and with increasing compression across, the Alpine Fault, was initiated after the end of the Oligocene (Carter & Norris, 1976; Walcott, 1978a). Oblique convergence became dominant and thick clastic sediments were deposited in irregular, often transient basins (Van der Lingen & Pettinga, 1980) as the present tectonic pattern emerged.

During the Miocene a calcalkaline andesitic volcanic arc was again established in New Zealand for the first time since the Rangitata Orogeny (Ballance, 1976; Wright & Black, 1979). This arc was restricted to northern New Zealand (Fig. 4e). Miocene volcanism in southern New Zealand produced alkaline and tholeiitic basalts and rhyolites (Gregg & Coombs, 1965). These are probably related to intracontinental rifting rather than subduction zone processes. In the Miocene, therefore, a transform structure, possibly subparallel to the present Alpine Fault (Fig. 4e) separated a part of New Zealand with active subduction from a part dominated by rifting. The vector of relative motion between the Indian and Pacific Plates was practically parallel to this transform.

Since about 10 m.y.BP oblique-slip with increasing compression across the Alpine Fault has prevailed. This has resulted in the destruction of some of the rifted basins formed in the Oligocene (Norris, Carter & Turnbull, 1978) and in the strong uplift of the Southern Alps and other mountain ranges during the last 2 m.y. (Walcott 1978a). With the establishment of two opposing subduction regimes north and south of New Zealand (Fig. 4f) back-arc rifting propagated south from the Havre trough into the North Island.

At present the vectors of convergence are at right angles to the plate boundary some distance north of New Zealand (Fig. 3) and become increasingly oblique towards the south.

CAINOZOIC TECTONIC PATTERN

The Cainozoic tectonic pattern is dominated by faulting (Figs 5, 6, 7). Where folds are present, they are always closely associated with fault zones. The fault grid has strongly influenced the shape of sedimentary basins. This is best shown by the facies maps and cross-sections in the offshore areas west of New Zealand (Pilaar & Wakefield, 1978). A smaller scale pattern of basement faults is shown in Fig. 8. The tectonic patterns in Figs 5 and 6 have been determined in part from the configuration of the widespread sub-Cretaceous and Tertiary erosion surface on the Mesozoic or Palaeozoic basement, and in part directly from strikes and dips of the Cainozoic strata.

South Island (Fig. 5)

The Alpine Fault and its branches are the major structures of the South Island. Uplift along the Alpine fault has amounted to about 10 mm/year (Lensen, 1975) during the last 2 m.y. In the central part of the fault, recent horizontal displacement approximately equals vertical displacement (Wellman, 1955). Mylonites parallel to the Alpine Fault are cut by Cretaceous lamprophyre dikes and may indicate early Cainozoic or even earlier phases of movement along the fault (Wellman & Cooper, 1971).

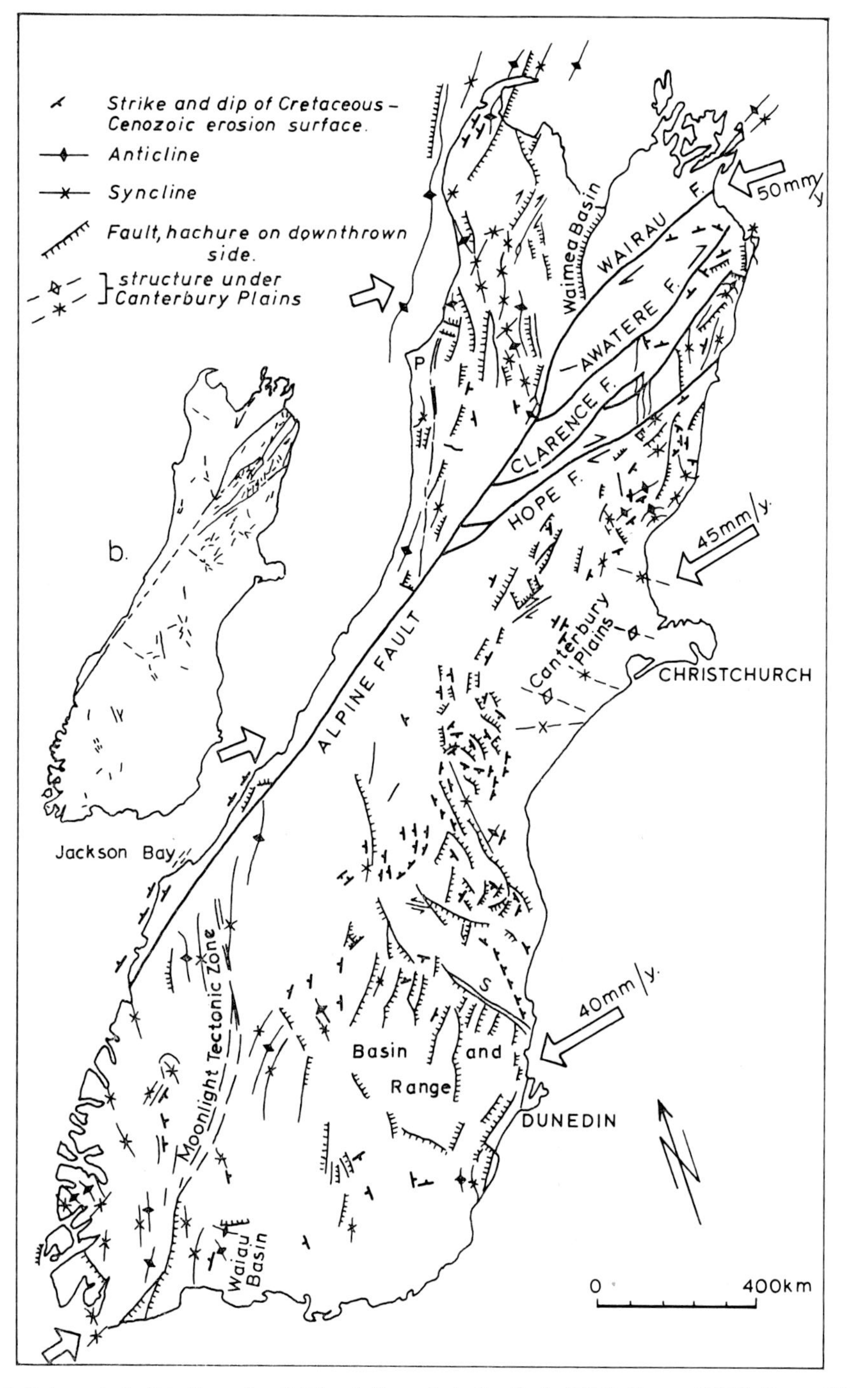

Fig. 5. Cainozoic fault pattern, South Island. Compiled after Beck (1964), Bowen (1964), Davey (1977), Gair (1967), Gregg (1964), Grindley (1961a, b), Laird (1968, 1972), Lensen (1962), McKellar (1966), Mutch (1963), Mutch & McKellar (1964), Norris *et al.* (1978), Oborn (1959), Warren (1967), Watters *et al.* (1968), Wilson, personal communication, 1978, Wood (1960, 1962, 1966). Open arrows indicate direction of present-day plate convergence (after Walcott, 1978a and personal communication). P: Paparoa Tectonic Zone. S: Shag River Fault Zone. Folds on either side of the Moonlight Tectonic Zone have been included under the assumption that the latest fold phases in the Otago Schists and in Fiordland are Cainozoic. Inset b: Late Quaternary faulting, after Lensen (1977).

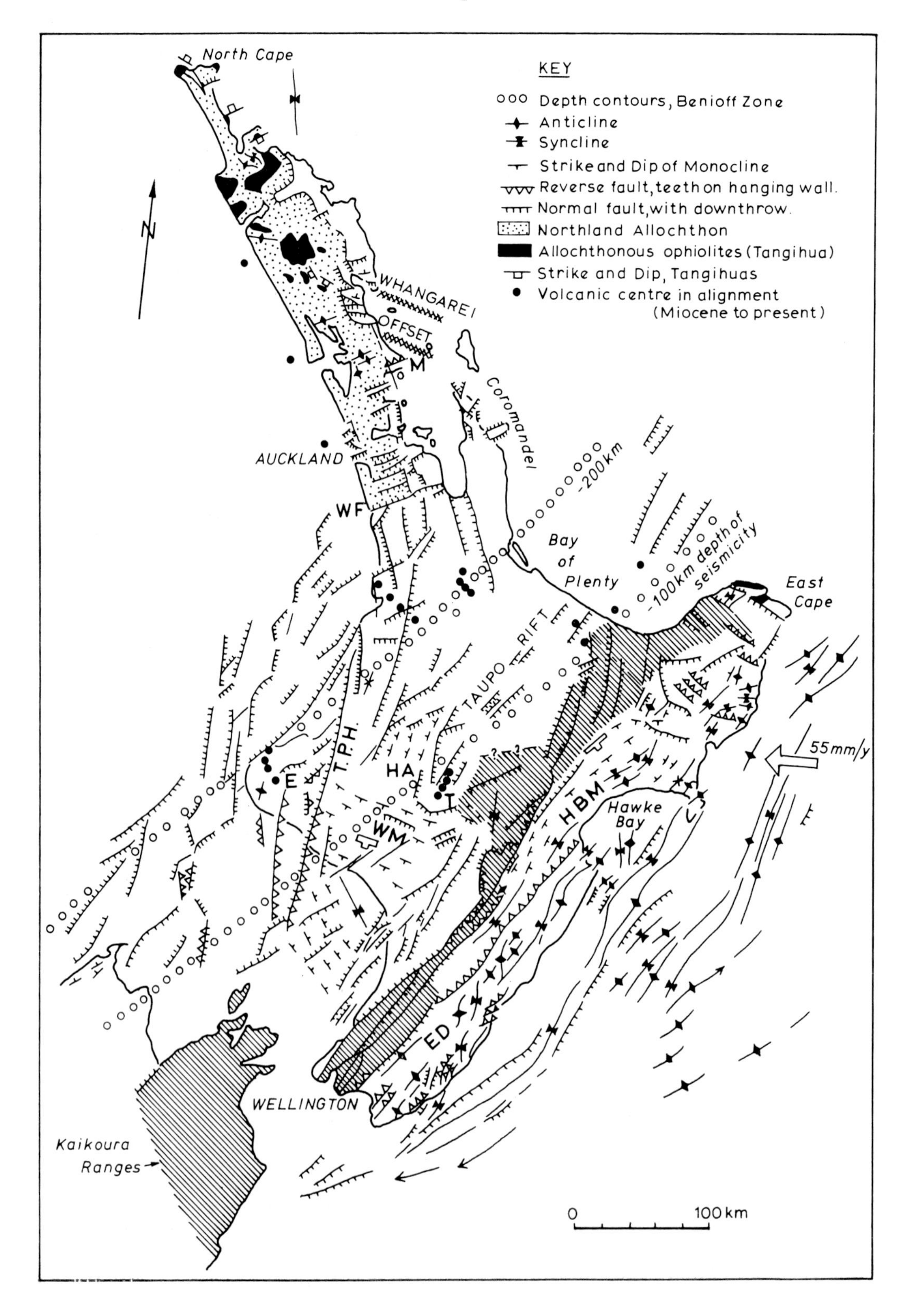
North Cape
KEY
Depth contours, Benioff Zone
Anticline
Syncline
Strike and Dip of Monocline
Reverse fault, teeth on hanging wall.
Normal fault, with downthrow.
Northland Allochthon
Allochthonous ophiolites (Tangihua)
Strike and Dip, Tangihuas
Volcanic centre in alignment
(Miocene to present)
N
WHANGAREI
OFFSET
M
Coromandel
AUCKLAND
-200 km
WF
Bay
of
Plenty
-100 km depth of seismicity
East
Cape
TAUPO RIFT
T.P.H.
HA
E
T
WM
HBM
Hawke
Bay
55 mm/y
ED
WELLINGTON
Kaikoura
Ranges
0
100 km

The northern branches of the Alpine Fault (Wairau, Awatere, Clarence and Hope faults) define a field of dominantly northeast-trending structure associated with dextral strike slip (e.g. Lensen, 1968; Freund, 1971). Microseismic studies indicate that this field may be slowly propagating towards the south (Rynn & Scholz, 1978). Elsewhere an oblique rhombic pattern prevails, except in the 'basin and range' province of eastern Otago (Fig. 5) where the pattern is rectangular. Trends similar to those of the rectangular pattern are also present in the subsurface of the Canterbury Plains (D. Wilson, N.Z. Geological Survey, Christchurch, personal communication) and in the Chatham Rise to the east (Austin, Sprigg & Braithwaite, 1973). At present most of the NNE and northeast-trending faults in the South Island have reverse components of movement. Because of this, there are folds parallel to both fault directions, giving rise to complex non-cylindrical interference structures in some of the Cainozoic basins (Bradshaw, 1975).

Reactivation of Cretaceous faults, followed by reversal of the sense of movement has been recorded from a number of localities (Bishop, 1974, p. 331). Usually the change is from a normal to a reverse component of movement. It commonly can be timed to have to have taken place during the Late Miocene (Carter & Norris, 1976).

Prominent master shear zones of the South Island besides the Alpine Fault and its branches are the Paparoa Tectonic Zone (Laird, 1968, 1972), the Moonlight Tectonic Zone (Norris *et al.*, 1978) and the yet little known Shag River Fault zone (Fig. 5). All of these faults have histories of movement dating back to the Cretaceous and, in the case of the Paparoa Tectonic Zone, possibly earlier. The Moonlight Tectonic Zone may be the compressed onland continuation of the Oligocene Emerald–Solander trough (Fig. 2) (Norris & Carter, 1980).

North Island (Figs 6 and 7a)

The influence of the Alpine Fault becomes diffuse in the North Island and the fault pattern is more complex than in the South Island. The zone of most intense Cainozoic deformation lies along the east coast (East Coast Deformed Belt). Together with the adjacent Hikurangi trough (Katz, 1974a; Eade & Carter, 1975) it represents the surface expression of subduction under the North Island (Lewis, 1980) outlined by earthquake foci (Adams & Ware, 1977).

From east of Wellington almost to Hawke Bay the structure is dominated by thrusts dipping away from the Hikurangi Trough, the thrusts becoming steeper landwards and changing into strike-slip faults adjacent to the axial ranges (Lewis, 1980). Actively growing anticlines are known on the coast and continental shelf of Hawke Bay (Lewis, 1971) and on the coast to the south (Ghani, 1978). Some distinct, north–south trending *en echelon* structures, basement cored in the south, but only affecting younger strata in the north, can be recognized within the belt (Kingma, 1967;

Fig. 6. Cainozoic fault pattern, North Island. Compiled after Adams & Ware (1977), Ballance & Spörli (1979), Davey (1974, 1977), Grindley (1960), Hay (1967), Healy *et al.* (1964), Katz (1974a, b), Kear (1960), Kear & Hay (1961), Kingma (1962, 1964, 1966, 1967), Lensen *et al.* (1959), Lewis (1971), Pilaar & Wakefield (1978), Ridd (1964), Schofield (1967), Speden (1976), Stonely (1968), Thompson (1961). Open arrow indicates direction of present day plate convergence. Shaded: uplifted axial basement ranges. Faults with both teeth and hachure are normal faults reactivated by reverse movement. ED: East Coast Deformed Belt. M: thrust locality, Mathesons Bay, North Auckland. WF: Waikato Fault. E: Mount Egmont. TPH: Tongapurutu–Patea High. HA: Hauhungaroa Block. T: Tongariro Volcanoes. WM: Wanganui Monocline. HBM: Hawke's Bay Monocline.

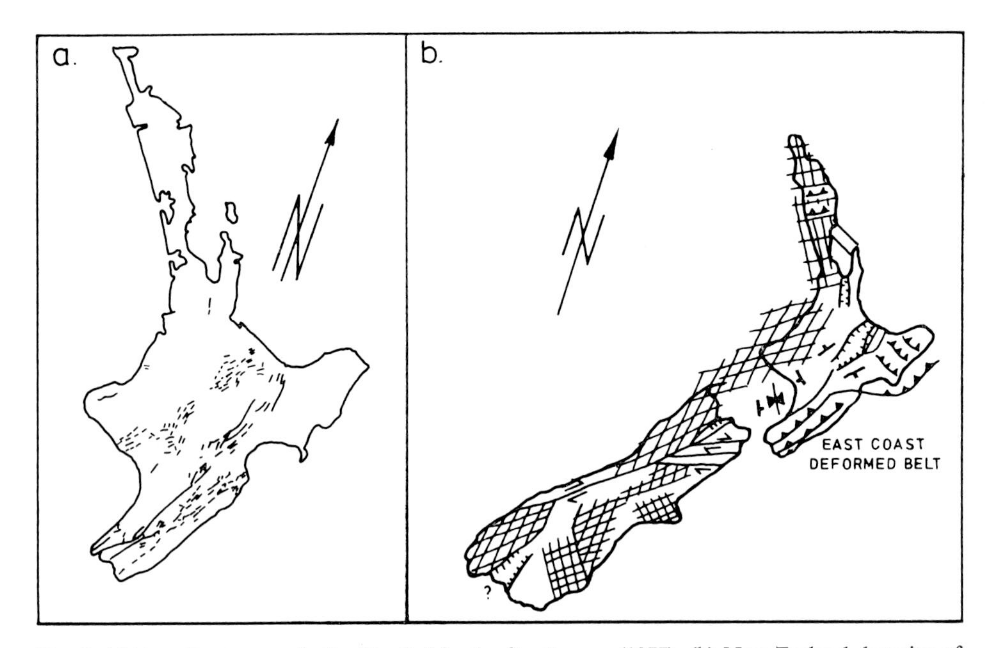

Fig. 7. (a) Late Quaternary faults, North Island, after Lensen (1977). (b) New Zealand domains of Cainozoic structure, showing rhombic and rectangular fault patterns, thrust zones (toothed lines), major folds, monoclines (strike and dip symbols) and rifts (lines with hachures).

Johnston, 1975; Fig. 6, this paper). The orientation of these oblique folds is consistent with dextral strike-slip along the northeast direction. Unconformities show that changes from thrust imbrication on northeast-striking faults to *en echelon* folding on north–south trending axes have taken place, while imbrication has persisted in adjacent belts. Thus an alternation between complete décollement of the sedimentary sequence and more 'autochthonous' *en echelon* folding, directly recording the oblique-slip vector, is indicated.

The East Coast Deformed Belt is bounded in the west by the uplifted, axial greywacke ranges (Figs 6, 8). For about 200 km northeast of Wellington they form a narrow block bounded both in the east and the west by steep reverse faults (Kingma, 1962, 1967). Oblique, very open folds and warps again indicate that the strain during uplift included dextral shear along northeast trending faults (Fig. 8).

Near the centre of the North Island, the zone of uplifted basement blocks widens and instead of being thrust-bound in the east it adjoins the actively tilting Hawkes Bay Monocline (Fig. 6). Faults at the northern end of the axial ranges originated in the Cretaceous and have experienced reversal of movement (Moore, 1978). A swing into a north–south trend of the tectonic grain corresponds to the sigmoidal bend on the west side of the North Island.

North of Hawkes Bay and east of the axial ranges, the structural pattern is strongly influenced by northwest trends. Some of these are inherited from a phase of southward thrusting or gravity sliding in the Late Oligocene (Ridd, 1964; Stonely, 1968; Speden, 1976) which involved emplacement of the Matakoa ophiolites (Brothers, 1974).

From the Bay of Plenty to the centre of the North Island, the Taupo Volcanic Zone (Cole, 1978) lies to the west of the uplifted axial ranges. It coincides with a rift-like

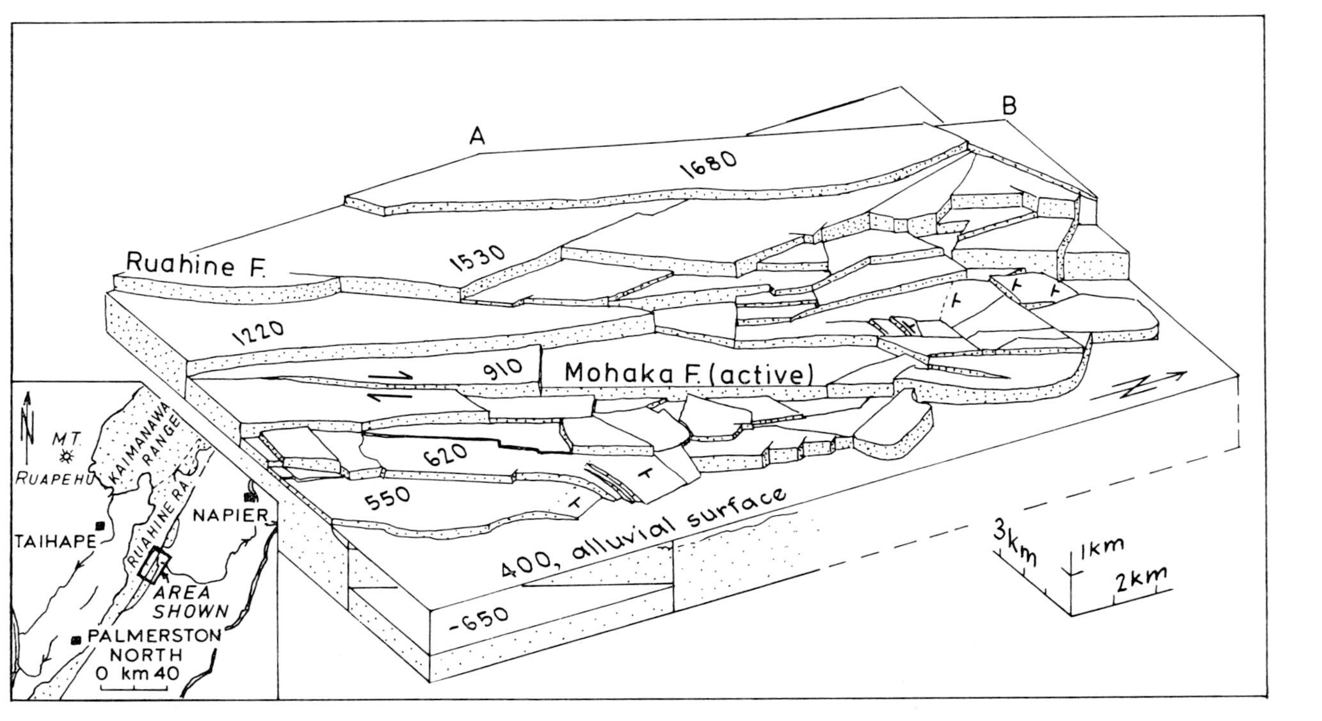

Fig. 8. Block Faults, northeastern Ruahine Range. Illustrates structure in strike-slip zone behind the accretion zone of the East Coast Deformed Belt. Similar basement block patterns control the subsurface geometry of some Cainozoic basins. Reconstructed from remnants of a sub-Waitotaran (Pliocene) erosion surface. Tectonic relief is entirely due to Quaternary movements. Numbers on the reconstructed surface give height above or below sea level in m. Line **AB** is an arbitrary northwestern boundary of the area studied, not a fault. Note northwest trending monoclinal warps, indicating dextral strike-slip on the major faults. Throw on the southeastern-most fault is estimated after data from Leslie & Hollingsworth (1972). Inset shows location of the area in the North Island; axial ranges are stippled.

depression bounded by a rhombic pattern of normal faults. Outward tilting of adjacent blocks is evident from the Hawke's Bay Monocline and the Hauhungaroa Block (Fig. 6). The southern end of the rift is marked by the Tongariro volcanoes.

South of the Taupo Volcanic Zone, a major WNW-trending downwarp, the Wanganui Monocline (Fig. 6) affects strata as young as Pleistocene (Fig. 2). The monocline is the northern limb of a northwest-trending syncline (Katz, 1974b) in which the Mesozoic basement is downwarped to about 4000 m depth. The junction of this structure with the uplifted ranges to the east is marked by a zone of northeast-trending, recent faults and actively growing anticlines (Te Punga, 1957; Stevens, 1974). The tectonic significance of this major WNW-trending syncline is by no means clear. It may be one of the structures associated with the northern termination of the Alpine Fault.

In the southwestern parts of the North Island and in the adjacent offshore, the fault pattern is distinctly rhombic and is the result of NNE directed transcurrent dextral shear applied to normal faults inherited from Cretaceous phases of deformation (Pilaar & Wakefield, 1978). Strike-slip and reverse movement on NNE trending faults is also evident from earthquake data (Robinson, Caelham & Thompson, 1976).

Many volcanic alignments in the west coast area are oriented NW–SE, approximately perpendicular to the trend of the Taupo rift and to the strike of the depth contours on the Benioff zone (Fig. 6). The line including Mount Egmont shows a southeastward progression from older to younger (Neall, 1974) at a rate of about 1·5 cm/year during the last 1·75 m.y. Kear (1964) interpreted the trends of the alignments to be representative of the direction of 'principal horizontal stress'.

Faults in Northland Peninsula north of the Waikato Fault (Fig. 6) are arranged in an almost rectangular grid. Some of the fractures are hinge faults with maximum throws over 1 km (Hochstein & Nunns, 1976).

The Northland Allochthon (Ballance & Spörli, 1979) was emplaced in late Oligocene time, possibly by gravity slide movement over several hundred kilometres, and consists of Cretaceous and Cainozoic lithologies mixed in an apparently chaotic manner. It can be correlated with overthrust masses in the East Cape Region, along the east coast of the North Island and possibly as far as the Kaikoura area of the northern South Island.

During the Miocene, Waitemata Group clastic sediments (Ballance, 1974) were deposited onto the Allochthon from the north and overstepped onto already existing fault line scarps along upthrown basement blocks to the east. Andesitic arcs were formed on either side of the Northland Peninsula (Ballance, 1976). Volcanism continued into the Pliocene in the eastern arc. In the area of Whangarei, the western boundary of the eastern arc is displaced westward to continue through the northern part of Northland Peninsula (Fig. 6). Since the Pleistocene, volcanicity in this area has been mostly alkali–basaltic (Kear, 1964; Brothers, 1965) possibly of intra-plate or behind-arc origin.

THE PRESENT TECTONIC REGIME

Fault plane solutions for earthquakes and geodetic data provide information about the present day strain field (Walcott, 1978a, b). A 70–100 km wide belt of dextral shear connecting the Hikurangi trough with the Puysegur trench can be recognized in onland New Zealand. The shear strain and orientations of the principal axes of compression correspond closely with those derived from the instantaneous pole of plate rotation. Two zones of predominant compression occur to the west of the shear belt (axial tectonic belt) in Nelson and off Fiordland. Major rifting is taking place in the Taupo Volcanic Zone of the North Island and, perhaps, the Waiau Basin of the South Island.

Strain deduced from plate tectonics and from geodetic data is two to three times greater than the slip summed along known active faults. This indicates that beside the elastic deformation accompanying fault movement a considerable amount of aseismic and anelastic strain is taking place.

The East Coast Deformed Belt of the North Island corresponds to a subduction thrust zone. The following model, similar to that proposed by Walcott (1978b), can account for the complex effects of oblique slip within this belt: since the shear strength on a strike slip zone is probably considerably lower than that on a reverse fault, the component of motion parallel to a plate boundary with oblique convergence can be readily taken up by strike-slip on a zone usually located some distance back from the trench, while the component normal to the plate boundary is accumulated as elastic strain which is relieved less frequently by rupturing on dip-slip thrusts located between the strike slip zone and the trench. Only a limited amount of the convergent movement can be taken up by thrust-thickening of this accretionary wedge. Therefore from time to time the subduction thrust itself must be ruptured during earthquakes leading to extensional relaxation in the overlying accretionary wedge. On the other hand the lithosphere overlying the subduction thrust occasionally becomes so completely locked to the subducting slab for short periods that the oblique vector of convergence is recorded directly in the form of *en echelon* folds.

CAINOZOIC SEDIMENTATION

Early Tertiary

With the structural outline developed in the previous paragraphs, an attempt can be made to assess the influence of oblique-slip processes on sedimentation. However it must be realized that sedimentation was during some periods controlled by other events. During the Oligocene for example a break in deposition occurred because of a change in ocean currents due to the separation of Australia and Antarctica (Carter & Landis, 1972; Kennett, 1977) and in the Pleistocene sea level fluctuated. This last effect will be especially difficult to distinguish from the tectonic influences on sedimentation (Lewis, 1973).

Oblique-slip processes were unimportant during the Late Cretaceous and early Tertiary. As the New Zealand block moved away from terrestrial source areas on Gondwanaland and became increasingly peneplaned, grain size generally became finer (e.g. Carter & Norris, 1976; Norris *et al.*, 1978). Deepwater marls and mudstones are very common in the Early Tertiary and indicate that large parts of New Zealand eventually subsided to considerable depths below sea level. Calcareous turbidites, some of which are rich in glauconite (e.g. Stonely, 1968; Nelson, 1968) indicate a considerable submarine relief. Coal deposits, crystalline limestones and reworked, residual lag conglomerates are typical for the central parts of the continental mass, which formed a shallow marine or very low level emergent platform with little relief.

The palaeogeographic situation may have been similar to that prevailing today on the Lord Howe Rise and the Campbell Plateau. Deformation during this time was gentle, with vertical block movements dominating. The general subsidence was probably caused by the cooling of the lithosphere as New Zealand moved away from the spreading centres which had separated it from Gondwanaland. The rifting was accompanied by extrusion of basaltic lavas which persisted into the early Tertiary on the Chatham Rise (Grindley *et al.*, 1977), in the southwestern South Island (Nathan, 1977), the northeastern South Island (Lensen, 1962), in the southern part of the East Coast Deformed Belt (Kingma, 1967) and in the source area of the early Tertiary sediments of the Northland Allochthon (Ballance & Spörli, 1979).

From Oligocene time strike-slip and oblique-slip regimes have influenced the sedimentation in New Zealand. Broad generalized facies patterns gave way to deposition in localized basins (Fig. 2). Major features of sedimentation in the various basins are discussed below.

Northland

In the Northland basin the Late Oligocene to Miocene Akarana Supergroup, including the Waitemata Group (Ballance, 1974; Ballance, Hayward & Wakefield, 1977) was laid down over the Northland Allochthon mainly by a system of turbidity currents flowing from the north along the peninsula. After initial subsidence which led to the formation of the central turbidite facies trough, the entire area was affected by a pulse of rapid subsidence. Complex, in part nappe-like structures have been interpreted as penecontemporaneous slumping of poorly consolidated sediments in response to seafloor tilting (Gregory, 1969). However, relatively deep-seated thrusting is indicated by faults which pass from the Mesozoic basement into the basal Waitemata sequence at Mathesons Bay (Fig. 6, and Spörli, in preparation). A significant Miocene or post-Miocene deformation is also recorded by the Onerahi Chaos Breccia (Ballance & Spörli, 1979), a deposit of reworked Northland Allochthon which either is intercalated with or overlies the Waitemata Group. This mélange-like unit contains large inclusions of serpentinite derived from the belt giving rise to the Junction Magnetic Anomaly (Hatherton & Sibson, 1970), indicating a major reactivation of the ultramafic-bearing suture formed in the Mesozoic. Similar reactivation is known from the Wairere Serpentinite further south (O'Brien & Rodgers, 1973). The movements may have been contemporaneous with renewed thrusting or gravity sliding in the eastern North Island (Ridd, 1964) and with the emplacement of the highly deformed, possibly allochthonous Jackson Bay Formation (Nathan, 1978) on the west side of the Alpine Fault. These events most likely mark the onset of compressive movement across the Alpine Fault around 10 m.y.BP.

West Coast North Island

Oligocene–Miocene basins probably formed a semi-continuous belt, interrupted by cross-trending thresholds, through the western North Island (Wanganui–Taranaki Basin) (Fig. 2) and into the northern South Island (Carter & Norris, 1976). In the Wanganui and Taranaki Basins, Late Eocene to Oligocene regional submergence, with widespread deposition of limestone ended the Lower Tertiary phase of transgression, which had been interrupted by local regression in the Eocene (Pilaar & Wakefield, 1978). In the Late Oligocene Mahoenui Formation (Glennie, 1959) turbidite sequences deposited in a N–S trending trough indicate transport parallel to tectonic grain, down a south-dipping palaeoslope; a situation similar to that described from the Waitemata Group to the north. In the Miocene, influx of clastic material including turbidites from easterly sources became important and the Taranaki Graben complex became established (Pilaar & Wakefield, 1978). Building out of the present day continental shelf began in the late Miocene. Initial downwarping on the Wanganui monocline occurred some time in the Pliocene. Conglomerates from easterly sources also became more prominent during this time (Hay, 1967).

The intensity of tectonic events as recorded in sedimentary sequences in the North Wanganui Basin (Nelson & Hume, 1977) reached a peak in the early Lower Miocene (24–20 m.y.BP) possibly due to the initiation of dextral movement on the Alpine Fault. Activity then decreased, to increase again from Late Miocene (11 m.y.BP) to present time, with the onset of compressional tectonics and uplift along the Alpine Fault.

West Coast Basin, South Island

Whereas the basins so far discussed lie some distance away from the Alpine Fault, the West Coast Basin is located in the immediate vicinity of this transform boundary. The Cainozoic succession correspondingly shows greater irregularity. Conglomerates, local unconformities, and contrasting facies are more common. Development of local troughs, some of which were later everted, is typical (Paparoa Geosyncline, Grey Valley Geosyncline and Murchison Geosyncline of Bowen, 1964). Coal measures in the Miocene pass laterally into flysch-like sediments (Nathan, 1974). Major reversal of movement took place in the early Miocene. Subsidence of some of the basins continues to the present. The earliest schist debris from the rising Southern Alps was deposited in the early Pleistocene Old Man Gravels (Bowen, 1964).

The situation at the southern end of the West Coast Basin is distinctively different, in that the early Tertiary sequence contains voluminous basalts (Nathan, 1977), and the Miocene rests unconformably on older rocks and begins with a shallow water limestone overlain by mudstones, flysch-type sandstones and granitic conglomerates, indicating rapid subsidence and deposition by currents flowing from NE to SW (Nathan, 1978). High in the sequence is the possibly allochthonous Jackson Bay Formation. Another unconformity occurs at the base of Pliocene conglomerates and Pleistocene glacial and fluvioglacial deposits.

East Coast, North Island

The East Coast Deformed Belt of the North Island is another area where oblique-slip, in this case associated with accretion in a subduction zone, has had a strong influence on sedimentation. Interference of *en echelon* domes and basins with NE-trending accretion thrusts has been described in an earlier section.

Offshore Pleistocene sediments are involved in complex deformations (Lewis, 1980). In the eastern onland part of the belt, Miocene clastic sequences generally lie unconformably on the highly folded and sliced older rocks. To the west, Pliocene unconformities cut down into the early Tertiary, indicating an uplifted zone where subsequently a prominent NE-trending trough was formed along the eastern foot of the axial ranges (Kingma, 1962; 1967; Leslie & Hollingsworth, 1972; Ballance, 1980). Discrete basins with turbidite sedimentation were formed in the early Miocene (Johnston, 1975), late Miocene (van der Lingen & Pettinga, 1980) and in the early

Pliocene (Neef, 1974). In many cases transport directions of turbidites are from north to south, while the palaeoslope indicated by facies changes was parallel to the present NE strike of the East Coast Deformed Belt. Conglomerates are only present near onlaps onto basement rocks. A prominent emergence of basement is recorded by mid-Miocene conglomerate in the eastern part of the belt (Lillie, 1953; Kingma, 1971; Johnston, 1975). Local unconformities occur throughout the Miocene (Lillie, 1953; Johnston, 1975). An unconformity at the base of the Pliocene may, however, be of more regional significance (Katz, 1973). Widespread shallow water limestones above it may indicate a general change in sea-level and/or a period of tectonic quiescence. Molasse-type conglomerates shed from the uprising axial ranges first appear some distance away from the ranges in mid-Pleistocene time (Kingma, 1971).

In the area north of Hawke's Bay, early Tertiary sediments are highly disturbed by southward gravity sliding and/or thrusting in the Oligocene and Late Miocene (Ridd, 1964, Stonely, 1968). Turbidites were deposited both in mid and Late Miocene time (Grindley, 1960; Ridd, 1964; Stonely, 1968). During this period a transgression from NE to SW covered the area of the future Hawkes Bay Monocline.

East Coast, South Island

The East Coast Deformed Belt of the North Island continues southwards into the Kaikoura Ranges (Fig. 4), where complex structure (Lensen, 1962; Prebble, 1980) marks the transition from the subduction complex in the north to the transform regime of the Alpine Fault. The early Tertiary sequence is dominated by fine-grained limestones (e.g. Amuri Limestone; Lensen, 1962). Basaltic volcanic rocks are very common. An important Miocene unit is the Great Marlborough Conglomerate (Lensen, 1962; Prebble, 1980), a product of mass flow processes, indicating a peak of tectonic unrest. Terrestrial, fluviatile sedimentation is presently taking place in active fault angle depressions (Ballance, 1980).

South of the zone of active faulting, Cainozoic stratigraphy and structure become less complex (Fig. 5; Wilson, 1963; Gregg, 1964; Gair, 1967). Shore lines generally trended parallel to the NE structural trends in the Mesozoic basement, and migrated from SE to NW during the early Tertiary. Large tracts of the present Southern Alps, however, remained above sea-level as low-lying land. After the Oligocene the shore line receded towards the present coast line. Folding and some uplift preceded deposition of Pleistocene sediments, when large volumes of gravel shed from the rising Alps began to be transported east to form the Canterbury Plains.

Southern South Island

The palaeogeographic pattern in the southern South Island is strongly influenced by the Moonlight Tectonic Zone (Carter & Norris, 1976; Norris *et al.*, 1978) which separates an eastern platform area with initial terrestrial conditions and later shallow water deposition from a western area with a complex Cretaceo-Tertiary history of transgressive sedimentation disturbed by block faulting. In a trough along the Moonlight Tectonic Zone, deposition of mass flow-emplaced breccias and sandy flysch from the sides of the trough in the Oligocene gave way in the Miocene to schist debris fed from the north together with muddy turbidites. A major tectonic pulse is recorded by a Late Miocene unconformity approximately marking the change from rifting to compression across the basin. The flysch basins of this zone were continuous with the

Solander trough to the south (Norris & Carter, 1980). In the Plio-Pleistocene, slightly differently oriented, gravel-filled basins were superimposed on a part of the Moonlight Tectonic Zone.

SUMMARY

New Zealand has experienced oblique-slip combined with compression possibly during the Palaeozoic, in the Mesozoic and definitely in the late Cainozoic. Oblique-slip is at present taking place along the northeast trending plate boundary and is accompanied by high rates of uplift. Oblique extension prevailed in the early Tertiary.

The Cainozoic field of faulting can be subdivided into domains with rhombic, rectangular, rift and thrust fault patterns (Fig. 7b). The oblique rhombic domains are situated along both sides of the Alpine Fault and may mark zones of flow which have accommodated the sigmoidal bending of New Zealand. (A similar suggestion has been put forward for the South Island by Norris, 1979). If this bending is Cainozoic, parts of New Zealand have deviated from rigid plate movement during this time, as is also indicated by the studies of Walcott (1978a, b). As a result, sedimentary basins in this zone of flow are likely to have been deformed during sedimentation, experiencing rotation. Source areas may have been displaced by transcurrent movement (Ballance, 1980). In the zones of thrusting, cycles of downwarping, eversion and extensional relaxation must have followed each other in rather rapid succession.

It is interesting to note that two major Cainozoic tectonic pulses can be recognized throughout New Zealand, one in the Late Oligocene, the other in the Late Miocene. In the North Island the Oligocene event was more pronounced, in the South Island the Miocene event, perhaps indicating a migration of peak deformation towards the South Island. A yet unexplained feature is the southward inclination of palaeoslope, parallel to the dominant tectonic grain, with resulting north to south turbidity currents in a number of Mio-Pliocene basins on both sides of the Alpine Fault (Glennie, 1959; Ballance, 1974; Neef, 1974; Nathan, 1978; Norris *et al.*, 1978; van der Lingen & Pettinga, 1980).

ACKNOWLEDGMENTS

I am indebted to P. F. Ballance, R. M. Carter, K. B. Lewis, R. J. Norris, H. G. Reading and R. I. Walcott for helpful comments on the manuscript. R. Harris draughted the figures.

REFERENCES

ADAMS, R.D. & WARE, D.E. (1977) Subcrustal earthquakes beneath New Zealand; locations determined with laterally inhomogeneous velocity model. *N.Z. J. Geol. Geophys.* **20,** 59–83.

ARONSON, J.L. (1968) Regional geochronology of New Zealand. *Geochim. Cosmochim. Acta,* **32,** 669–697.

AUSTIN, P.M., SPRIGG, R.C. & BRAITHWAITE, J.C. (1973) Structural development of the eastern Chatham Rise of the New Zealand region. In: *Oceanography of the South Pacific* (comp. by R. Fraser) N.Z. National Commission for UNESCO. 201–215.

BALLANCE, P.F. (1974) An interarc flysch basin in northern New Zealand: Waitemata Group (Upper Oligocene-Lower Miocene). *J. Geol.* **82,** 439–471.

BALLANCE, P.F. (1976) Evolution of the upper Cenozoic magmatic arc and plate boundary in northern New Zealand. *Earth Planet. Sci. Lett.* **28,** 356–370.

BALLANCE, P.F. (1980) Models of sediment distribution in non-marine oblique slip fault zones. In: *Sedimentation in oblique-slip mobile zones* (Ed. by P. F. Ballance & H. G. Reading). *Spec. Publ. int. Ass. Sediment.,* **4,** 229–236.

BALLANCE, P.F., HAYWARD, B.W. & WAKEFIELD, L.L. (1977) Group nomenclature of Late Oligocene and Early Miocene rocks in Auckland and Northland, New Zealand; and an Akarana Supergroup. *N.Z. J. Geol. Geophys.* **20,** 673–686.

BALLANCE, P.F. & SPÖRLI, K.B. (1979) Northland Allochthon. *J. R. Soc. N.Z.* **9,** 259–275.

BECK, A.C. (1964) Sheet 14 Marlborough Sounds (1st edn). *Geological Map of New Zealand* 1 : 250 000. D.S.I.R. Wellington, N.Z.

BISHOP, D.G. (1974) Stratigraphic, structural and metamorphic relationships in the Dansey Pass area, Otago, New Zealand. *N.Z. J. Geol. Geophys.* **17,** 301–335.

BOWEN, F.E. (1964) Sheet 15 Buller (1st edn). *Geological Map of New Zealand* 1 : 250 000 D.S.I.R. Wellington.

BRADSHAW, J.D. (1975) The folds at Castle Hill (Canterbury) and their bearing on Kaikouran deformation style in the Canterbury Basin. *J. R. Soc. N.Z.* **5,** 209–217.

BROTHERS, R.N. (1965) Northland igneous geology. In: *New Zealand volcanology, Northland, Coromandel, Auckland* (Ed. by B. N. Thompson & L. O. Kermode), *N.Z. D.S.I.R. Inform. Ser.* **49,** 17–21.

BROTHERS, R.N. (1974) Kaikoura Orogeny in Northland. *N.Z. J. Geol. Geophys.* **17,** 1–18.

CARTER, R.M., & LANDIS, C.A. (1972) Correlative Oligocene unconformities in Southern Australasia. *Nature* (*Phys. Sci.*) **237,** 12–13.

CARTER, R.M. LANDIS, C.A., NORRIS, R.J. & BISHOP, D.G. (1974) Suggestions towards a high level nomenclature for New Zealand rocks. *J. R. Soc. N.Z.* **4,** 5–18.

CARTER, R.M. & NORRIS, R.J. (1976) Cainozoic history of southern New Zealand; an accord between geological observations and plate tectonic predictions. *Earth Planet Sci. Lett.* **31,** 85–94.

CHRISTOFFEL, D.A. (1978) Interpretation of magnetic anomalies across the Campbell Plateau, south of New Zealand. *Bull. Australian Soc. Explor. Geoph.* **9,** 143–145.

COLE, J.W. (1978) Andesites of the Tongariro Volcanic Centre, North Island, New Zealand. *J. Volc. Geoth. Res.* **3,** 121–153.

COOMBS, D.S., LANDIS, C.A., NORRIS, R.J., SINTON, J.M., BORNS, D.J. & CRAW, D. (1976) The Dun Mountain ophiolite belt, New Zealand, its tectonic setting, constitution and origin, with special reference to the southern portion. *Am. J. Sci.* **276,** 561–603.

COOPER, R.A. (1975) New Zealand and South East Australia in the Early Paleozoic. *N.Z. J. Geol. Geophys.* **18,** 1–20.

CRADDOCK, C. (1975) Tectonic evolution of the Pacific margin of Gondwanaland. In: *Gondwana Geology* (Ed. by K. S. W. Campbell), pp. 609–618. Australian National University Press.

CULLEN, D.J. (1970) A tectonic analysis of the southwest Pacific. *N.Z. J. Geol. Geophys.* **13,** 7–20.

DAVEY, J.J. (1974) Magnetic anomalies off the west coast of Northland, New Zealand. *J. R. Soc. N.Z.* **4,** 203–216.

DAVEY, F.J. (1977) Marine seismic measurements in the New Zealand region. *N.Z. J. Geol. Geophys.* **20** 719–777.

EADE, J.V. & CARTER, L. (1975) Definitions and code of nomenclature for naming of morphological features on the New Zealand sea floor. *N.Z. Oceanogr. Inst. Records,* **2,** 129–140.

FLEMING, C.A. (1970) The Mesozoic of New Zealand: chapters in the history of the circum-Pacific mobile belt. *Q. J. geol. Soc. Lond.* **125,** 125–170.

FREUND, R. (1971) The Hope Fault: a strike-slip fault in New Zealand. *Bull. N.Z. Geol. Surv.* n.s. **86,** 49pp.

GAIR, H.S. (1967) Sheet 20 Mt. Cook (1st edn) *Geological Map of New Zealand* 1 : 250 000. D.S.I.R. Wellington N.Z.

GHANI, M.A. (1978) Late Cenozoic movements in the southern North Island, New Zealand, *N.Z. J. Geol. Geophys.* **21,** 117–125.

GLENNIE, K.W. (1959) The graded sediments of the Mahoenui Formation (King Country, North Island) *N.Z. J. Geol. Geophys.* **2,** 613–621.

Grapes, R.H. (1975) Petrology of the Blue Mountain Complex, Marlborough, New Zealand. *J. Petrology*, **16,** 371–428.

Gregg, D.R. (1964) Sheet 18 Hurunui (1st edn). *Geological Map of New Zealand* 1 : 250 000. D.S.I.R. Wellington, New Zealand.

Gregg, D.R. & Coombs, D.S. (1965) Volcanism in the South Island. In: *New Zealand Volcanology* (Ed. by B. N. Thomposn & L. O. Kermode). *D.S.I.R. Information Ser.* **51,** 5–8.

Gregory, M.R. (1969) Sedimentary features and penecontemporaneous slumping in the Waitemata Group, Whangaparaoa Peninsula, North Auckland. *N.Z. J. Geol. Geophys.* **12,** 248–282.

Griffiths, J.R. (1974) Revised continental fit of Australia and Antarctica. *Nature,* **249,** 336–338.

Grindley, G.W. (1960) Sheet 8 Taupo (1st edn). *Geological Map of New Zealand* 1 : 250 000. D.S.I.R. Wellington, N.Z.

Grindley, G.W. (1961a) Sheet 13 Golden Bay (1st edn). *Geological Map of New Zealand* 1 : 250 000. D.S.I.R. Wellington, N.Z.

Grindley, G.W. (1961b) Sheet S8 Takaka, *Geological Map of New Zealand* 1 : 63 360. D.S.I.R. Wellington, N.Z.

Grindley, G.W. (1974) New Zealand. In: *Mesozoic-Cenozoic Orogenic Belts.* (Ed. by A. M. Spencer), pp. 387–416. Geological Society of London.

Grindley, G.W., Adams, C.J.D., Lumb, J.T. & Watters, W.A. (1977) Paleomagnetism, K-Ar dating and tectonic interpretation of Upper Cretaceous and Cenozoic volcanic rocks of the Chatham Islands, New Zealand. *N.Z. J. Geol. Geophys.* **20,** 425–467.

Harrington, H.J., Burns, K.L. & Thompson, B.R. (1973) The Gambier-Beaconsfield and Gambier-Sorell fracture zones and the movement of plates in the Australia-Antarctica-New Zealand region. *Nature,* **245,** 109–112.

Hatherton, T. (1975) Stokes magnetic anomaly—magnetic system of magnetic supergroup? *N.Z. J. Geol. Geophys.* **18,** 519–521.

Hatherton, T. & Sibson, R.H. (1970) Junction magnetic anomaly north of Waikato River. *N.Z. J. Geol. Geophys.* **13,** 655–662.

Hay, R.F. (1967) Sheet 7 Taranaki (1st edn). *Geological Map of New Zealand* 1 : 250 000 D.S.I.R. Wellington, N.Z.

Hayes, D.E. & Ringis, J. (1973) Sea floor spreading in the Tasman Sea. *Nature,* **243,** 454–458.

Healy, J., Schofield, J.C. & Thompson, B.N. (1964) Sheet 5 Rotorua (1st edn) *Geological Map of New Zealand* 1 : 250 000 D.S.I.R. Wellington, N.Z.

Hochstein, M.P. & Nunns, A.G. (1976) Gravity measurements across the Waikato Fault North Island, New Zealand. *N.Z. J. Geol. Geophys.* **19,** 347–358.

Hunt, T. & Nathan, S. (1976) Inangahua Magnetic Anomaly, New Zealand. *N.Z. J. Geol. Geophys.* **19,** 395–406.

Johnston, M.R. (1975) Sheet N 159 and pt. N 158 Tinui-Awatoitoi (1st edn). *Geological Map of New Zealand* 1 : 63 360. D.S.I.R. Wellington, N.Z.

Katz, H.R. (1973) Pliocene unconformity at Opau stream, Hawke's Bay, New Zealand, *N.Z. J. Geol. Geophys.* **16,** 917–925.

Katz, H.R. (1974a) Margins of the Southwest Pacific. In: *The Geology of Continental Margins* (Ed. by C. A. Burk & C. L. Drake), pp. 549–565. Springer Verlag, New York.

Katz, H.R. (1974b) Offshore petroleum potential in New Zealand. *Aust. Petrol. Expl. Ass. J.* 1974, 3–13.

Katz, H.R. (1976) Sedimentary basins and petroleum prospects, onshore and offshore New Zealand. In: *Circum-Pacific Energy and Mineral Resources.* (Ed. by M. T. Halbouty, J. C. Maher & H. M. Lian). *Am. Assoc. Petrol. Geol. Mem.* **25,** 217–288.

Kear, D. (1960) Sheet 4 Hamilton (1st edn). *Geological Map of New Zealand* 1 : 250 000. D.S.I.R. Wellington, N.Z.

Kear, D. (1964) Volcanic alignments north and west of New Zealand's central volcanic region. *N.Z. J. Geol. Geophys.* **7,** 24–44.

Kear, D. & Hay, R.F. (1961) Sheet 1 North Cape. *Geological Map of New Zealand* 1 : 250 000. D.S.I.R. Wellington, N.Z.

Kennett, J.P. (1977) Cenozoic evolution of Antarctic glaciation, the Circum-Antarctic Ocean and their impact on global paleooceanography. *J. geoph. Res.* **82,** 3843–3860.

Kingma, J.T. (1962) Sheet 11 Dannevirke (1st edn). *Geological Map of New Zealand* 1 : 250 000. D.S.I.R. Wellington, N.Z.

Kingma, J.T. (1964) Sheet 9 Gisborne (1st edn). *Geological Map of New Zealand* 1 : 250 000. D.S.I.R. Wellington, N.Z.

Kingma, J.T. (1966) Sheet 6 East Cape (1st edn) *Geological Map of New Zealand* 1 : 250 000 D.S.I.R. Wellington, N.Z.

Kingma, J.T. (1967) Sheet 12 Wellington (1st edn). *Geological Map of New Zealand* 1 : 250 000. D.S.I.R. Wellington, N.Z.

Kingma, J.T. (1971) Geology of the Te Aute Subdivision. *Bull. N.Z. Geol. Surv.* **70,** 173pp.

Laird, M.G. (1968) The Paparoa Tectonic Zone. *N.Z. J. Geol. Geophys.* **11,** 435–454.

Laird, M.G. (1972) Sedimentology of the Greenland Group in the Paparoa Range, West Coast, South Island. *N.Z. J. Geol. Geoph.* **15,** 372–393.

Landis, C.A. & Coombs, D.S. (1967) Metamorphic belts and orogenesis in southern New Zealand. *Tectonophysics,* **4,** 501–518.

Lensen, G.J. (1962) Sheet 16 Kaikoura (1st edn). *Geological Map of New Zealand* 1 : 250,000. D.S.I.R. Wellington, N.Z.

Lensen, G.J. (1968) Analysis of progressive fault displacement during downcutting at the Branch River terraces, South Island, New Zealand. *Bull. geol. Soc. Am.* **79,** 545–566.

Lensen, G.J. (1975) Earth deformation studies in New Zealand. *Tectonophysics,* **29,** 541–551.

Lensen, G.J. (1977) Late Quaternary tectonic map of New Zealand 1 : 2 000 000 (1st edn). *N.Z. Geol. Surv. Misc. Series Map* 12. D.S.I.R. Wellington, N.Z.

Lensen, G.J., Fleming, C.A. & Kingma, J.T. (1959) Sheet 10 Wanganui (1st edn). *Geological Map of New Zealand* 1 : 250 000. D.S.I.R. Wellington, N.Z.

Le Pichon, X. (1968) Seafloor spreading and continental drift. *J. geophys. Res.* **73,** 3661–3697.

Leslie, W.C. & Hollingsworth, R.J.S. (1972) Exploration in the East Coast Basin, New Zealand. *Aust. Petrol. Expl. Ass. J.* **12,** 39–44.

Lewis, K.B. (1971) Growth rate of folds using tilted wave-planed surfaces: coast and continental shelf, Hawkes Bay, New Zealand. In: *Recent Crustal Movements* (Ed. by B. W. Collins & R. Frazer). *R. Soc. N.Z. Bull.* **9,** 225–231.

Lewis, K.B. (1973) Erosion and deposition on a tilting continental shelf during Quaternary oscillations of sea level. *N.Z. J. Geol. Geophys.* **16,** 281–301.

Lewis, K.B. (1980) Quaternary sedimentation in the Hikurangi oblique subduction and transform margin, New Zealand. In: *Sedimentation in oblique-slip mobile zones* (Ed. by P. F. Ballance & H. G. Reading). *Spec. Publ. int. Ass. Sediment.*, **4,** 171–189.

Lillie, A.R. (1953) The Geology of the Dannervirke Subdivision. *Bull. N.Z. Geological Survey* **46,** 152 p.

Lingen, G.J. van der & Pettinga, J.R. (1980) Miocene slope-basin sedimentation along the New Zealand sector of the Australian-Pacific convergent plate boundary. In: *Sedimentation in oblique-slip mobile zones* (Ed. by P. F. Ballance & H. G. Reading). *Spec. Publ. int. Ass. Sediment.* **4,** 191–215.

McKellar, I.C. (1966) Sheet 25 Dunedin. *Geological Map of New Zealand* 1 : 250 000. D.S.I.R. Wellington, N.Z.

Molnar, P., Atwater, T., Mammerickx, J. & Smith, S.M. (1975) Magnetic anomalies, bathymetry and tectonic evolution of the South Pacific since the Late Cretaceous. *Geophys. J. R. Astr. Soc.* **60,** 383–420.

Moore, P.R. (1978) Geology of western Koranga Valley, Raukumara Peninsula. *N.Z. J. Geol. Geophys.* **21,** 1–20.

Mutch, A.R. (1963) Sheet 23 Oamaru (1st edn). *Geological Map of New Zealand* 1 : 250 000. D.S.I.R. Wellington, N.Z.

Mutch, A.R. & McKellar, I.C. (1964) Sheet 19 Haast (1st edn). *Geological Map of New Zealand* 1 : 250 000. D.S.I.R. Wellington, N.Z.

Nathan, S. (1974) Stratigraphic nomenclature for the Cretaceous-Lower Quaternary rocks of Buller and North Westland, West Coast, South Island, New Zealand. *N.Z. J. Geol. Geophys.* **17,** 423–445.

Nathan, S. (1977) Cretaceous and Lower Tertiary stratigraphy of the coastal strip between Buttress Point and Ship Creek, South Westland, New Zealand. *N.Z. J. Geol. Geophys.* **20,** 615–654.

Nathan, S. (1978) Upper Cenozoic stratigraphy of South Westland, New Zealand. *N.Z. J. Geol. Geophys.* **21,** 329–361.

Neall, V.E. (1974) *The Volcanic history of Taranaki.* Egmont National Park Board, 14 pp.

NEEF, G. (1974) Sheet N 153 Eketahuna (1st edn). *Geological Map of New Zealand* 1 : 63 360 D.S.I.R. Wellington, N.Z.

NELSON, C.S. (1968) Sedimentology of redeposited calcareous and glauconitic beds at Pahaoa, south-east Wellington. *Trans. R. Soc. N.Z.* **6,** 45–62.

NELSON, C.S. & HUME, T.M. (1977) Relative intensity of tectonic events revealed by the Tertiary sedimentary record in the North Wanganui Basin and adjacent areas, New Zealand. *N.Z. J. Geol. Geophys.* **20,** 369–392.

NORRIS, R.J. (1979) A geometrical study of finite strain and bending in the South Island. In: *Origin of the Southern Alps* (Ed. by R. I. Walcott & M. M. Cresswell) *Roy. Soc. N.Z. Bull.* **18,** 21–28.

NORRIS, R.J., CARTER, R.M. & TURNBULL, I.M. (1978) Cainozoic sedimentation in basins adjacent to a major continental boundary in southern New Zealand. *J. geol. Soc. Lond.* **135,** 191–205.

NORRIS, R.J. & CARTER, R.M. (1980) Offshore sedimentary basins at the southern end of the Alpine Fault, New Zealand. In: *Sedimentation in oblique-slip mobile zones* (Ed. by P. F. Ballance & H. G. Reading). *Spec. Publ. int. Ass. Sediment.,* **4,** 237–265.

OBORN, L.E. (1959) Sheet 21 Christchurch (1st edn). *Geological Map of New Zealand* 1 : 250 000. D.S.I.R. Wellington, N.Z.

O'BRIEN, J.P. & RODGERS, K.A. (1973) Alpine-type serpentinites from the Auckland Province-1. The Wairere Serpentinite. *J. R. Soc. N.Z.* **3,** 169–190.

PACKHAM, G.H. & ANDREWS, J.E. (1975) Results of Leg 30 and the geologic history of the southwest Pacific arc and marginal sea complex. In: *Initial Reports of Deep Sea Drilling Project* (Ed. by J. E. Andrews & G.H. Packham *et al.*), **30,** 691–705. Washington D.C.

PACKHAM, G.H. & TERRILL, A. (1975) Submarine Geology of the South Fiji Basin. In: *Initial Reports of the Deep Sea Drilling Project* (Ed. by J.E. Andrews & G. H. Packham *et al.*), **30,** 617–633. Washington D.C.

PILAAR, W.F.H. & WAKEFIELD, L.L. (1978) Structural and stratigraphic evolution of the Taranaki Basin, offshore New Zealand. *APEA J.* 1978, 93–102.

PREBBLE, W. (1980) Late Cainozoic sedimentation and tectonics of the East Coast Deformed Belt in Marlborough, New Zealand. In: *Sedimentation in oblique-slip mobile* zones (Ed. by P.F. Ballance & H. G. Reading) *Spec. Publ. int. Ass. Sediment.* **4,** 217–228.

RIDD, M.F. (1964) Succession and structural interpretation of the Whangara-Waimate area. *N.Z. J. Geol. Geophys.* **7,** 279–298.

ROBINSON, R., CALHAEM, I.M. & THOMSON, A.A. (1976) The Opunake, New Zealand earthquake of 5 Nov. 1974. *N.Z. J. Geol. Geophys.* **19,** 335–345.

RYNN, J.M.W. & SCHOLZ, C.H. (1978) Seismotectonics of Arthur's Pass region, South Island, New Zealand. *Geol. Soc. Am. Bull.* **89,** 1373–1388.

SCHOFIELD, J.C. (1967) Sheet 3 Auckland (1st edn) *Geological Map of New Zealand* 1 : 250 000· D.S.I.R. Wellington, N.Z.

SCLATER, J.G., J.W. Jr, MAMMERICKX, J. & CHASE, C. G. (1972) Crustal extension between the Tonga and Lau Ridges: Petrologic and geophysical evidence. *Bull. geol. Soc. Am.* **83,** 505–518.

SPEDEN, I.G. (1976) Geology of Mt. Taitai, Tapuaeroa Valley, Raukumara Peninsula. *N.Z. J. Geol. Geoph.* **19,** 71–119.

SPÖRLI, K.B. (1978) Mesozoic tectonics, North Island, New Zealand. *Geol. Soc. Am. Bull.* **89,** 415–425.

STEVENS, G.R. (1974) *Rugged Landscape, the Geology of Central New Zealand,* 286 pp. A. H. & A. W. Reed, Wellington.

STONELEY, R. (1968) A Lower Tertiary décollement on the east coast, North Island, New Zealand. *N.Z. J. Geol. Geophys.* **11,** 128–156.

SUGGATE, R.P. (1963) The Alpine Fault. *Trans. R. Soc. N.Z.* **2,** 105–129.

SUGGATE, R.P. (1972) Mesozoic-Cenozoic development of the New Zealand region. *Pacific Geology,* **4,** 113–120.

TE PUNGA, M.T. (1957) Live anticlines in western Wellington. *N.Z. J. Science and Technology,* **38B,** 433–436.

THOMPSON, B.N. (1961) Sheet 2A Whangarei. *Geological Map of New Zealand* 1 : 250 000 D.S.I.R. Wellington, N.Z.

WALCOTT, R.I. (1978a) Present tectonics and Late Cenozoic evolution of New Zealand. *Geophys. J. R. astr. Soc.* **52,** 137–164.

WALCOTT, R.I. (1978b) Geodetic strains and large earthquakes in the Axial Tectonic Belt, of North Island, New Zealand. *J. geophys. Res.,* **83,** 4419–4429.

Warren, G. (1967) Sheet 17 Hokitika (1st edn). *Geological Map of New Zealand* 1 : 250 000. D.S.I.R. Wellington, N.Z.

Watters, W.A., Speden, I.G. & Wood, B.L. (1968) Sheet 26 Stewart Island (1st edn). *Geological Map of New Zealand* 1 : 250 000. D.S.I.R. Wellington, N.Z.

Watts, A.B., Weissel, J.K. & Davey, F.J. (1977) Tectonic evolution of the South Fiji marginal basin. In: *Island Arcs, Deep Sea Trenches and Back Arc Basins* (Ed. by M. Talwani & W. C. Pitman III). *Am. Geophys. Union, Maurice Ewing Ser.* **1,** 419–427.

Weissel, J.K. (1977) Evolution of the Lau Basin by the growth of small plates. In: *Island Arcs, Deep Sea Trenches and Back Arc Basins* (Ed. by M. Talwani & W. C. Pitman III). *Am. Geophys. Union Maurice Ewing Ser.* **1,** 429–436.

Weissel, J. K., Hayes, D.E. & Herron, E.M. (1977) Plate tectonic synthesis: the displacements between Australia, New Zealand and Antarctica since the Late Cretaceous. *Mar. Geol.* **25,** 231–277.

Wellmann, H.W. (1955) New Zealand Quaternary tectonics. *Geol. Rundschau,* **43,** 248–257.

Wellman, H.W. (1973) The Stokes Magnetic Anomaly. *Geol. Mag.* **110,** 419–429.

Wellman, H.W. (1974) Recent crustal movements in New Zealand. *Tectonophysics,* **23,** 423–424.

Wellman, P. & Cooper, A. (1971) Potassium argon age of some New Zealand lamprophyre dikes near the Alpine Fault. *N.Z. J. Geol. Geoph.* **14,** 341–350.

Wilson, D.D. (1953) The geology of the Waipara subdivision. *N.Z. Geol. Survey Bull.* **64,** 122pp.

Wood, B.L. (1960) Sheet 27 Fiord (1st edn). *Geological Map of New Zealand* 1 : 250 000. D.S.I.R. Wellington, N.Z.

Wood, B.L. (1962) Sheet 22 Wakatipu (1st edn). *Geological Map of New Zealand* 1 : 250 000. D.S.I.R. Wellington, N.Z.

Wood, B.L. (1966) Sheet 24 Invercargill (1st edn). *Geological Map of New Zealand* 1 : 250 000. D.S.I.R. Wellington, N.Z.

Wright, A.C. & Black, P.M. (1979) Geochemistry and petrology of the Waitakere Group, North Auckland, New Zealand. *49th ANZAAS Congress*, Auckland, Abstracts 1,181.

Spec. Publ. inst. Ass. Sediment. (1980) **4,** 171–189

Quaternary sedimentation on the Hikurangi oblique-subduction and transform margin, New Zealand

K. B. LEWIS

New Zealand Oceanographic Institute, Wellington North, New Zealand

ABSTRACT

The Hikurangi Margin/Trough system on the northeastern side of New Zealand is mainly an extension of the Tonga–Kermadec Arc/Trench subduction system into a continental environment. Structural elements become progressively more elevated and subduction progressively more oblique towards the south until the whole system is truncated at a strike-slip, transform boundary that extends along the south-western part of the Hikurangi Trough and the Hope Fault to the Alpine Fault.

Subduction, combined with rapid detrital sedimentation, has led to the development of a 150 km wide, imbricate-thrust controlled, accretionary borderland of seaward-faulted, anticlinal ridges and landward-tilting basins. The basins are generally 5–30 km wide, 10–60 km long and contain fill 200–2000 m thick. The borderland continues on land where the highest accretionary ridges (normally at mid-bathyal depths) form a line of coastal hills in front of a line of strike-slip faulted, highest accretionary basins and volcano-backed, frontal ranges.

Turbidites, hemipelagic muds and volcanic ash layers fill the tilting slope basins with wedge-shaped layers, at rates ranging from about 0·1 m/1000 years in lower slope basins to about 0·3 m/1000 years in upper slope basins. Prisms of sandy mud build the shelf upwards at maximum rates of 3·0 m/1000 years during interglacial periods of high sea level, and the upper slope upwards and outwards at similar rates during glacially lowered sea level.

On the upper slope, sheets of sediment more than 10 km in length but only a few tens of metres thick have slumped on slopes of as little as 1°, probably after dewatering of surface mud has trapped excess pore water in underlying coarse silty layers.

Along the steep, southwestern transform margin much of the influx of muddy sediment is fed direct to the Hikurangi Trough along canyons.

INTRODUCTION

The plate boundary through New Zealand

Quaternary sedimentation in New Zealand is dominated by its position across the boundary between the Indian and the Pacific Plates. A plate boundary has passed through the continental mass of New Zealand for the last 40 m.y., with the Indian

0141-3600/80/0904-0171$02.00

Plate moving more rapidly northwards away from the spreading Pacific–Antarctic Ridge than the Pacific Plate (Molnar *et al.*, 1975). The two plates have sheared some 1000 km past one another, the relative motion having been taken up by fault movements and by rock deformation. However, the movement has rarely been simple shear. Various changes in the configuration of the boundary and of the spreading centre have meant that, from time to time and from place to place, relative motion has been either oblique extension ('transtension') or oblique compression ('transpression') (Carter & Norris, 1976; Ballance, 1976; Spörli, 1980).

For the last 10 m.y., the basic movement on most sectors of the boundary has tended more and more towards oblique compression. The compressional element appears to have migrated southwards from the Kermadec Trench area and to have increased in intensity about 2–3 m.y. ago. It was then that the Southern Alps began to rise rapidly towards the climax of the Kaikoura Orogeny (Walcott, 1978a).

To the north of New Zealand along the Tonga–Kermadec Margin, it is the Pacific plate that is subducted. To the south of New Zealand, along parts of the Macquarie–Puysegur Margin, it is the Indian Plate that underrides (Hayes & Talwani, 1972). In each case oceanic crust is subducted beneath oceanic crust (Fig. 1).

Because of the long continued strike-slip movement along the boundary, there are mirror image zones in the northeast and southwest where oceanic crust is subducted beneath continental crust. The floor of the southwest Pacific Basin disappears beneath

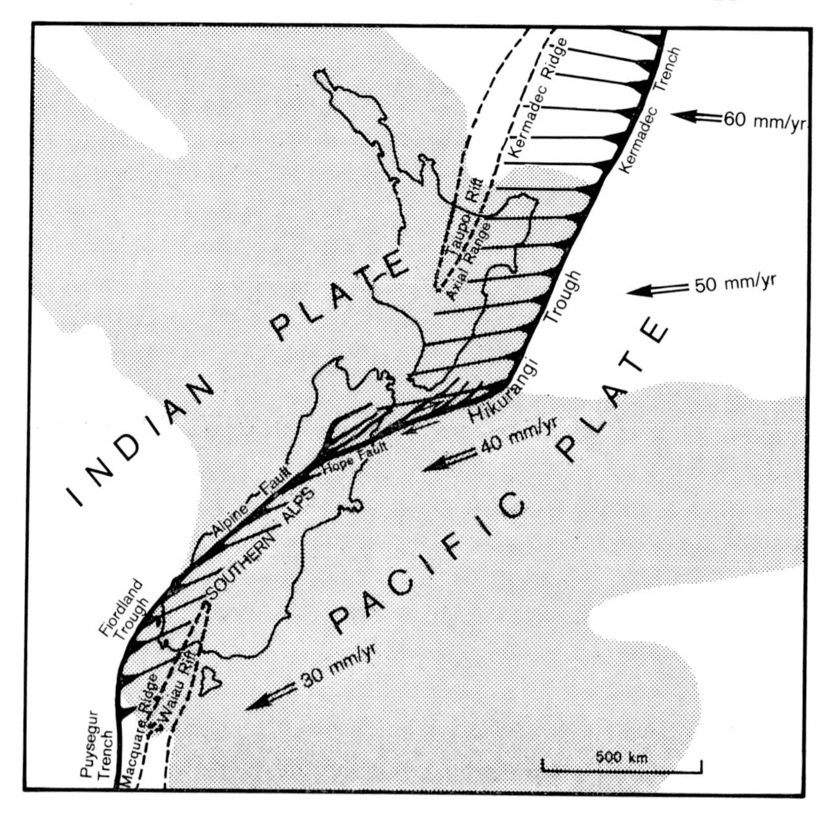

Fig. 1. Major elements of the Indian–Pacific Plate boundary in the New Zealand Region. Stippling represents continental crust. Arrows show relative motion of Pacific Plate with respect to the Indian Plate. Lines represent direction of motion of underthrusting plate (Adapted from Walcott, 1978a, b).

the 500 km long, eastern side of the North Island and the floor of the Tasman Basin dives under the Fiordland corner of South Island (Cristoffel & van der Linden, 1972).

Along the west coast of South Island the obliquely compressional boundary cuts through continental crust. Since continental crust is not easily subducted, the edge of the Pacific Plate rides up onto the Indian Plate to form the Southern Alps. Some of the motion is taken up by thrusting on the Alpine Fault but much of it takes place over a much wider zone of deformation (Walcott, 1978b).

The Hikurangi Trough

In the Cook Strait Area, central New Zealand, the boundary is not so obvious. The Alpine Fault divides into numerous smaller faults that continue northwards into the ranges of North Island. One or several of these faults have been regarded as the plate boundary but it is now considered (Walcott 1978b) that the real boundary lies offshore to the east along the elongate depression known as the Hikurangi Trough, or 'Hikurangi Trench' (Brodie & Hatherton, 1958). The feature is only 3 km deep and barely deeper than the seafloor to the east until it merges southwards into the V between central New Zealand and the Chatham Rise (Fig. 3). In accepted bathymetric terminology the feature is a 'trough' (Eade & Carter, 1975).

Most authors (Brodie, 1959; Pantin, 1963; Houtz *et al.*, 1967; Hatherton, 1970; Ballance, 1976; Walcott, 1978a) have regarded the Hikurangi Margin as a continuation of the Kermadec Trench/Rise compressional system. However, Katz (1974a), citing a gravity anomaly well to the west of its typical trench position (Hamilton & Gale, 1968) together with some of the profiles presented here, concluded that the whole of the broad trough between eastern North Island and the Chatham Rise is simply a downwarped continental basin. The alternative explanation is that the displaced gravity anomaly coincides with an abrupt increase in steepness of the downgoing plate and the Hikurangi Trough represents the site of low angle underthrusting (Walcott, 1978b).

It is considered here that the Hikurangi Trough is two separate structures; a northern, sediment flooded, volcanic-knoll studded SSW trending structural continuation of the Kermadec Trench and a southern, more east–west trending, transform boundary between central New Zealand and the downwarped continental crust of the Chatham Rise (Fig. 1). The northern boundary is at an acute angle to the direction of relative plate motion (Minster *et al.*, 1974) so that Pacific oceanic crust is obliquely subducted beneath North Island. The southern boundary is almost parallel with the direction of relative plate motion and is more nearly a simple, strike-slip transform although Central New Zealand rises up slightly onto the Chatham Rise crustal block (Bennett, in prep.). In fact, relative plate motion is taken up non-uniformly across a 1–250 km wide zone of deformation to landward. Much of the strike-slip movement is in the most landward part of the zone so that relative movement at the boundary itself may be nearly at right angles to the boundary (Walcott, 1978b).

ANATOMY OF AN OBLIQUE-SUBDUCTION MARGIN

The model

Structural and sedimentary mechanisms on the slopes adjacent to trenches have been proposed by Karig & Sharman (1975) and Moore & Karig (1976). Structural

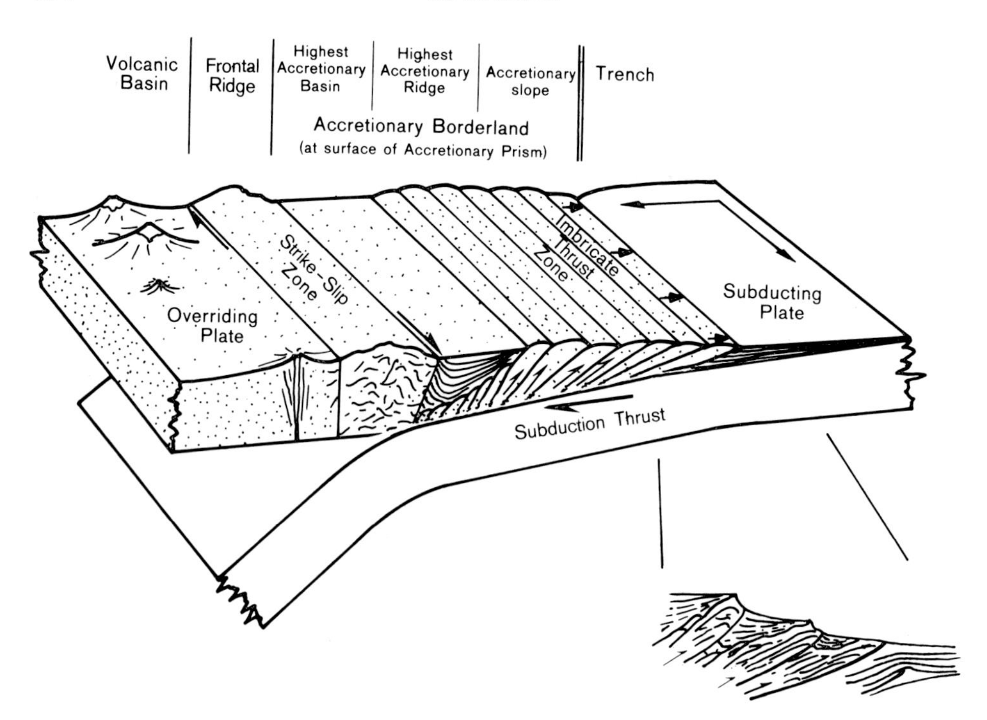

Fig. 2. Model of an obliquely subducting margin. Inset is detail of toe of accretionary slope. Stippling indicates over-riding plate: (Adapted from Walcott, 1978b; Karig & Sharman, 1975.)

mechanisms at an obliquely convergent boundary involving subduction have been modelled by Walcott (1978b). A combination of these is illustrated in Fig. 2.

An 'inner trench' zone of thrust faults may be a more or less general phenomenon at subducting margins (Karig & Sharman, 1975; Moore & Karig, 1976). It is believed to originate by landward-thinning wedges of sediment, mainly scraped from the surface of the subducting plate, being forced underneath the feather edge of the over-riding plate and accreted to it. As each new wedge is accreted, older wedges and the active thrust-faults between them are pushed upwards and landwards and rotated towards the vertical. Each wedge can then form a topographic ridge that dams sediment to landward in accretionary slope (or 'trench-slope') basins. The imbricate stack of wedges as a whole, known as the accretionary prism, and the highest accretionary ridge in particular (the so-called 'trench-slope break') commonly form a major, trench-flank ridge that dams a relatively large, highest accretionary basin (also termed the 'upper slope basin' or 'forearc basin') between it and a frontal ridge of older sediments.

At the Hikurangi Margin, the subducting Pacific Plate dips at 12° for some 250 km before diving down at about 45° beneath the Taupo Volcanic Zone (Walcott, 1978b). Immediately landward (or arcward) from the trench, landward-dipping thrust faults overlie the gently dipping, subduction thrust. Arcward of this accretionary borderland and overlying the zone of rapidly increasing dip of the subduction thrust there is a zone of high-angle strike-slip faulting.

The names adopted here are overtly topographic and generic and hopefully unambiguous in terms of the concept. They deliberately avoid reference to a bathymetric position, e.g. 'trench-slope break', 'upper slope basin' and 'upper slope discontinuity'

(Karig & Sharman, 1975), because terminology based on bathymetric position is impossibly confusing, for reasons discussed later, when used on the Hikurangi Margin.

The Hikurangi accretionary borderland

The Hikurangi continental slope (Fig. 3) is characterized by sedimentary basins and ridges aligned more or less parallel to the slope (Pantin, 1963; Lewis, 1976). The

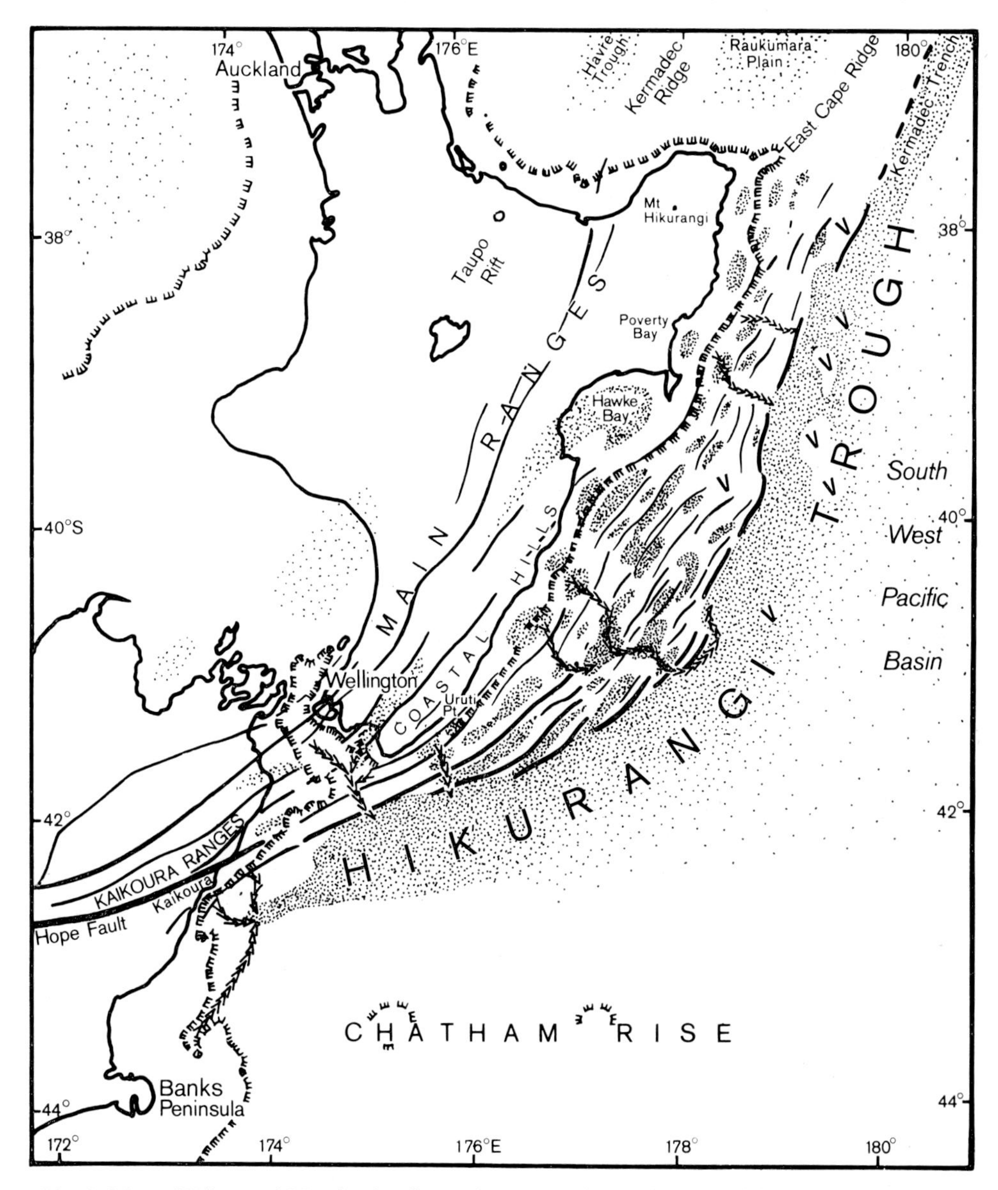

Fig. 3. Map of Hikurangi Margin showing main topographic and structural features. Stippled areas are sedimentary basins. Lines represent faults either evident in seismic profiles or inferred (on seaward side of ridges) from bathymetry. Line of 'E's represents shelf edge. Line of 'V's represents major canyons or seavalley systems. Large 'V' indicates possible volcanic knolls in Hikurangi Trough (Adapted from Katz, 1974a).

basins range from 5 to 30 km wide (ridge crest to ridge crest) and 10–60 km long. A typical size is 15 km wide by 30 km long. Early seismic profiles showed that, in general, the basins are synclinal and the ridges anticlinal (Figs 4, 5 profile E-16) and that there is a boundary at the foot of the slope between the folded slope strata and undeformed, parallel-bedded trough strata (Houtz *et al.*, 1967). Later high-resolution, seismic profiles from the shelf and upper slope showed that the folds there are actively growing structures, the difference between the rate of uplift of the ridge crests and the rate of subsidence of the basin axes being about 3 m/1000 years (Lewis, 1971a). Many of the anticlines are steeply dipping or faulted on their seaward side (Lewis, 1973a; Katz, 1974b). Deep water foraminiferal faunas in the rocks in the cores of some upper slope anticlines have been uplifted at least 1000 m since Pliocene times (Lewis, 1974a). Multichannel seismic profiles collected by the Mobil Oil Corporation in 1972, and now on open file at the New Zealand Oceanographic Institute, Wellington, show that growing monoclinal or seaward-faulted anticlines and landward-tilting basins occur right down the slope (Figs 4, 5).

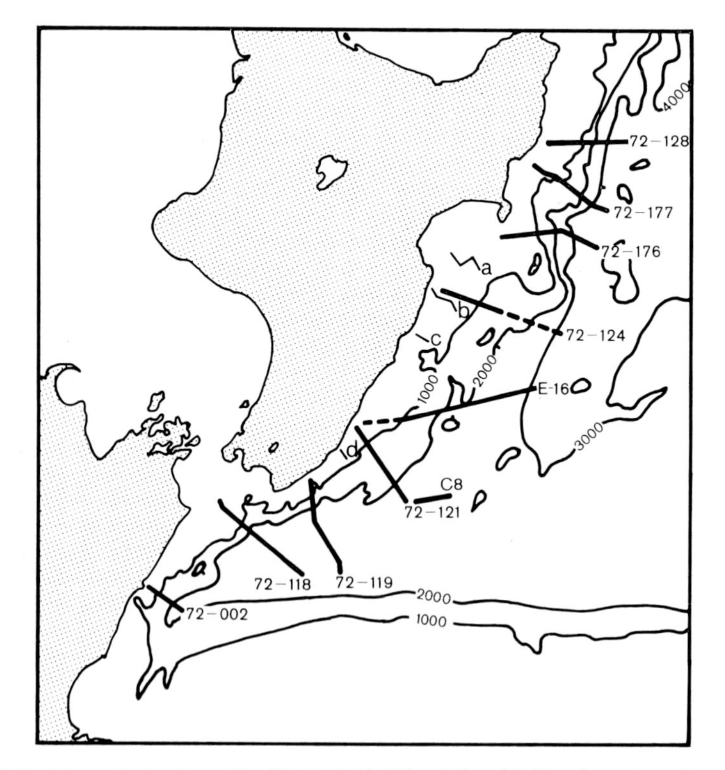

Fig. 4. Positions of seismic profiles illustrated in Figs 5, 6 and 8. Depth contours in metres.

In the far north, between Hawke's Bay and East Cape, the slope is steep and sediment-filled basins are small and comparatively rare (Fig. 5 profiles 72–177, 72–128). From Hawke's Bay south to Uruti Point, seaward-faulted anticlines and flat-floored, sediment-filled basins are spectacularly developed. Away from the prograding shelf edges, strata in virtually every basin are thickest at their landward limit and then wedge out towards the seaward anticline. Thus, the oldest beds in any basin

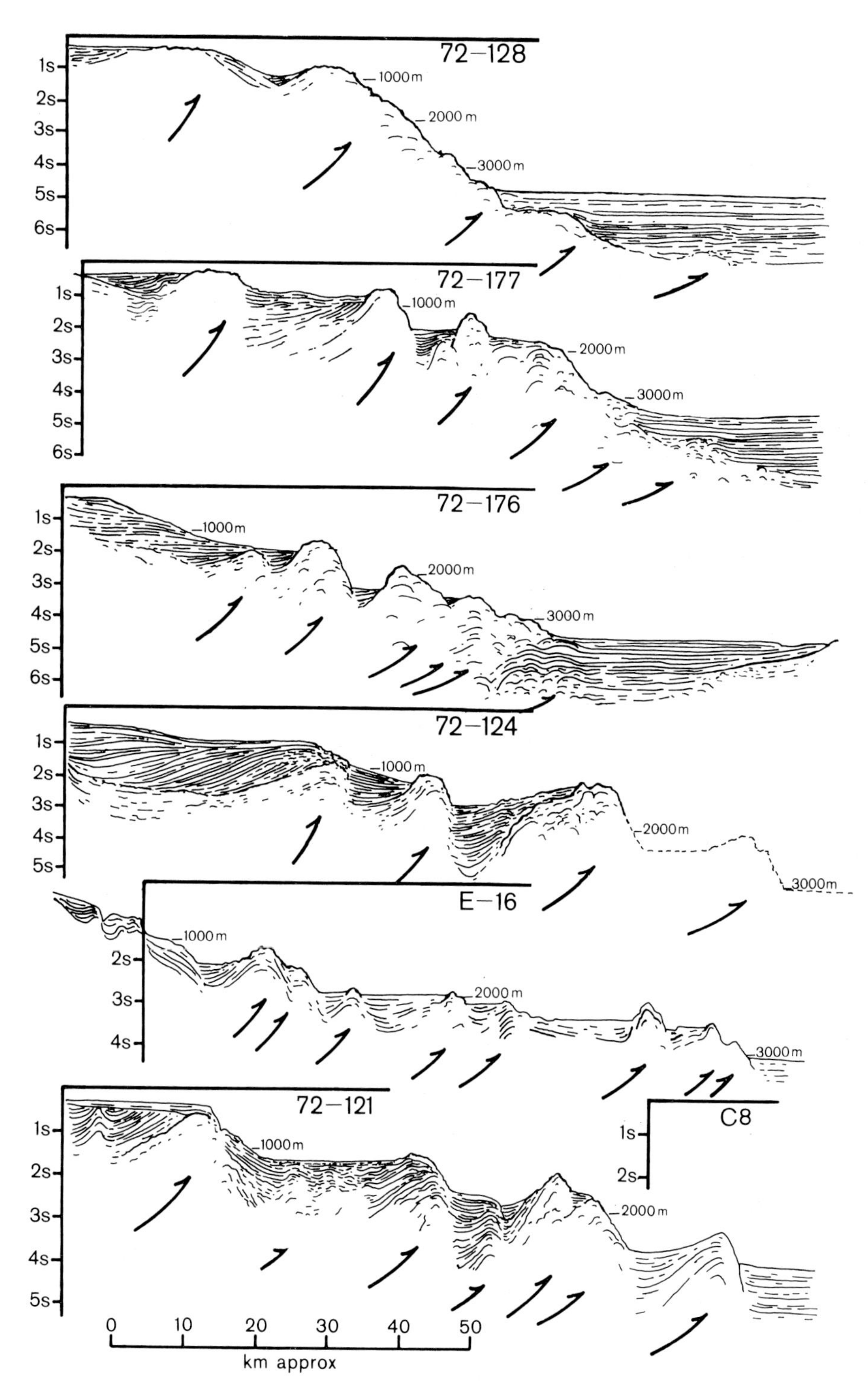

Fig. 5. Tracings of seismic profiles from the Hikurangi Oblique-subduction Margin, continental shelf (left) to Hikurangi Trough (right). Arrows represent probable position of thrust faults. At left, two-way travel time in seconds. On slope, depth to seabed in metres. Horizontal scale approximate only. A line sloping at 45° on the profiles represents a surface with a dip of about 6° in the line of the profile. Note underthrust toe of slope in 72–176 and 72–177. Basin-fill sediments mainly Quaternary in age. Ridge-crest strata range from Quaternary to Early Tertiary but are mainly Late Tertiary.

dip most steeply landward and it is inferred that they have been tilted from an original near-horizontal inclination displayed by the youngest beds. At some places, there appears to be a sharp distinction between the ridge-forming rocks, which are probably mainly Upper Tertiary in age, and the basin-filling sediments, which are probably mainly Quaternary. At other places there is no sharp distinction. Basin fill ranges from 200 to 2000 m thick.

In some of the profiles that continue eastward as far as the Hikurangi Trough (Fig. 5, 72–176, 72–177) flat, trough sediments appear to continue beneath the deformed base of the slope. This may indicate low-angle thrusting (cf. Moore & Karig, 1976).

In the Hikurangi Trough sediments some 2 km thick overlie strong reflectors emanating from the flanks of knolls that are thought to be volcanic (Katz, 1974a).

The Hikurangi transform margin

Along the southern part of the Hikurangi Margin, from Uruti Point south to Kaikoura, the slope narrows and steepens, and the imbricate pattern of seaward-faulted anticlines and landward-tilted basins fades out. Beneath the southwestern Hikurangi Trough, flat-lying, near parallel-bedded, Upper Cainozoic strata, up to 4 km thick, unconformably onlap the eroded and downwarped edge of Miocene to

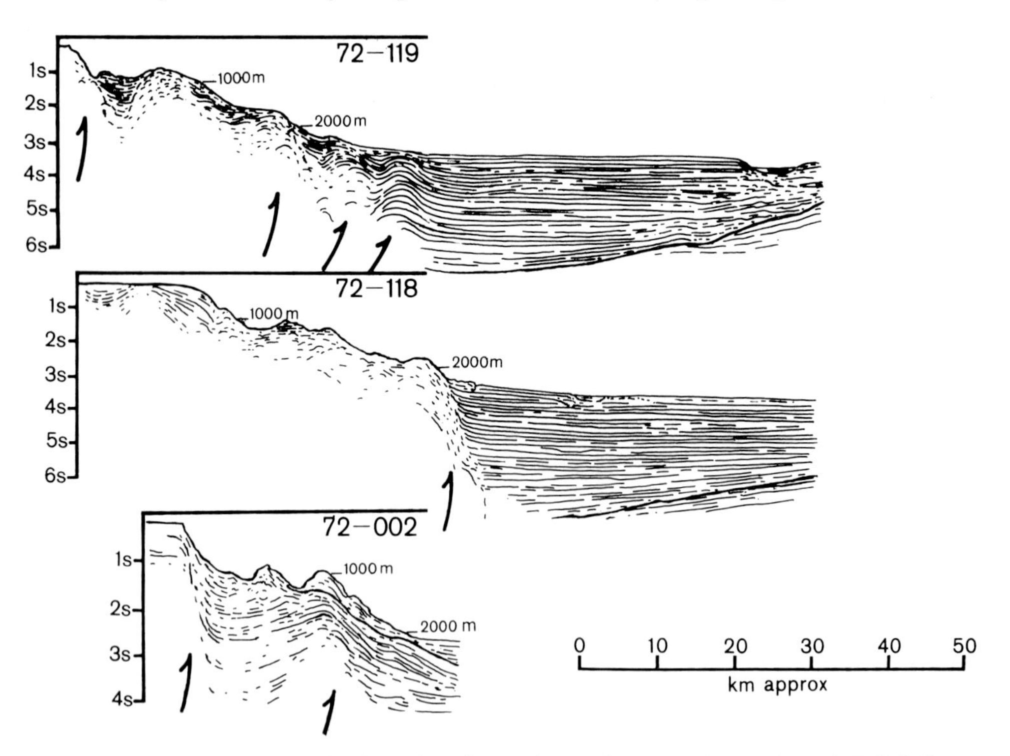

Fig. 6. Tracing of seismic profiles from the Hikurangi Transform Margin, continental shelf (left) to Hikurangi Trough (centre and right). Arrows represent probable position of reverse faults with major strike-slip component of motion. At left, two-way travel time in seconds. On slope, depth to seabed in metres. Horizontal scale approximate only. A line sloping at 45° on the profile represents a surface with a dip of about 6° in the line of the profile. Slope strata on the left mainly Late Mesozoic and Early Tertiary. Trough-fill sediments on right are Plio-Pleistocene.

mid-Mesozoic sediments that underlie the Chatham Rise (Fig. 6) (Katz, 1974a; Bennett, in prep.). A series of faults separates these trough strata from more intensely deformed strata on the New Zealand continental slope. The faults appear to be the extremities of thrust-faults from further north that change direction and character as they reach their southern limits (Fig. 3). If the imbricate-thrust concept is correct, then progressively older, steeper, faults form the boundary towards the southwestern limit of the Hikurangi Trough; they may also show increasing evidence of strike-slip movement. From Cape Palliser on the southern tip of North Island to the southern limit of the Hikurangi Trough off Kaikoura, only one major fault is apparent on the slope. Off Cook Strait this fault is at the foot of the slope and very sharply divides intensely crumpled slope beds from completely undeformed, parallel-bedded, trough beds (Fig. 6, 72–118). The fault then trends south-westward across the slope, and off Kaikoura is on the upper slope where it forms a steep, faulted edge to the shelf (Fig. 6 72–002). Here all that remains of the Hikurangi Trough is a broad syncline. Although the evidence is bathymetric and by no means conclusive, this major fault in the southern Hikurangi Trough appears to pass south of the Hope Fault, rather than link directly with it. The Hope Fault is probably the most conspicuous feature of the next part of the transform boundary zone from Kaikoura to the Alpine Fault.

An emergent accretionary borderland

The proximity of the Hikurangi Margin to rising mountain ranges gives rise to

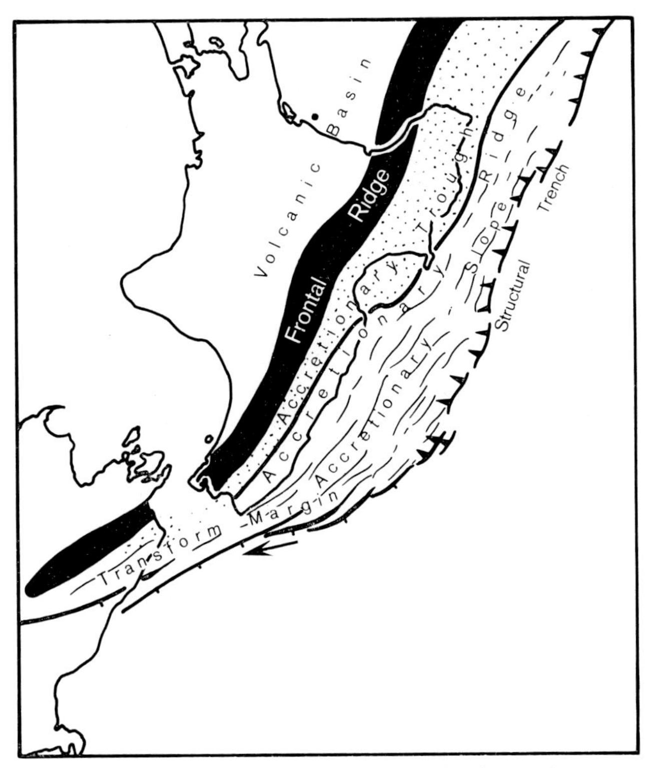

Fig. 7. Major structural elements of the Hikurangi Margin showing onshore propagation of some elements from Kermadec system to the north and termination of oblique-subduction system at the transform boundary to the south.

abundant sedimentation. This allows a wide, continental terrace to outbuild rapidly in front of a pre-existing margin at the frontal ridge (Fig. 7). Thus, the accretionary prism is wide and sediments near its upper surface are sufficiently gently deformed for their structure to be evident on seismic profiles.

Older strata in the prism probably formed in a similar oblique subduction system that curved around to the northwest before the Tonga–Kermadec system spread south (Cole, in press). Thus, with transcurrent movement, older sediments could have been derived from what is now the Auckland Peninsula.

Because of their position at the continent-limited end of a subduction zone and next to a transcontinental transform system, all structural and geomorphic features are relatively elevated, even allowing for the influx of sediment. The Hikurangi Trough is only 3 km deep, in contrast to normal trench depths of 6–10 km. The sediment thickness is at least 2 km and overlies strong, possibly volcanogenic, reflectors that may conceal deeper sediments. The accretionary borderland of ridges and basins is also shallower than normal and the highest accretionary ridges which normally form a mid-slope, trench-side bank or terrace, form the shallow East Cape Ridge in the far north, the shelf edge of Poverty Bay and Hawke's Bay and, most surprising of all, a line of coastal hills south of Hawke's Bay (Figs 3, 7; van der Lingen & Pettinga, 1980). The highest accretionary basins, between highest accretionary ridges and frontal ridge or arc, may be represented by the massive Raukumara Plain north of East Cape, by a series of slump and diapirically distorted, coastal basins around East Cape and Poverty Bay, by Hawke's Bay itself and by the long Quaternary sediment-filled valley that runs, on land, all the way from Hawke's Bay between main ranges and coastal hills to the shores of Cook Strait. The volcano-backed frontal ridge is represented by the Kermadec Ridge and by the main axial ranges of North Island. However, there is a disturbance of gravity anomalies just north of East Cape (Hatherton & Syms, 1975) close to the position where the suboceanic Tonga–Kermadec subduction system intersects the subcontinental Hikurangi system. To the south, all of the structures end at the South Hikurangi–Hope Fault Transform Margin in the vicinity of Cook Strait.

SEDIMENTATION ON THE HIKURANGI MARGIN

Quaternary basin filling

Rapid sedimentation has undoubtedly affected the structure of the Hikurangi Margin, but structure, in turn, controls the pattern of sedimentation.

Immediately adjacent to the main frontal ranges, faults are steep and strike-slip movement probably plays an important part in fluviatile and lacustrine deposition (Ballance, 1980). On the accretionary slope, thrust faulting apparently controls basin formation and filling.

Erosion of rising coastal hills and axial ranges, together with showers of ash from the volcanoes to windward, brings a flood of sediment to the continental shelf and slope. The slope has filled basins and in some places is now draped over the growing, seaward-faulted anticlines (Figs 5, 8). The flat tops of the wedge-shaped, landward-faulted, basin-fill sediments suggest deposition from some sort of dense, turbid flow. The increasing landward dip of progressively older, more deeply buried strata suggests growth of the basins (and anticlines) synchronous with deposition. In the north, near East Cape, and in the south, near Cook Strait, it appears that much of the sediment is by-passing the slope through canyons and being deposited in the Hikurangi Trough (Fig. 3).

Quaternary shelf progradation

Deposition on and adjacent to the continental shelf is more complex than on the slope basins to seaward because of the particularly abundant supply of sediment and the effects of Quaternary oscillations of sea-level. However, depositional patterns are more easily interpretable here than on many other continental margins because rapid tectonism has facilitated the preservation of datable structures. Sediment deposited on subsiding parts of the shelf during interglacial periods of sea-level as high as the present, was only partly removed during the subsequent glacial period of low sea-level,

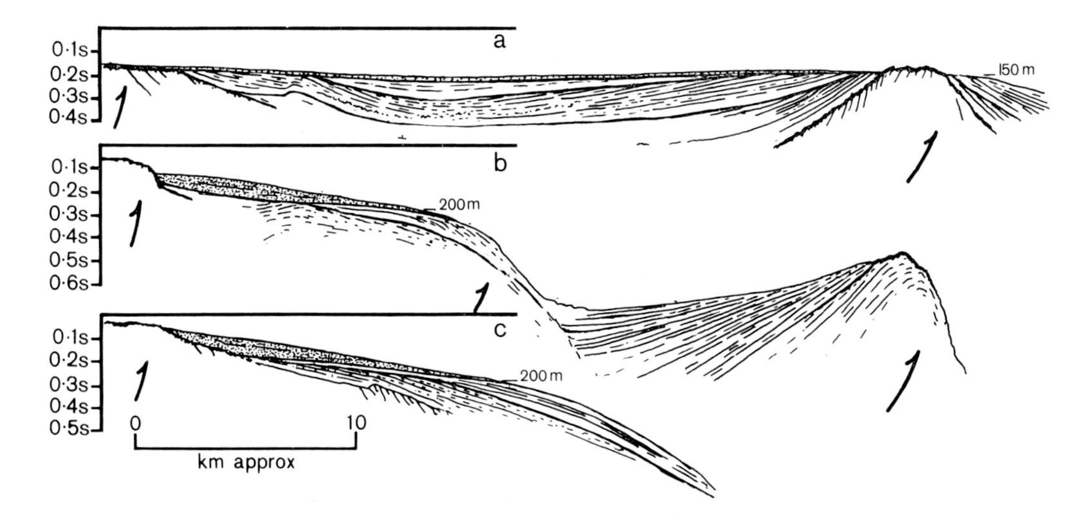

Fig. 8. High-resolution profiles of the continental shelf and upper slope illustrating development of continental shelf in continually folding basins during Quaternary oscillations of sea level. Stippling denotes post-20 000 year old sediments. Below this sediments of the Last Glacial Age overlie an Early Last Glacial Age unconformity and are themselves truncated at a 20 000 year old unconformity. Deeper still are Last Interglacial Age sediments. At left two-way travel time in seconds. On profile, depth to seabed in metres. Horizontal scale approximate only. A line sloping at 45° on the profiles represents a reflective surface dipping at 5° (modified from Lewis, 1973a).

thereby producing conspicuous unconformities within the Quaternary strata (Fig. 8; Lewis, 1971, 1973a). These unconformities can be tentatively correlated with dated maximum glacial extensions, just as raised beaches on the adjacent coastal hills can be correlated with interglacial periods of particularly high sea-level. On the basis of these correlations, rates of uplift on the coastal hills have been estimated to range up to 1·7 m/1000 years, rates of subsidence in the shelf basins has been estimated to range up to 1·5 m/1000 years and rates of tilting reach a maximum of about 0·03°/1000 years (30°/m years; Lewis, 1971a). In general, the shoreline should stabilize in a position close to a line of no vertical movement, the zero isobase, but this equilibrium is difficult to attain during rapid glacio-eustatic changes of sea level (Lewis, 1974b).

Because of the rapid tectonic and sedimentary processes, the basic morphology of the continental shelf along the Hikurangi Margin is primarily a result of wave-planation during the last glacial lowering of sea-level. So, at any place, the depth of the shelf break depends largely on the degree of induration of the rock or sediment at the shelf edge. In the south around Kaikoura and Cook Strait, where hard Lower Cainozoic or Mesozoic rocks form the outer shelf and upper slope, the shelf break is around 120–130 m deep, that is, the depth to which sea-level fell during the last glacial age (Fig. 6). At many places from Hawke's Bay northwards, the shelf is mainly a Quaternary sediment-filled basin with a Lower Cainozoic-cored, unconformity-draped, growing anticline at the shelf edge (Fig. 8a). The depth of the shelf break here depends on the degree of sediment build-out beyond the core of the anticline but it generally ranges from 140 to 160 m deep. Really deep shelf breaks occur for 150 km south from Hawke's Bay, where there is no fold or fault-lifted rock at the shelf edge (Fig. 8b, c) and all, except the inner shelf, is up and out built onto a subsiding substrate, the next anticline to seaward being too deep to have been affected by lowered sea level (Lewis, 1973a). During the last glacial age, waves planed soft sediments deposited during the preceding interglacial age to a depth of 50–70 m, which means that the shelf break now ranges from 170 to 200 m deep.

It is clear from high resolution profiles of the area south of Hawke's Bay that sediment is deposited most rapidly in a coast-parallel prism (Lewis, 1973a). In this area, prisms of sediment have built the continental shelf upwards during interglacial high sea-levels and the upper continental slope upwards and outwards during glacial low sea-levels (Fig. 8b, c). The zone of most rapid deposition has migrated back and forth as the level of the sea fell and rose in response to waxing and waning of ice caps. Since sea level reached its present level after the last glacial age, a prism has formed on the mid shelf. Its thickest part, a line of most rapid deposition, ranges from 4 to 10 km from shore where the water depth presently ranges from 30 to 100 m. Estimates of the age of reflectors suggests that sediment is deposited along this line at rates ranging from 1·5 to 3·0 m/1000 years. Estimates of the age of reflectors on the outer shelf and upper slope suggest that rates along the line of most rapid deposition were essentially similar during the last period of low sea level and during the subsequent rise to present sea level.

Present sedimentation

Present sedimentation on the Hikurangi Margin is for the most part rapid and muddy.

Except on offshore banks and ridges, the detrital grains of surface sediments are predominantly terrigenous and become finer offshore (Lewis, 1973b; McDougall, 1975). Where they are backed by Mesozoic 'greywacke' Ranges, for instance the Kaikoura Ranges, the southern tip of the Coastal Hills and the Main Ranges behind Hawke's Bay, beaches are composed of gravel and dark lithic sand. Elsewhere beaches are generally felsic sand. Offshore, the modal grain size ranges from fine sand on the inner shelf, to medium silt on the outer shelf to fine silt in the Hikurangi Trough. Almost all surface sediments have a 'tail' of fine silt and clay grains from disaggregated flocs. The zone of most rapid deposition on the mid shelf is characterized by muddy sand and sandy mud.

At the crests of banks and ridges, where the rate of detrital deposition is low, surface sediments are predominantly biogenic (mainly molluscan, foraminiferal and colonial coral debris) in the south, and predominantly volcanic in the north. Some shelf banks have locally derived, (residual) lithic fragments and some shelf and slope banks have glauconite in their veneer of sediment.

The source, transport and mode of deposition of coarse detrital grains is reasonably clear. Mesozoic 'greywacke' ranges, supply coarse lithic debris to rivers and ultimately to beaches. Cainozoic hills supply much of the felsic beach sand. Beach deposits move northwards under the influence of southerly swell until trapped by a headland. Only at Kaikoura does a canyon approach the surf zone and perhaps funnel coarse sediment down the slope.

The sedimentary history of fine sediment is much more complex. Coastal hills and valleys, composed predominantly of soft Cainozoic mudstones, are probably the major source, although the main axial ranges of both islands, particularly where slopes are more gentle and deeply weathered, also contribute. There is some contribution to the southern end of the Hikurangi Trough by the large eastern rivers of South Island, sediment moving northwards around Banks Peninsula under the influence of storm-intensified, coastal currents (Herzer, 1977). Some of this sediment may move down the Pegasus and Kaikoura Canyons in diffuse suspension but these ducts are probably most active during periods of low sea level. There is also some contribution to the same area from the west coast of New Zealand. Fine sediment from the large rivers flowing into the shallow western Cook Strait is resuspended on the storm-dominated shelf (Lewis, 1979) having little chance to settle before being swept by tidal currents through the narrows to the Hikurangi Margin. Off Hawke's Bay, fine sediment from unstable hills and muddy rivers tends to move south (rather than north like the beach sediments), under the influence of weak coastal currents (Lewis, 1973a, b). Hence various sediment migration paths converge independently on the sediment-swamped borderland south of Hawke's Bay.

The general increasing fineness offshore implies that the surface sediments have been deposited by settling through the water column as sediment-laden water dispersed away from the coast.

However, the reduced deposition on slope highs might also suggest that 'hemipelagic' mud settling through the water column, together with resuspended sediment, forms a diffuse turbid cloud near the sea floor (Moore, 1969). The cloud drifts down slopes, across basins and around banks that project above it.

Sediment influxes: turbidites and ashes

Gravity cores and piston cores from the Hikurangi Margin show that, in the basins at least, the regular pattern of fine hemipelagic sedimentation is spasmodically interrupted by 'catastrophic' influxes of coarse sediment. Below the seabed there are layers of sand, white ash and dark mud, none of which are represented in the top few centimetres of pale grey or brownish (oxidized) mud (Lewis, 1973c).

A recognizable cycle is repeated, with various members periodically left out, in most of the basins (Fig. 9). The basal member is a 1–50 mm thick, laminated, very fine sand or very coarse silt. Except for a few coarser grains at the base there is no obvious sign of grading but there are examples of current ripple laminations and convolute laminations sandwiched between parallel laminations. This layer, although thin, is probably equivalent to the B, C and D divisions of Bouma (1962) recognized on the

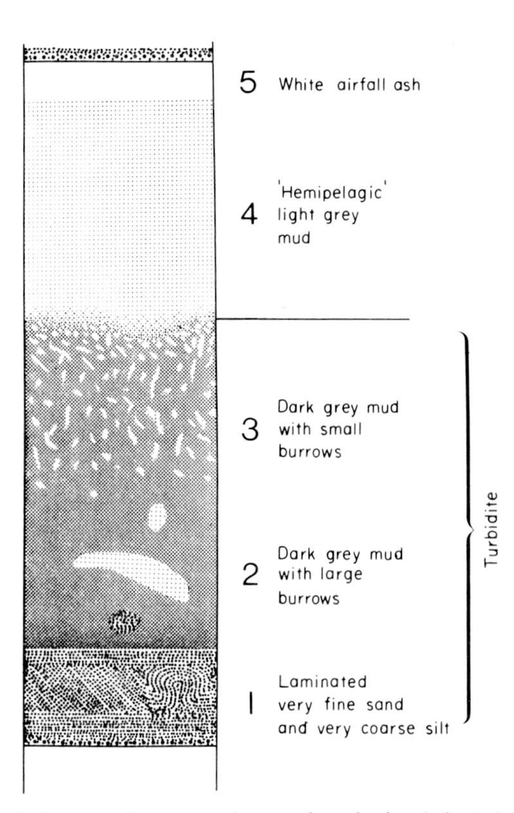

Fig. 9. Sedimentary cycle in cores from accretionary slope basins (adapted from Lewis, 1973c; for interpretation, see text).

adjacent land (van der Lingen, 1969; van der Lingen & Pettinga, 1980). The sand is considered to have been deposited from a turbidity current. Above the laminated coarse layer, there is always a dark grey mud with light grey, mud-filled burrows. In the lower part of thick layers, the dark grey mud is massive or with only occasional large burrows 5–15 mm in diameter. In the upper part of thick layers and the whole of thin layers the dark mud is riddled with small burrows 1–3 mm in diameter. The dark mud is probably equivalent to the pelitic interval (E) (van der Lingen & Pettinga, 1980) and represents the muddy 'tail' of a turbidity current. The smaller burrowing organisms can only penetrate the top 60–80 mm of a newly deposited turbidite but the rarer larger organisms can penetrate 150 mm or more.

The dark mud grades up into a light grey mud, which is the 'hemipelagic' layer sampled at the seabed, and is equivalent to the F interval in uplifted basin sediment on the adjacent land (van der Lingen & Pettinga, 1980).

Overlying, or within the hemipelagic layer, there is commonly a conspicuous white layer of airfall ash. Where the ash immediately underlies sand, the ash shower may have triggered the next turbidity current. Locally too, the ash appears to have been artificially thickened by some sort of sediment gravity flow from the adjacent ridges. These ash layers have been correlated with dated ash layers in the Taupo Area using the trace element composition of their titanomagnetite content (Lewis & Kohn, 1973). One ash, the 3400 year old Waimihia Ash, is recognized in all cores from the borderland south of Hawke's Bay. In most cores, from ridges as well as troughs, it ranges from

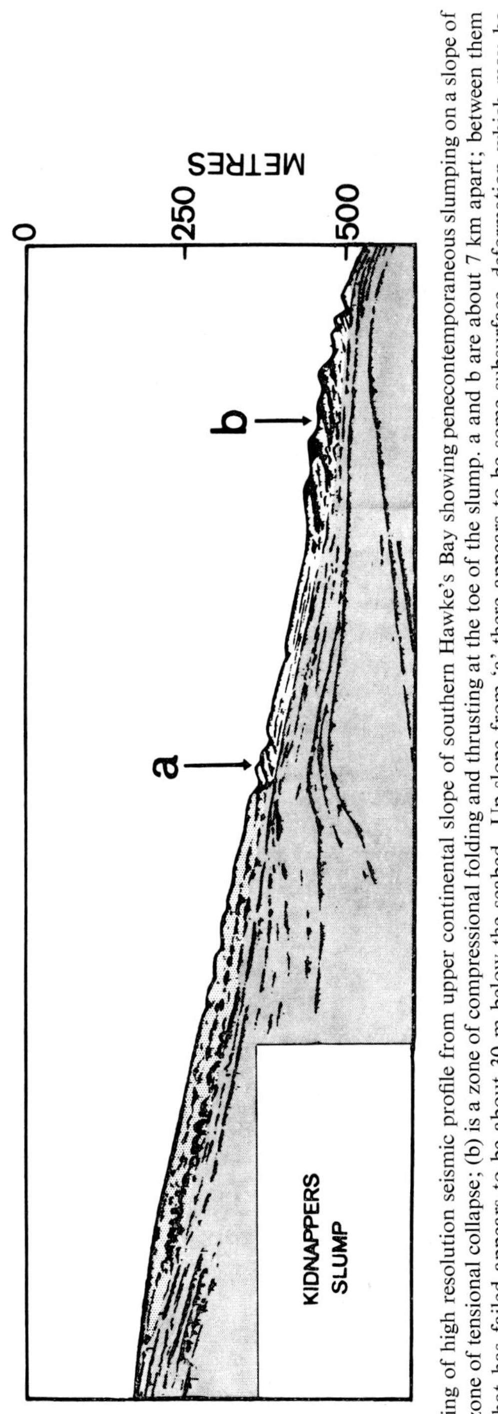

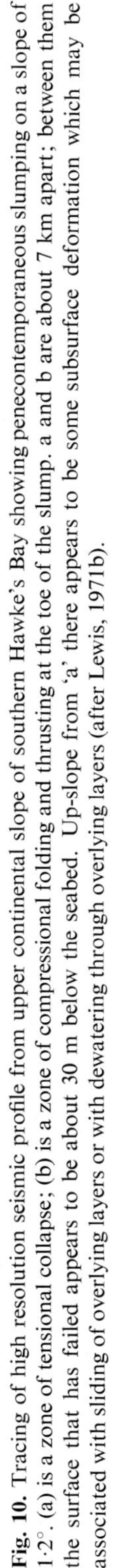

Fig. 10. Tracing of high resolution seismic profile from upper continental slope of southern Hawke's Bay showing penecontemporaneous slumping on a slope of 1·2°. (a) is a zone of tensional collapse; (b) is a zone of compressional folding and thrusting at the toe of the slump. a and b are about 7 km apart; between them the surface that has failed appears to be about 30 m below the seabed. Up-slope from 'a' there appears to be some subsurface deformation which may be associated with sliding of overlying layers or with dewatering through overlying layers (after Lewis, 1971b).

20 to 50 mm thick. However, if the ash is locally less than a critical thickness, of perhaps 20 mm, then it did not destroy the local infauna. Instead the fauna destroyed the ash as a discrete layer leaving white ash present only in burrows. This destruction of thin ashes contrasts with the preservation of very thin sands which are protected by their overlying 'tail' of mud.

The Waimihia Ash is an invaluable marker horizon. A layer of pumice halfway between it and the seabed is probably the 1800 year old Taupo eruption. Another ash that ranges from three to seven times deeper than the Waimihia Ash has a ferromagnesian assemblage typical of the 20 500 year old Oruanui Ash. Overall rates of deposition, based on the Waimihia Ash correlations, range from a high of 0·36 m/1000 years in an upper slope basin where 9 turbidites overlie the ash, to 0·05–0·15 m/1000 years on flat-floored, mid and lower slope basins where only one or two turbidite layers overlie the same ash. Rates are less than 0·01 m/1000 years on most of the ridges.

All along the Hikurangi Margin canyons and sea valleys lead from the shelf to slope basins and the Hikurangi Trough (Fig. 3). Locally levéed channels cross the basins and the trough. All presumably act as conduits for various types of dense, sediment-laden water. At many places, this moving water appears to have inhibited deposition in the channels while deposition occurred on either side of these channels. However, some profiles show clear evidence of erosion and in others there is evidence of channel infilling.

Penecontemporaneous deformation

There is abundant evidence in cores and high resolution profiles that recently deposited sediment on the Hikurangi Margin can be remobilized and redeposited.

Pebbly muds in Hawke's Bay shelf sediments appear to have been deposited from some sort of dense flow following the breaching of coastal shingle spits (Pantin, 1966, 1967).

Creeping and sliding of pebbly muds and soft surface sediment are evident in cores from the relatively steep slopes and narrow basins just north of Cook Strait (Pantin, 1972). Coarse silt and fine sand layers are contorted between layers of apparently undeformed mud.

Off southern Hawke's Bay, where deposition has been particularly rapid, profiles show that surface sediments can slump on slopes of only just over 1° (Lewis, 1971b). The slumps can be more than 10 km in length, up and down slope, but are generally only a few tens of metres thick. There is dislocation at the head, and thrusting and contortion at the toe of the slumps but no obvious signs of disturbance of the rafted sheet between (Fig. 10). Such structures are physically impossible on land and only occur in the submarine environment when compaction and dewatering of rapidly deposited, clay-rich sediments produces an impermeable layer overlying a coarse silt or fine sand-rich layer from which water can escape only slowly (Yagashita, 1977). The result is an anomalously high pore pressure in the more slowly compacting coarser silt or sand layer. The competent clay surface sheet then virtually floats on the underlying fluid coarser layer and any disturbance, such as earthquake stress, is sufficient to trigger motion downslope, no matter how gentle the slope. Except at the head and toe, sedimentary deformation only occurs within the coarse silt–fine sand layer that has failed. Slope failure that is the result of dewatering is always penecontemporaneous.

CONCLUSIONS

(1) An oblique-slip plate boundary traverses the full length of the Hikurangi Trough.

(2) The Hikurangi Trough is two, more or less distinct features; a northern structural trench involving oblique subduction beneath North Island and a southern fault-angle depression involving predominantly strike-slip movement along an intra-continental, transform zone of faulting.

(3) The northern oblique-subduction margin includes a wide, emergent, imbricate thrusted, accretionary prism of growing, seaward-faulted monoclinal ridges and landward-tilting basins: the structural effects are mainly those of thrusting although strike-slip motion may be more evident in progressively steeper faults adjacent to the frontal ridge.

(4) On the continental shelf, which is built up and out across one or more basins, unconformities and migrations of a rapidly deposited near-shore prism of sediment provide a clear record of Quaternary oscillations of sea level.

(5) On the upper slope, thin sheets of sediment, several kilometres across but only a few tens of metres thick, have slid down slopes as low as one degree, probably after excess water was trapped in rapidly buried sandy layers.

(6) Slope basins, 5–50 km wide by 10–60 km long are filled by recurring sequences of turbidites, hemipelagic mud and ash layers, the layers being near horizontal at the seafloor and tilted progressively more landward at depths below the seafloor. Basin fill ranges from 200–2000 m thick.

(7) The southern transform margin is a steep, fault-controlled slope with some mass movement of sediment and with turbidity currents flowing directly to the Hikurangi Trough rather than being trapped in slope basins. There are as yet no documented sedimentary effects of the strike-slip component of movement.

(8) The Hikurangi Margin is an excellent laboratory for the study of sedimentation at an oblique-slip plate boundary, firstly, because rapid sedimentation combined with rapid coastal movement have facilitated the preservation and seismic interpretation of sedimentary structure and secondly because effects at oblique subduction and transform plate boundaries can be studied both in the formational stage below the sea and on the adjacent land when uplifted and dissected.

ACKNOWLEDGMENTS

I should like to thank Dr P. F. Ballance for his encouragement and assistance in the formative stages of this paper and to Dr H. G. Reading and Dr R. M. Carter for their constructive criticism of the manuscript. I am also grateful to Mobil Oil Corporation and to Lamont Doherty Geological Observatory for permission to use their seismic profiles.

REFERENCES

Ballance, P.F. (1976) Evolution of the Upper Cenozoic Magmatic Arc and plate boundary in northern New Zealand. *Earth Planet. Sci. Lett.* **28,** 356–370.

BALLANCE, P.F. (1980) Models of sediment distribution in non-marine and shallow marine environments in oblique-slip fault zones. In: *Sedimentation in oblique-slip mobile zones* (Ed. by P. F. Ballance & H. G. Reading). *Spec. Publ. int. Ass. Sediment.*, **4**, 229–236.

BOUMA, A.H. (1962) *Sedimentology of Some Flysch Deposits: A Graphic Approach to Facies Interpretation.* Elsevier, Amsterdam, 168 pp.

BRODIE, J.W. (1959) Structural significance of sea floor features around New Zealand. *Geol. Rdsh.* **47,** 662–667.

BRODIE, J.W. & HATHERTON, T. (1958) The morphology of Kermadec and Hikurangi Trenches. *Deep Sea Res.* **5,** 18–28.

CARTER, R.M. & NORRIS, R.J. (1976) Cainozoic history of Southern New Zealand: an accord between geological observations and plate-tectonic predictions. *Earth Planet. Sci. Lett.* **31,** 85–94.

CHRISTOFFEL, D.A. & VAN DER LINGEN, W.J.M. (1972) Macquarie Ridge–New Zealand Alpine Fault transition. In: *Antarctic Oceanology II: The Australian–New Zealand Sector* (Ed. by D. E. Hayes). *Antarct. Res. Ser. Am. geophys. Union,* **19,** 235–242.

COLE, J.W. (in press) Structure petrology and genesis of Cainozoic vulcanism, Taupo Volcanic Zone, New Zealand. *N.Z. J. Geol. Geophys.* **22**

EADE, J.V. & CARTER, L. (1975) Definitions and code of nomenclature for the naming of morphologic features on the New Zealand seafloor. *N.Z. Ocean. Inst. Records* **2,** 129–140.

HAMILTON, R.M. & GALE, A.W. (1968) Seismicity and structure of North Island, New Zealand. *J. geophys. Res.* **73,** 3859–3876.

HATHERTON, T. (1970) Upper Mantle inhomogeneity beneath New Zealand: surface manifestations *J. geophys. Res.* **75,** 269–284.

HATHERTON, T. & SYMS, M. (1975) Junction of Kermadec and Hikurangi negative gravity anomalies (Note) *N.Z. J. Geol. Geophys.* **18,** 753–756.

HAYES, D.E. & TALWANI, M. (1972) Geophysical investigation of the Macquarie Ridge Complex. In: *Antarctic Oceanology II: The Australian–New Zealand Sector* (Ed. by D. E. Hayes) *Antarct. Res. Ser. Am. geophys. Union,* **19,** 211–234.

HERZER, R.H. (1977) *Late Quaternary geology of the Canterbury continental terrace.* Unpublished Ph.D. thesis, Victoria University of Wellington. 268 pp, (Chapter D, pp. 170–195).

HOUTZ, R., EWING, J., EWING, M. & LONARDI, A.G. (1967) Seismic reflection profiles of the New Zealand Plateau. *J. geophys. Res.* **72,** 4713–4729.

KARIG, D.E. & SHARMAN, G.F. (1975) Subduction and accretion in trenches. *Bull. geol. Soc. Am.* **86,** 377–389.

KATZ, H.R. (1974a) Margins of the Southwest Pacific. In: *The Geology of Continental Margins* (Ed. by C. A. Burk & C. L. Drake). Springer-Verlag, New York, 549–565.

KATZ, H.R. (1974b) Ariel Bank off Gisborne – an offshore late Cenozoic structure, and the problem of acoustic basement on the East Coast, North Island, New Zealand. *N.Z. J. Geol. Geophys.* **18,** 93–108.

LEWIS, K.B. (1971a) Growth rates of folds using tilted wave-planed surfaces: coast and continental shelf, Hawke's Bay, New Zealand. In: *Recent Crustal Movements* (Ed. by B. W. Collins & R. Fraser). *Bull. R. Soc. N.Z.* **9,** 225–231.

LEWIS, K.B. (1971b) Slumping on a continental slope inclined at 1°–4°. *Sedimentology* **16,** 97–110.

LEWIS, K.B. (1973a) Erosion and deposition on a tilting continental shelf during Quaternary oscillations of sea level. *N.Z. J. Geol. Geophys.* **16,** 281–301.

LEWIS, K.B. (1973b) Sediments on the continental shelf and slope between Napier and Castlepoint, New Zealand. *N.Z. J. Mar. Freshwater Res.* **7,** 183–208.

LEWIS, K.B. (1973c) Ashes, turbidites and rates of sedimentation on the continental slope off Hawke's Bay. *N.Z. J. Geol. Geophys.* **16,** 439–454.

LEWIS, K.B. (1974a) Upper Tertiary rocks from the continental shelf and slope of Southern Hawke's Bay. *N.Z. J. Mar. Freshwater Res.* **8,** 663–670.

LEWIS, K.B. (1974b) The Continental Terrace. *Earth Sci. Rev.* **10,** 37–71.

LEWIS, K.B. (1976) Turnagain Bathymetry. *N.Z. Ocean. Inst. Chart, Coastal Series* 1:200,000 2nd edn.

LEWIS, K.B. (1979) A storm dominated inner shelf, Western Cook Strait, New Zealand. *Mar. Geol.* **31,** 31–43.

LEWIS, K.B. & KOHN, B.P. (1973) Ashes, Turbidites and rates of sedimentation on the continental slope of Hawke's Bay. *N.Z. J. Geol. Geophys.* **16,** 439–454.

LINGEN, G.J. VAN DER (1969) The turbidite problem. *N.Z. J. Geol. Geophys.* **12,** 7–47.

LINGEN, G.J. VAN DER & PETTINGA, J.R. (1980) Miocene slope-basin sedimentation along the New Zealand sector of the Australian–Pacific convergent plate boundary. In: *Sedimentation in oblique-slip mobile zones* (Ed. by P. F. Ballance & H. G. Reading). *Spec. Publ. int. Ass. Sediment.* **4,** 191–215.

McDOUGALL, J.C. (1975) Cook sediments. *N.Z. Ocean. Inst. Chart, Oceanic Series* 1 : 1,000,000.

MINSTER, J.B., JORDAN, T.H., MOLNAR, P. & HAINES, E. (1974) Numerical modelling of instantaneous plate tectonics. *Geophys. J. R. astr. Soc.* **36,** 541–576.

MOLNAR, P., ATWATER, T., MAMMERICK, J. & SMITH, S.M. (1975) Magnetic anomalies, bathymetry and the tectonic evolution of the South Pacific since the late Cretaceous. *Geophys. J. R. astr. Soc.* **40,** 383–420.

MOORE, D.G. (1969) Reflection profiling studies of the California Continental Borderland: structure and Quaternary turbidite basins. *Spec. Pap. geol. Soc. Am.* **107,** 142 pp.

MOORE, G.F. & KARIG, D.E. (1976) Development of sedimentary basins on the lower trench slope. *Geology,* **4,** 693–697.

PANTIN, H.M. (1963) Submarine Morphology east of the North Island, New Zealand. *Bull. N.Z. Dep. sci. ind. Res.* **149,** 44 pp.

PANTIN, H.M. (1966) Sedimentation in Hawke's Bay. *Bull. N.Z. D.S.I.R.* **171,** 70 pp.

PANTIN, H.M. (1967) The origin of water-borne diamictons and their relation to turbidites. *N.Z. J. Mar. Freshwater Res.* **2,** 118–138.

PANTIN, H.M. (1972) Internal structure in marine shelf, slope and abyssal sediments east of New Zealand. *Bull. N.Z. D.S.I.R.* **208,** 56 pp.

SPÖRLI, K.B. (1980) New Zealand and oblique-slip margins: tectonic development up to and during the Cainozoic. In: *Sedimentation in oblique-slip mobile zones* (Ed. by P. F. Ballance & H. G. Reading). *Spec. Publ. int. Ass. Sediment.,* **4,** 147–170.

WALCOTT, R.I. (1978a) Present tectonics and Late Cenozoic evolution of New Zealand. *Geophys. J. R. astr. Soc.* **52,** 137–164.

WALCOTT, R.I. (178b) Geodetic strains and large earthquakes in the axial tectonic belt of North Island, New Zealand. *J. geophys. Res.* **83,** 4419–4429.

YAGASHITA, K. (1977) Possible mechanism of submarine sliding and its associated minor slump fold. *Earth Sci. J. Assoc. geol. Collaboration in Japan,* **31,** 179–192.

Spec. Publ. int. Ass. Sediment. (1980) **4**, 191–215

The Makara Basin: a Miocene slope-basin along the New Zealand sector of the Australian–Pacific obliquely convergent plate boundary

GERRIT J. VAN DER LINGEN *and* JARG R. PETTINGA*

Sedimentology Laboratory, New Zealand Geological Survey, University of Canterbury, Christchurch and Geology Department, University of Auckland, Auckland, New Zealand

ABSTRACT

The Miocene Makara Basin in the Hawke's Bay area is one of a series of small Neogene flysch basins within the North Island oblique subduction system. It forms part of the East Coast Deformed Belt, which, together with the Axial Ranges to the west, is situated between the Taupo Volcanic Zone (volcanic arc) and the Hikurangi Trough (subduction trench).

The basin is bounded by narrow zones of highly deformed Upper Cretaceous to Lower Tertiary strata, the Waimarama–Mangakuri Coastal High to the east, and the Otane Anticlinal Complex to the west. These zones were active structural highs during flysch sedimentation. A third structural high, the Elsthorpe Anticline, developed late in the history of the basin, and divided the basin sediments into eastern and western asymmetric synclines.

Basin sediments consist of flysch strata, pebbly mudstones, tuff beds, and hemipelagic mudstones, up to 2200 m thick. The flysch strata, subdivided into four subfacies, are frequency graded, display modified Bouma divisions of sedimentary structures, and have a constant maximum grain size. One subfacies, the Ponui Sandstone, consists of thick units of amalgamated, non-graded, mainly structureless sandstone beds. All these strata were probably deposited by a spectrum of sediment gravity flows, ranging from turbidity currents to fine-grained debris flows, which originated higher up the slope. The local occurrence of pebbly mudstones, deposited by gravelly debris flows and by slumps, reflects derivation from the structural highs. Rhyolitic tuff beds, intercalated in the flysch sequence, and forming the top of the basin fill, were deposited both by sediment gravity flows and ash-falls.

The Makara Basin (and similarly the other Neogene flysch basins in the East Coast Deformed Belt) is interpreted as a small basin (about 30 by 20 km) which originated on the inner slope of the Hikurangi subduction trench. The flysch sediments were ponded in between thrust ridges (the structural highs), formed by imbricate thrust faulting and deformation of the leading edge of the upper plate. The volcanogenic sediments were probably derived from the Coromandel Arc, which was active to the west in Miocene time. The hemipelagic mudstones represent normal slope sedimentation.

Although subduction is taking place obliquely to the trend of the East Coast Deformed Belt, deformation is by compressive folding and thrust faulting only. Strike-slip movements probably were located further to the west.

* Present address: Department of Geology, University of Canterbury, Christchurch, N.Z.

0141-3600/80/0904-0191$02.00

The setting, structure, and sediment types of the Makara Basin conform extremely well with models of inner trench-slope basins derived from subduction systems elsewhere.

INTRODUCTION

The Upper Miocene Makara Basin is located within the North Island Subduction System which lies between the Taupo Volcanic Zone and the Hikurangi Trough (Fig.

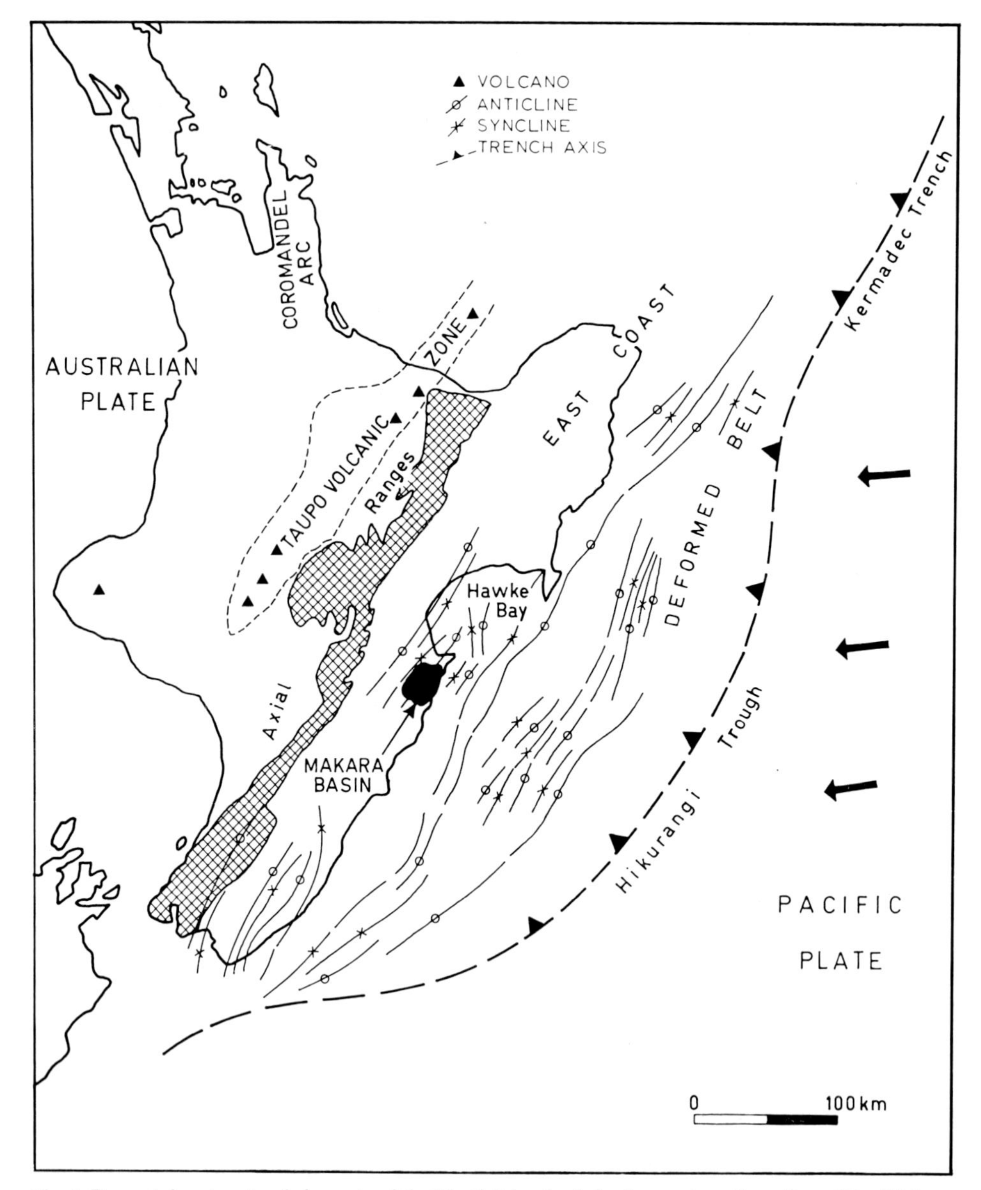

Fig. 1. Present-day structural elements of the North Island subduction system. Location of the Makara Basin is indicated.

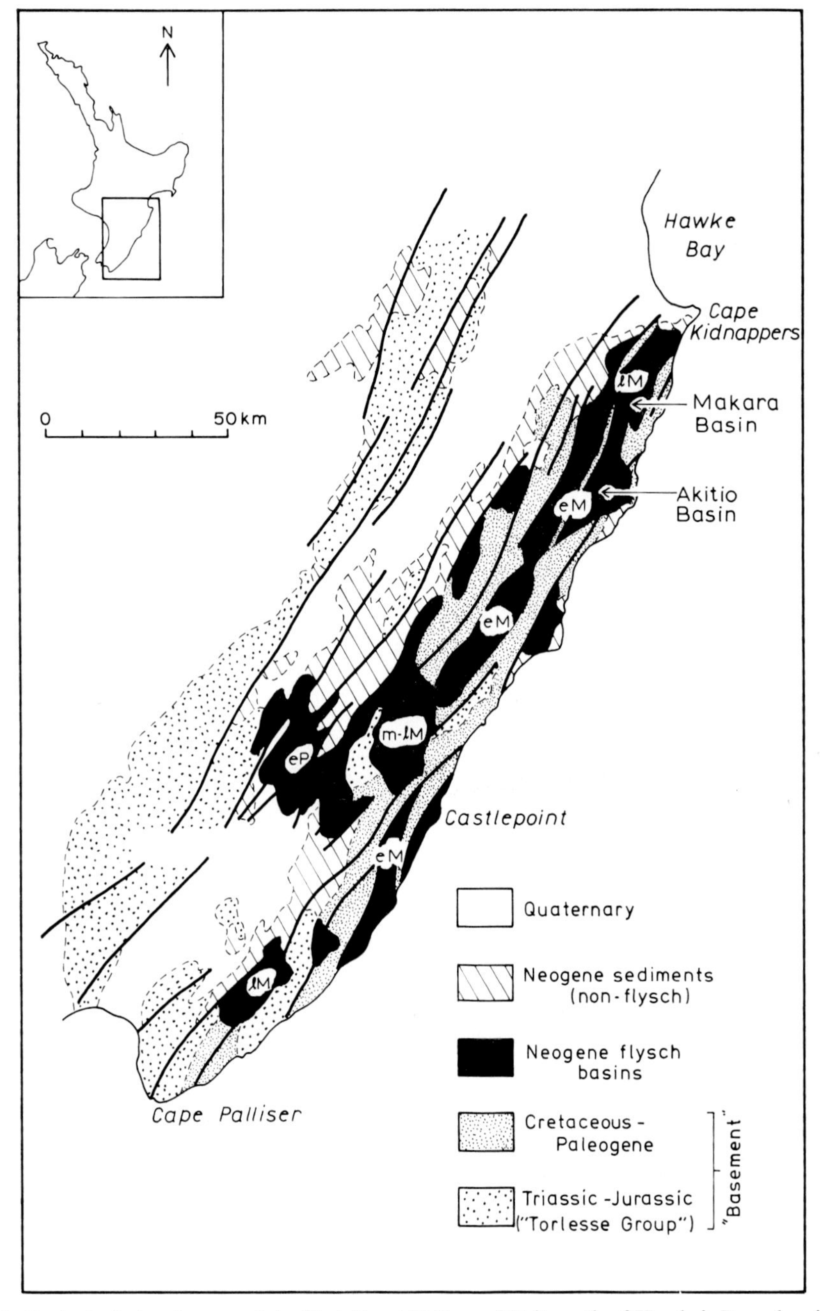

Fig. 2. Geological sketch map of the East Coast Deformed Belt south of Hawke's Bay, showing the present-day outcrop areas of Neogene flysch basins; (eM—early Miocene; mM—mid Miocene; lM—late Miocene; eP—early Pliocene).

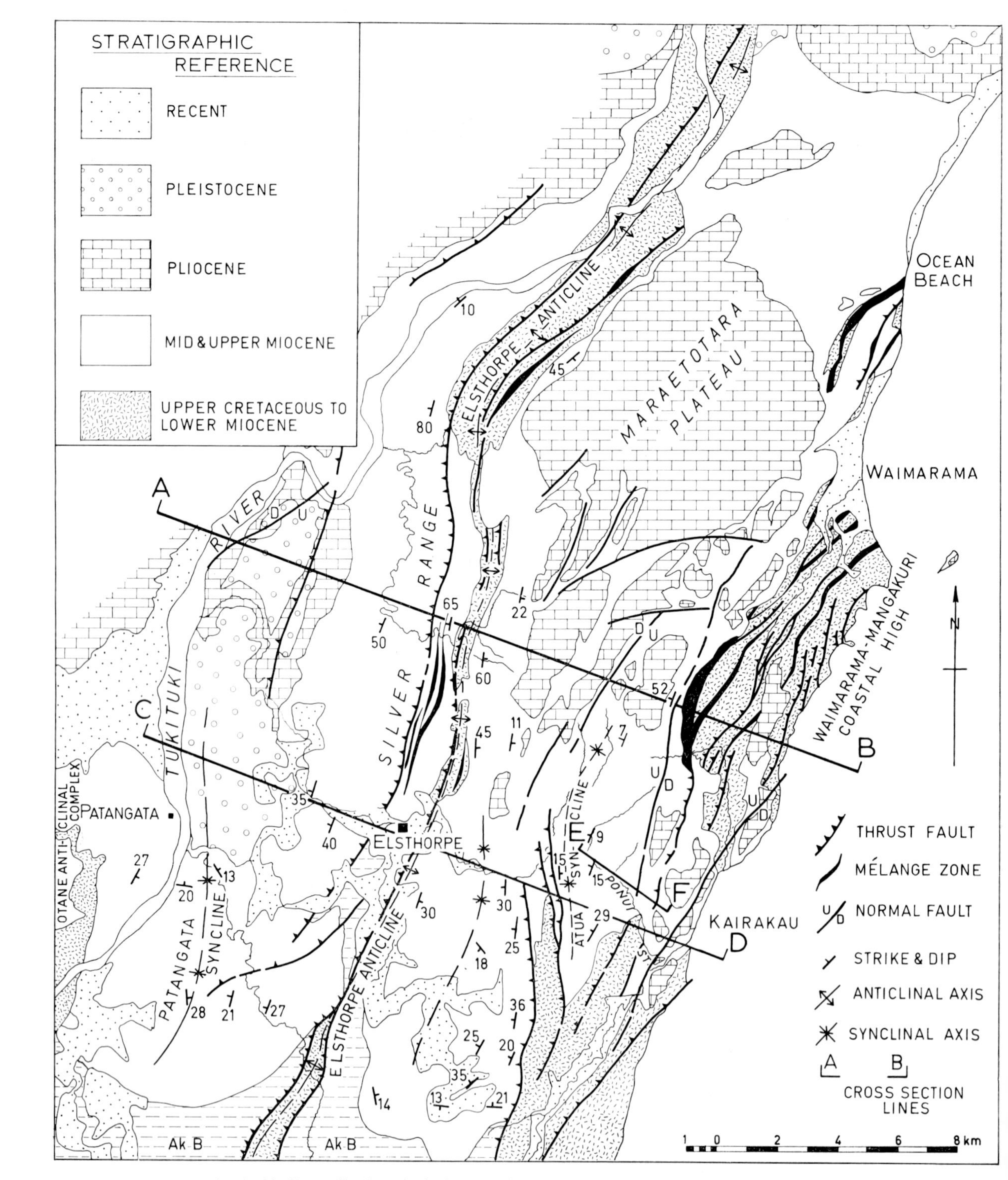

Fig. 3. (a) Generalized geological map of the Makara Basin area; Ak B-Akitio Basin.

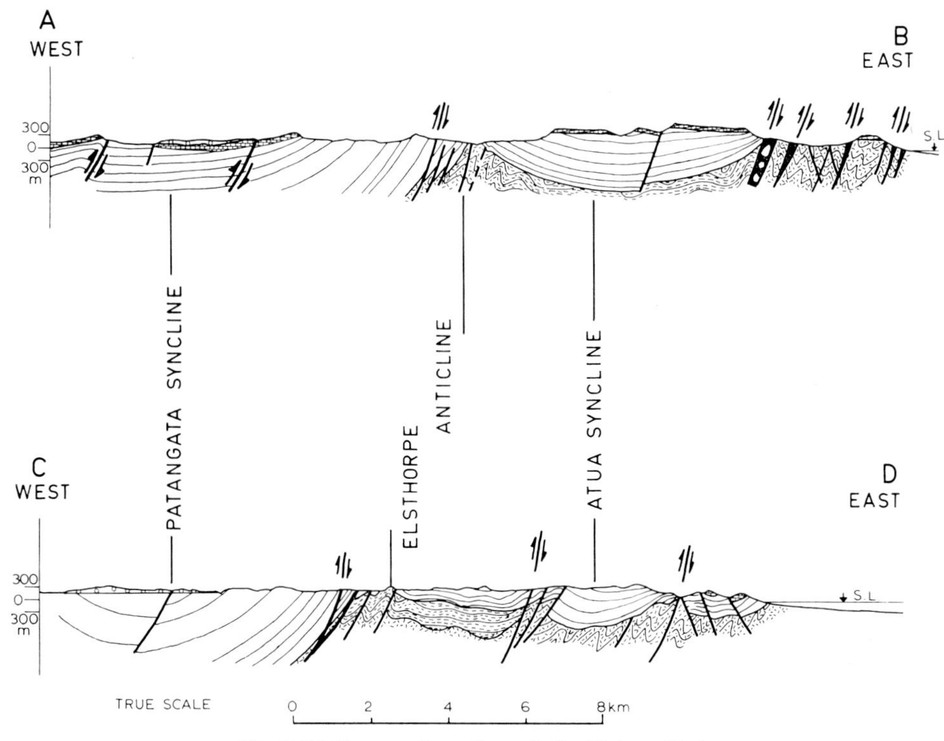

Fig. 3 (b) Cross-sections through the Makara Basin.

1) (Spörli, 1980; Lewis, 1980). The Subduction System is composed of two major elements, the western Axial Ranges and the East Coast Deformed Belt. The entire area is part of the boundary zone between the Australian and Pacific plates (Fig. 1) (Walcott, 1978a, b; Spörli, 1980).

The relative movement between the plates is obliquely compressive (Spörli, 1980, Fig. 3). It is taken up by underthrusting (subduction) of the Pacific plate and folding and thrust faulting of the overriding Australian plate. The strike-slip component of movement is concentrated in the Axial Ranges, while the compressive component is largely taken up in the East Coast Deformed Belt (Walcott, 1978a, b; Spörli, 1980; Lewis, 1980).

The Makara Basin is situated within the East Coast Deformed Belt, and is interpreted as a fossil slope basin of the inner trench-slope, in accordance with the model proposed by Moore & Karig (1976, Fig. 5). Lewis (1980) extends this interpretation to the offshore basins on the present-day inner trench-slope.

The area including the Makara Basin was first described by Kingma (1971). The region immediately to the south, including the Akitio Basin (Fig. 2), was first mapped by Lillie (1953). Kingma (1958a, 1960) did not accept that the flysch sediments of the Makara Basin were deposited by turbidity currents, and proposed an alternative model. Much discussion and controversy followed (see van der Lingen, 1969, 1970; Kuenen 1960, 1970).

The present paper presents a re-interpretation of the basin in the light of new work by the authors and of modern ideas about plate tectonics and sediment-gravity flow processes.

Specific localities are indicated on the maps by the following method: two figures are given, the first is a compass bearing (e.g.: 014) the second is the distance in kilometres from Elsthorpe (e.g.: 9·8, hence 014/9·8).

GEOLOGICAL SETTING OF THE MAKARA BASIN

The Makara Basin is one of a series of small Neogene flysch basins within the East Coast Deformed Belt south of Hawke's Bay (Fig. 2). These basins unconformably overlie, and are bounded by, highly deformed Mesozoic-Palaeogene 'basement' rocks. Neogene flysch sediments also occur in the East Coast Deformed Belt north of Hawke's Bay, but are little known.

In the region of the Makara Basin, the basement rocks crop out in three narrow

Fig. 4. West dipping thrust fault (top of hammer) in the Waimarama–Mangakuri Coastal High, near Ocean Beach, showing mélange rock (exotic clasts in sheared bentonitic mudstone) on top of sheared Tongaporutuan hemipelagic mudstone.

north-northeast trending structural highs (Fig. 3), the Waimarama–Mangakuri Coastal High (Pettinga, 1980) bounding the basin to the east, the Otane Anticlinal Complex (not studied in detail) to the west, and the Elsthorpe Anticline dividing the basin in half. The southern part of the Waimarama–Mangakuri Coastal High has a north-trending 'offshoot' extending into the basin sequence (Fig. 3). The three highs continue southwards where they have a similar structural relationship to the Akitio Basin (Fig. 2; Lillie, 1953).

The highs are composed of Upper Cretaceous flysch, massive mudstones and fine-grained siliceous sediments, isoclinally folded and tectonically imbricated with

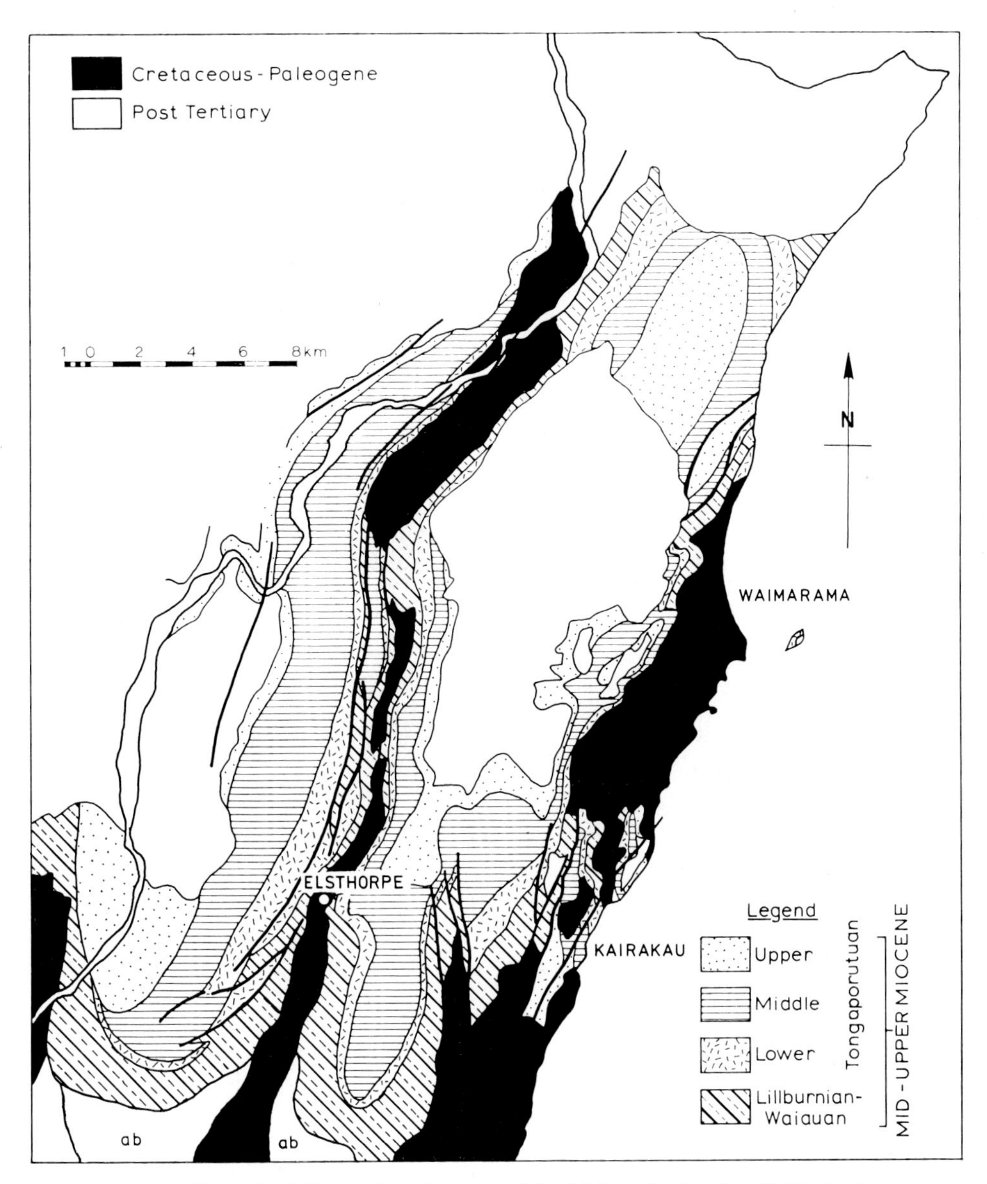

Fig. 5. Generalized biostratigraphic map of the Makara Basin. ab—Akitio Basin.

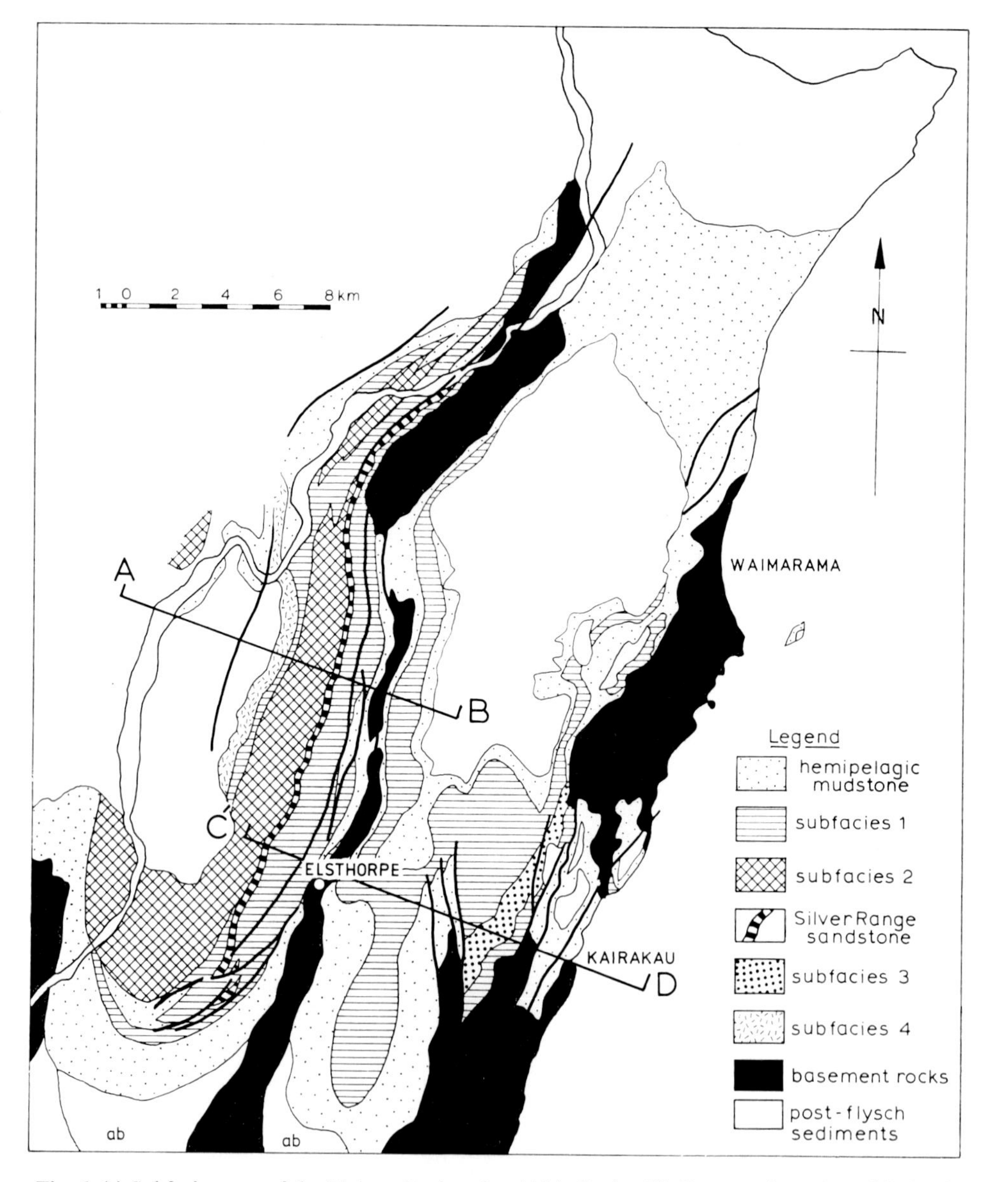

Fig. 6. (a) Subfacies map of the Makara Basin; ab—Akitio Basin. (b) Cross-sections. 1—subfacies 1; 2—subfacies 2; 3—Silverrange Sandstone (belonging to subfacies 2); 4—subfacies 3 (Ponui Sandstone); 5—subfacies 4; 6—hemipelagic mudstone; 7—tuff bed; 8—pebbly mudstone bed; 9—deformed basement rocks; 10—Waiauan limestone; 11—Te Aute Limestone (Pliocene); 12—Quaternary lake sediments.

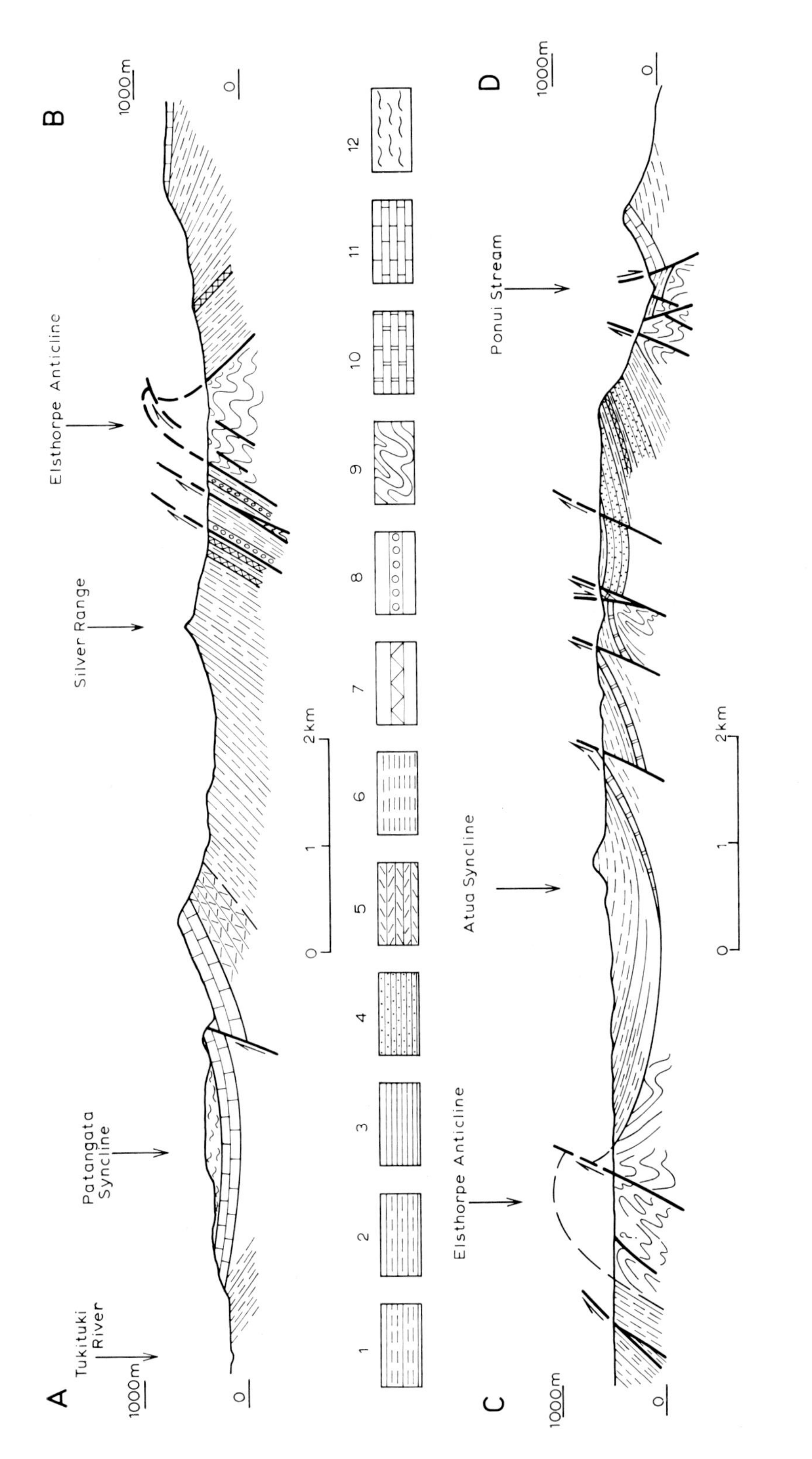
A
Tukituki River
1000m
0
Patangata Syncline
Silver Range
Elsthorpe Anticline
B
1000m
0
1
2
3
4
5
6
7
8
9
10
11
12
0
1
2km
C
1000m
0
Elsthorpe Anticline
Atua Syncline
Ponui Stream
D
1000m
0
0
1
2km

tightly folded Eocene and Oligocene massive mudstones and flysch. Westerly dipping thrust faults have sheared out the cores and limbs of the isoclinal folds and caused tectonic thickening (Pettinga, 1976, 1980). Commonly the thrust faults have associated mélange zones, often incorporating Lower Tertiary bentonitic clays (Fig. 4). The latest Oligocene to early Miocene sequence contains sediment gravity flow deposits and tuffs interpreted as recording the earliest phase of plate convergence and volcanism. Subsequently they were incorporated in the structural highs.

The Otane and Waimarama–Mangakuri highs have a continuous history of deformation from the late Oligocene to the present day. In the Neogene they formed submarine ridges on the inner trench-slope (the so-called 'thrust slices' in the Moore & Karig (1976) model), between which the flysch sediments of the Makara Basin were ponded. The Elsthorpe Anticline is of post-Miocene age and does not appear to have affected sedimentation of the flysch.

Kingma (1958b) interpreted the narrow structural highs as 'piercement structures', controlled by transcurrent fault movements. He also thought that the flysch basins were controlled by the same transcurrent movements. However, no field evidence has been found to substantiate this hypothesis (Pettinga, 1977, 1980).

The basin-fill flysch sequence of the Akitio Basin is slightly older (Early Miocene: Altonian–Clifdenian) than that of the Makara Basin. The Makara Basin flysch overlaps the northern portion of the Akitio Basin sediments, but the two sequences are separated stratigraphically and in part unconformably by a conglomeratic limestone and massive mudstone successions.

SEDIMENTS

Most of the sediments filling the Makara Basin are flysch: alternating sandstone and mudstone. Intercalated in the flysch are subordinate acidic tuff beds and pebbly mudstones. Surrounding the flysch are massive mudstones.

Flysch

The sediments can be ascribed to Ricci Lucchi's (1975) flysch facies A, B, C, D, F and G. To describe structures in the flysch strata a modified Bouma sequence (van der Lingen 1969, Fig. 2; Bouma 1972) in which Bouma's T divisions are replaced by M divisions will be used. The Ma division is poorly laminated as distinct from Bouma's massive graded Ta division. The remaining divisions are closely related in both schemes.

Based on sandstone–mudstone ratios the Makara flysch has been subdivided into two subfacies above and below a ratio of 0·8; this rather arbitrary choice is made as it aids in depicting the broad organization of strata in the basin (Figs 6, 7). Two further subfacies have been recognized: massive sandstone, and tuffaceous sediments (Fig. 6).

Subfacies 1

Sandstone–mudstone ratios less than 0·8, with sandstone thickness up to 85 cm (Fig. 8a).

The most common Bouma division sequence is Mbc. Thicker beds can have more

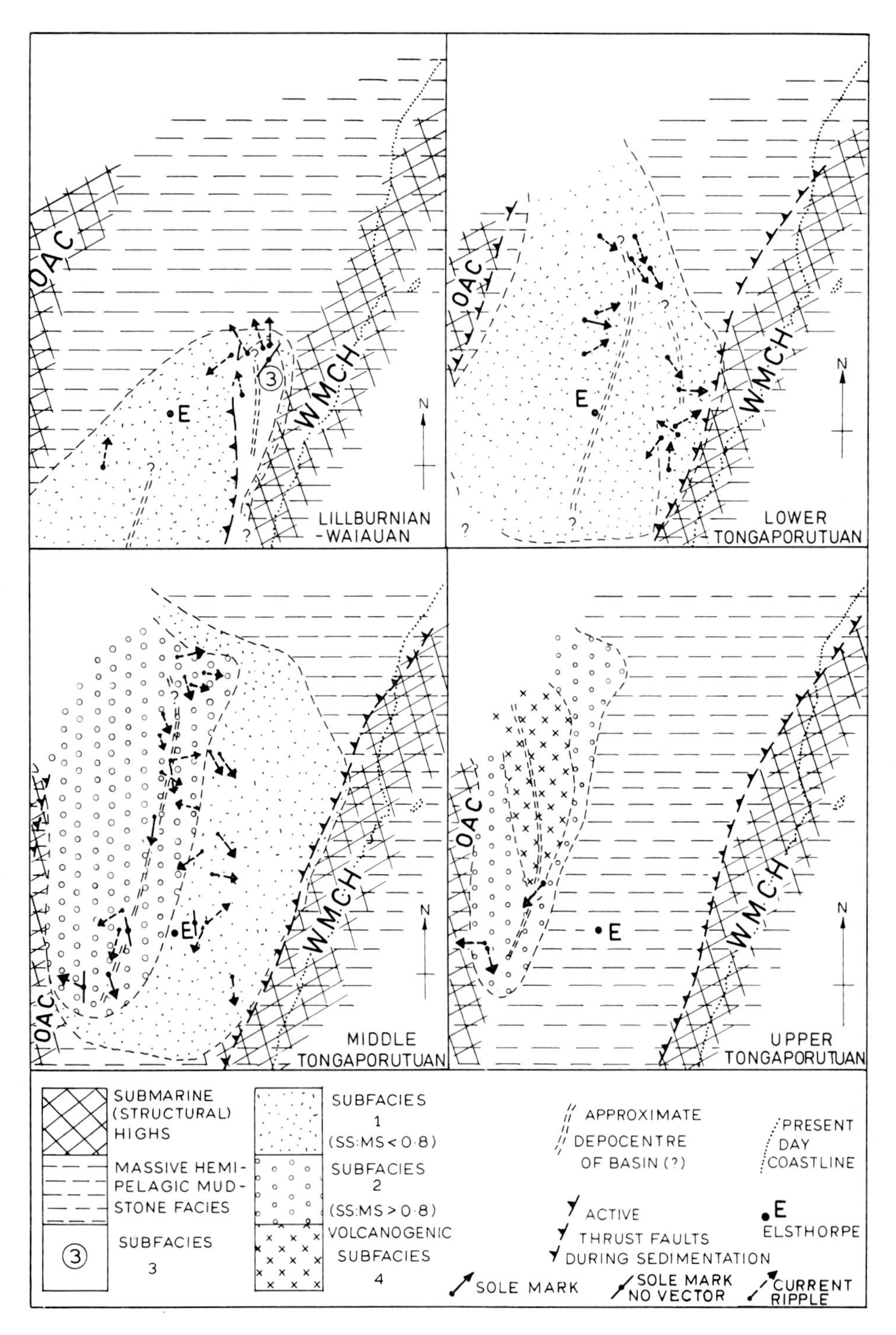

Fig. 7. Palaeogeographic maps depicting the development of the Makara Basin during Middle and Late Miocene. WMCH: Waimarama–Mangakuri Coastal High; OAC: Otane Anticlinal Complex; E: Elsthorpe. Palaeocurrent directions are indicated.

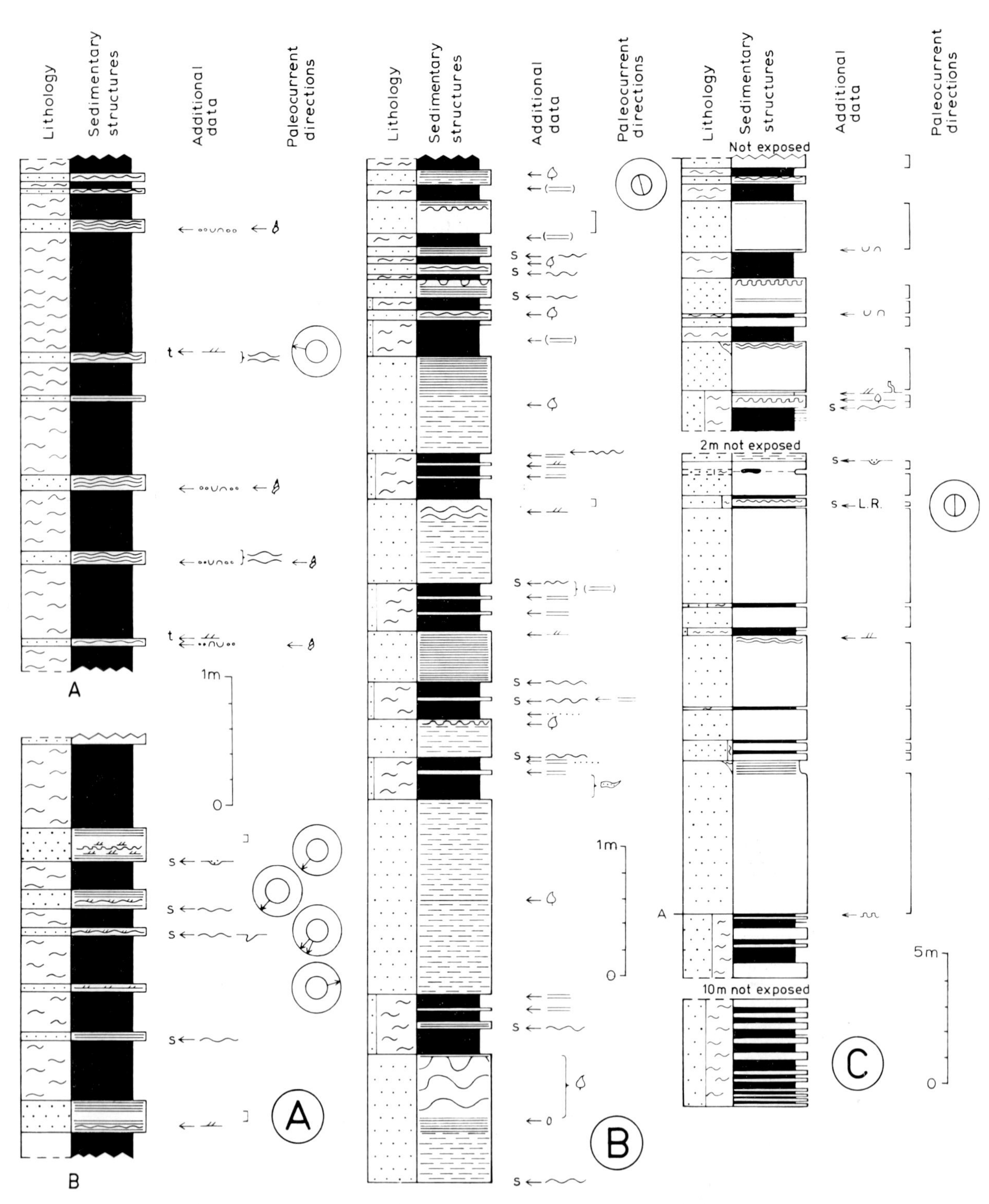

Fig. 8. (a) Representative sections of subfacies 1. A, loc. 014/6·8. B, loc. 099/6·5.

(b) Representative section of subfacies 2. Loc. 288/3·8.

(c) Representative section of the Silverrange Sandstone Unit (part of subfacies 2). The Silverrange Sandstone starts at A. Upwards this subfacies grades into 'normal' subfacies 2 strata. Below A, no sedimentary structures were recorded. Loc. 003/6·6.

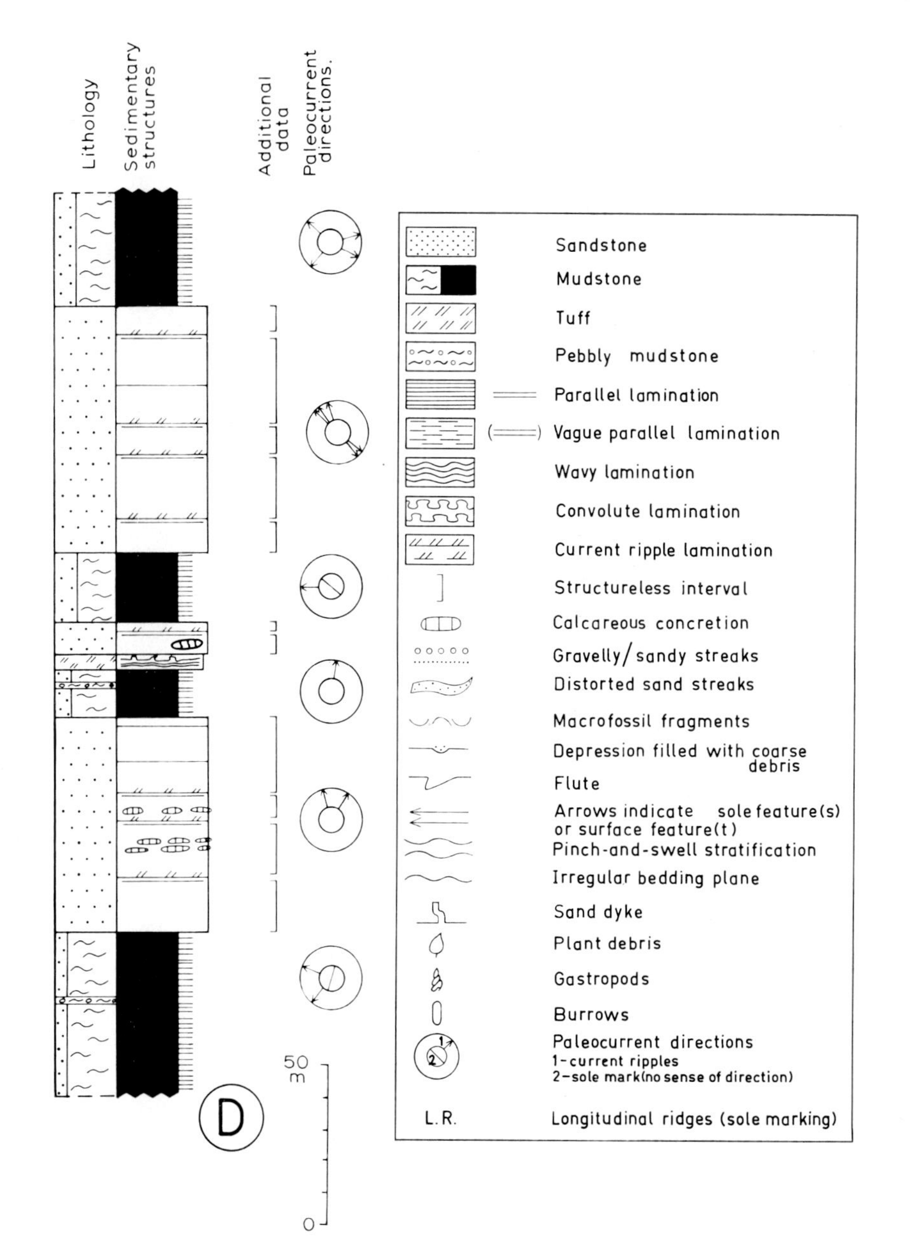

(d) Composite schematic section of subfacies 3 (Ponui Sandstone), near Kairakau, eastern Makara Basin. Plus Key.

complete Mb-e sequences. Most beds are frequency graded (van der Lingen, 1968a) and grading is often irregular. Occasionally strata have coarser grained particles and shale fragments concentrated in 'pockets' or thin laminae at their base.

Wedging of beds has never been observed. Lower boundaries of sandstone beds

Fig. 9. Flysch strata of subfacies 1. Western Makara Basin.

are sharp and can be straight or irregular. Upper boundaries are sharp or gradational. Sole markings are rarely observed, owing to the softness of the rocks and the absence of overturned strata. Upper surfaces may display ichnofossils. The sandstone beds do not differ in texture or sedimentary structures from sandstone beds of similar thickness in subfacies 2.

In the facies scheme of Ricci Lucchi (1975) most beds of this subfacies belong to facies D1. Systematic sequential arrangement of strata is not seen. At best there is an occasional suggestion of 'bundles' of strata (Fig. 9).

Apart from bioturbation the mudstone interbeds are structureless. They always contain a small percentage of sand, and a substantial percentage of silt. Maximum grain-size, sometimes only represented by a few clastic grains is commonly the same as for the underlying sandstone. There is no break in colour or texture to indicate pelitic and pelagic intervals.

Subfacies 2

Sandstone–mudstone ratios greater than 0·8. The highest ratio recorded is 8·5 (loc. 353/7·5). The sandstone thickness ranges up to 600 cm (Fig. 8b, c).

The thicker beds sometimes display the complete Ma-e sequence (Fig. 8b) (see also van der Lingen 1969, Figs 3, 4, 10–18; 1970, Figs 2, 3). Some beds are parallel laminated throughout, or have apparently structureless (but not coarse-grained) intervals (Fig. 8b, c). Most beds are frequency graded. Beds are continuous in outcrop. Lower boundaries of sandstone beds are sharp, and can be straight or irregular (Fig. 10).

Fig. 10. Flysch strata of subfacies 2. Western Makara Basin. Thickest bed is about 50 cm thick.

This subfacies includes a sequence of exceptionally thick sandstone beds named the Silverrange Sandstone by Kingma (1958a). The basin reached its maximum extent during deposition of this unit, which can be followed along strike for at least 29 km (Fig. 6). In the centre of the basin a substantial thickness of flysch strata is exposed below it but at its northern end it rests directly on massive hemipelagic mudstone (loc. 014/17·5). The Silverrange Sandstone can be considered as an imperfect thinning-upward sequence (Fig. 8c).

In the facies scheme of Ricci Lucchi (1975) most beds of this subfacies belong to either facies C, D1 or D2.

Subfacies 3

Thick sandstone beds, generally without mudstone interbeds. This subfacies is restricted to the eastern Makara Basin (Fig. 5, 6) and is near the base of the basin sequence. Three thick units of amalgamated sandstone beds are intercalated in thin-bedded sequences of subfacies 1 (Fig. 8d).

The 'Ponui Sandstone' is given subfacies status because of the general lack of mudstone interbeds and the distinctive texture and sedimentary structures. Individual beds range in thickness from 4 to 15 m and are structureless apart from parallel and current-ripple lamination in the upper 10–30 cm. Grading is absent, and maximum grain-size is the same as for all other subfacies, but the sandstones have a larger average grain-size, and are slightly better sorted, than 'normal' flysch sands. They contain about 80% sand, which is slightly higher than for sandstones in subfacies 1 and 2.

The lower boundary of each amalgamated unit is sharp and planar. At their northern limit the sandstone beds are separated by thin mud laminae and wedge into thin bedded flysch. In the Ricci Lucchi (1975) scheme, these units could be ascribed to facies A.

Subfacies 4

Volcaniclastic sediments. This subfacies is restricted, stratigraphically, to the uppermost part of the sequence in the western Makara Basin (Figs 6, 7). It is characterized by abundant acidic volcaniclastics, and consists of glass shards, with subordinate quartz, feldspar and biotite. Compositionally, the sediment is similar to the tuff beds intercalated in the underlying flysch sediments.

The subfacies consists of sandstone beds up to 60 cm thick, and thick massive volcanogenic mudstone units. The lowest parts of the sandstone beds show parallel, wavy, and current-ripple lamination, but not in a systematic arrangement. Large burrows are abundant.

The current structures, and the fact that the pyroclastics have been mixed with glauconite, forams, radiolaria, and probably non-volcanic quartz, indicate that the tuffs were reworked.

Composition, provenance and palaeocurrents

The composition of sandstones and mudstones is constant over the entire basin. Most sandstones are either feldspathic litharenites or lithic feldsarenites (classification by Folk, Andrews & Lewis, 1970), containing subordinate glauconite, mica, heavy and opaque minerals. Recognizable rock fragments are fine-grained quartzose sedimentary, acid and intermediate volcanic, and mudstone.

The mudstones contain fine-grained detrital particles (mainly quartz and feldspar), coccoliths, and clay minerals.

Several possible sources can be considered for the non-biogenic component:

(a) Triassic–Jurassic (Torlesse Group) rocks as exposed in the Axial Ranges (Fig. 2).

(b) Cretaceous and Palaeogene basement rocks as exposed in the structural highs (Figs 2, 3).

(c) Volcanic ash from contemporary volcanism.

A large percentage of the feldspar grains are fresh. Some quartz grains are also fresh and angular with little or no undulose extinction. Feldspar percentages are much higher than in either Torlesse rocks or the local Cretaceous and Lower Tertiary strata. Accordingly a volcanic source for a substantial portion of the sediment is likely. This argument is strengthened by the numerous intercalated tuff beds.

Coarse grains and granules of 'Whangai' argillite (Lillie, 1953) of Upper Cretaceous age and more altered and worn grains of feldspar and quartz, suggest sediment was also derived locally from the submarine highs and possibly also from the Torlesse terrain to the west.

Palaeocurrent directions are known mostly from current ripples, either internal or on top of bedding surfaces.

During the Mid-Miocene (Waiauan Stage) current directions are commonly from the SSE and SSW. During the late Miocene (Tongaporutuan Stage) the majority are from the north and west (Fig. 7).

Palaeontology

Foraminifera in the flysch have been studied by Dr N. de B. Hornibrook, Mr. G. H· Scott and Dr R. H. Hoskins of the New Zealand Geological Survey and Dr G. W. Gibson of the University of Auckland. Three Miocene stages occur: Lillburnian, Waiauan and Tongaporutuan. The latter can be subdivided into three (Fig. 5).

Two depth ranges are recognized, a relatively shallow water (inner to mid neritic) and a relatively deep water (outer neritic to bathyal). The sandstones contain shallow-water benthic assemblages, and few planktic species. Size sorting is common, and improves from base to top of individual beds. Breakage is common. Ostracods of shelf origin and abundant plant debris are also present. Some deeper-water species and reworked Cretaceous foraminifera can occur. The mudstones generally contain shallow water and/or deep water microfossils. In outcrop no clear distinction can be made between 'pelitic' and 'pelagic' mudstone intervals. If a distinction can be made at all, it is purely on faunal differences. In rare cases, pelagic mudstone rests directly on sandstone, without an intervening pelitic interval.

A limited number of ichnofossils has been identified in the flysch sediments, including *Helminthoida crassa*, *Zoophycos*, and *Chondrites*. None, however, are diagnostic for palaeobathymetry (see Häntzschel, 1975).

The upper parts of the mudstones are commonly strongly bioturbated; this may in part account for the lack of recognized 'pelagic' intervals—other possible explanations for the latter include strong bottom currents, and frequency of sediment gravity flows. Occasionally, entire molluscs occur in the mudstones.

Mechanisms of flysch deposition

Flysch beds of the Makara Basin were deposited by a variety of sediment gravity flows, including turbidity currents, fine-grained debris flows, debris flows, slumps, grain flows and/or modified grain flows.

Many sandstones of subfacies 1 and 2 show the modified sequence of Bouma divisions (van der Lingen, 1969). Systematic variations of division-combinations in space and time do not occur. The more complete Mabcd divisions occur in the thicker beds of subfacies 2. The thinner beds of subfacies 1 and 2 commonly consist of Mc and Mcd sequences. Many beds of these subfacies were probably deposited by turbidity currents, especially those with clear Bouma sequence divisions.

Beds with an Mabcd sequence of which the Mc division is parallel laminated, show a combination of frequency and coarse-tail grading and may have been deposited by a fine-grained debris-flow with some turbulence. The scarcity of scour sole markings (e.g. flutes, see Enos, 1977), and absence of pelitic divisions support this interpretation.

Grain-size distributions show a fairly constant maximum grain-size (van der Lingen, 1971). This may be attributable to several causes. The source area may not have contributed coarser grains, if the source was pyroclastic for instance. Alternatively, size sorting may have occurred higher up the continental margin before sediments were redeposited into the Makara (slope) Basin. Another possibility is the restricted competence of flows to carry sediment up to a maximum grain-size, as is theoretically the case with fine-grained debris flows (Middleton & Hampton, 1976). An exception to the cut-off maximum grain-size is the occurrence of pockets or laminae of coarser grains and granules at the base of sandstone beds in both subfacies 1 and 2. This coarser debris consists of fossil and mudstone fragments and glauconite, never of quartz or feldspar grains. These may have been deposited from the nose of a turbidity current, having been derived from the source area of the flow; or they may represent redistributed debris which slumped or was reworked off the submarine highs and was subsequently carried over the basin floor by bottom currents.

The Ponui Sandstone strata have somewhat smaller amounts of mud than sandstones from the other subfacies. The lack of grading, sedimentary structures, and higher porosity suggest a grain flow or modified grain flow mechanism of emplacement for this subfacies.

Tuff beds in the flysch

These beds are distinct from reworked tuffaceous beds in subfacies 4, and are intercalated in the flysch facies at various levels (Fig. 6; van der Lingen, 1968b). The thickest bed recorded, below the second Ponui Sandstone unit (Figs 6, 8d), is 450 cm. Most are fine-grained and have a light-grey or white colour. Grading is absent; vague current structures (lamination) and small water-escape structures are more frequent. Compositionally these tuffs resemble those of subfacies 4 and are considered pure ash-fall deposits. A few coarser-grained, graded, darker-grey tuff beds also occur (e.g. loc. 030/5·7). Current structures are abundant, and these beds represent tuffaceous sediment reworked by sediment gravity flows. Their darker colour is due to mixing with non-volcanic components.

Pebbly mudstones

These occur either as beds intercalated between flysch strata or as irregular masses associated with slumped flysch strata. The first type has been found from the (Waiauan) base of the flysch sequence in the eastern Makara Basin to the Middle Tongaporutuan strata in the western Makara Basin (Figs 6, 8d). Exotic clasts of up to 30 cm are distributed at random through a muddy matrix.

The second type forms irregular bodies. Exotic clasts include large blocks of up to several metres, and these occur with slumped flysch strata. Exotic clast lithologies in both types include most rock types found in the structural highs bordering the basin.

The first type displays characteristics of debris flow deposits (Middleton & Hampton, 1976) while the second probably represents slumps. The latter were penecontemporaneous, as shown by undisturbed flysch strata overlying the slumped horizons. In Ricci Lucchi's facies scheme, all these lithotypes would belong to facies F.

Hemipelagic mudstones

Massive mudstones surround the flysch sediments (Figs 6, 7) and represent normal hemipelagic slope sedimentation (see also Lewis, 1980).

Generally the sand percentage is less than 1%, but, in the southeastern part of the area studied, mudstones can contain up to 20% sand.

Based on faunal content, the massive mudstones can be subdivided into two subfacies. One is characterized by the presence of common to abundant macro- and microfauna indicative of an outer-neritic palaeobathymetry. The other subfacies lacks macrofossils and contains a bathyal microfauna.

The upper-Middle Miocene (Waiauan) sandy mudstones to the south and southeast of the area generally belong to the shallow-water subfacies. Deepening occurred towards the top of the Waiauan, and most Tongaporutuan massive mudstones comprise the deep-water subfacies.

In Ricci Lucchi's facies scheme the hemipelagic mudstones would belong to facies G.

BASIN DEVELOPMENT

During flysch sedimentation in the adjacent Akitio Basin in the Early to Middle Miocene, the area of the future Makara Basin was blanketed by fossiliferous massive mudstones and muddy sandstones (Kingma, 1971; Pettinga, 1976, 1980). Continued growth (thrusting) of the Waimarama–Mangakuri Coastal High and the Otane Anticlinal Complex created the Makara Basin during the Middle Miocene, and shifted flysch sedimentation from the Akitio Basin to the Makara Basin.

We assume that the extent of flysch strata in the Makara Basin indicates the extent of ponding of sediment, and hence the area of the basin (Fig. 6). This now measures 30 km north to south, and 20 km east to west. Post depositional folding and thrusting have caused some east to west shortening, but the extent of this is difficult to reconstruct; it is probably not more than 15%.

The ponding of the flysch between actively rising highs is clearly demonstrated by the onlap relationships with the Waimarama–Mangakuri Coastal High to the east (Figs 5, 7, 11). The upper parts of the eastern high are overlain by a condensed mudstone sequence. Insufficient detailed work has been done on the western high to be certain that the same is the case there, but in the southwest corner of the basin (loc. 250/9·8) Latest Miocene (Upper Tongaporutuan) flysch strata unconformably overlie mid-Miocene (Waiauan) limestone in a depositional contact (Fig. 5; Kingma, 1971).

No onlap relationship exists between the flysch strata and the Elsthorpe Anticline. The strata have been deformed into an anticline (thus dividing the basin into two broad synclines) by the upward (thrust) movements of the 'basement' rocks (Fig. 6). This high must therefore have become active near the end of, or after, flysch sedimentation.

The oldest flysch sediments of the Makara Basin are exposed in the southeast (Figs 5, 7). They are thin-bedded strata of subfacies 1. In the eastern part of the basin, subfacies 1 sedimentation continued through the Middle Tongaporutuan (Figs 5, 7). Localized in the area between the Waimarama–Mangakuri Coastal High and the north trending offshoot near Kairakau the thick massive units of the Ponui Sandstone subfacies 3 are intercalated in subfacies 1 sequences (Figs 6, 7, 8d).

The Silverrange Sandstone unit of subfacies 2, exposed in the western part of the basin, represents a sudden increase in sand supply. The unit can be followed along

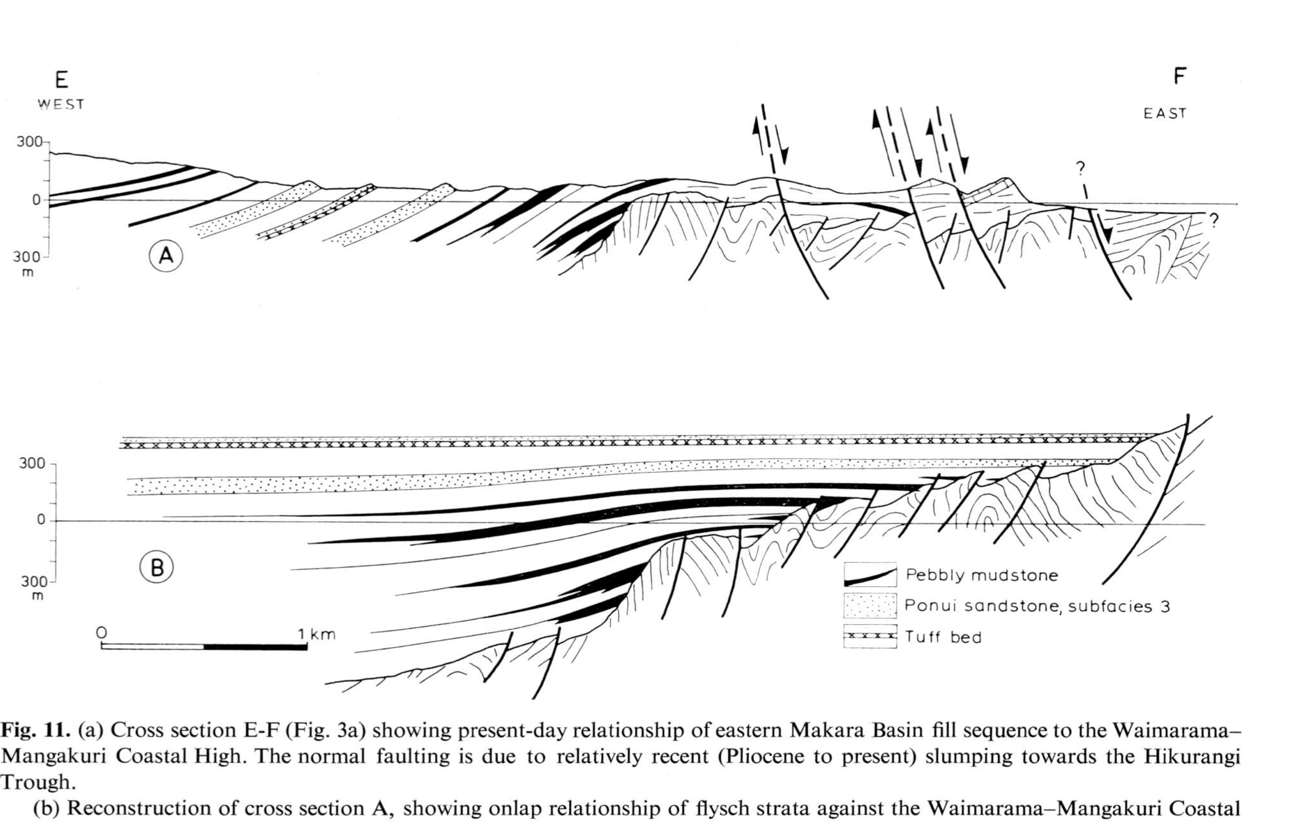

Fig. 11. (a) Cross section E-F (Fig. 3a) showing present-day relationship of eastern Makara Basin fill sequence to the Waimarama–Mangakuri Coastal High. The normal faulting is due to relatively recent (Pliocene to present) slumping towards the Hikurangi Trough.

(b) Reconstruction of cross section A, showing onlap relationship of flysch strata against the Waimarama–Mangakuri Coastal High, during deposition of the Ponui Sandstone. The contemporary origin of the reverse faults is considered probable, but not proven.

strike for almost the entire length of the basin (Fig. 6) and extends over a slightly larger area than the preceding flysch strata, but is nevertheless confined to the western part of the basin. The thinning-upward unit may represent a channelized sequence, but other explanations are possible (see Martini & Sagri, 1977).

A westward shift of the basin depocentre occurred during the Late Miocene (Figs 5, 7). This was due to continued thrust movement and uplift of the Waimarama-Mangakuri Coastal High and possibly the early development of the Elsthorpe Anticline (Fig. 7). During Upper Tongaporutuan time only massive hemipelagic mudstones accumulated over the eastern part of the basin, while to the west flysch sedimentation continued well into the Upper Tongaporutuan (Fig. 7).

During the entire history of the basin, rhyolitic volcanism was active. Tuff beds, both ash-fall deposits and layers of ash redeposited by sediment gravity flows, are intercalated in the flysch sediments. A substantial component of the flysch sediments themselves consists of volcanic particles. In the Upper Tongaporutuan, towards the top of the basin sequence, rhyolitic volcanism increased, and current-deposited sandstone beds almost exclusively consist of volcanic ash, while the mudstones also have a high ash content (subfacies 4).

Intensification of tectonic activity at the close of the Miocene terminated the flysch sedimentation in the Makara Basin and caused an unconformity to develop in the vicinity of the Otane Anticlinal Complex and the Waimarama–Mangakuri Coastal High (Fig. 3). Substantial uplift occurred, and the Te Aute limestone, representing widespread shallow marine deposition, was deposited during Pliocene time (Kingma, 1971; Pettinga, 1976, 1977, 1980). The Elsthorpe Anticline probably developed fully during the late Pliocene and Pleistocene times, a period during which the region was also uplifted above sea level. The coastal district has since been affected by gravitational tectonic gliding manifested onshore by the presence of numerous normal faults with substantial downthrows to the east, and westward tilting of large fault-bound blocks (Pettinga, 1980) (Figs 3, 6).

Flysch sedimentation continued in other parts of the East Coast Deformed Belt well into the Pliocene (Fig. 2), and is still active in the present-day offshore area (Lewis, 1980).

SUMMARY AND CONCLUSIONS

The Miocene Makara Basin in the Hawke's Bay area is one of a series of Neogene flysch basins which form part of the East Coast Deformed Belt which is situated within the present-day North Island oblique subduction system.

The Makara Basin (and the same holds for the other Neogene flysch basins) is thought to have formed on the inner slope of the subduction trench (Hikurangi Trough). It is now uplifted above sea level by imbricate thrust faulting of the leading edge of the upper plate (Fig. 12). The interpretation of the basin as a 'fossil slope basin' is based on comparison with other slope basins described in the literature (Grow, 1973; Karig & Sharman, 1975; Moore & Karig, 1976; Smith, Howell & Ingersoll, 1979), and by analogy to the present-day setting of basins in the inner trench-slope offshore Hawke's Bay (Lewis & Kohn, 1973; Lewis, 1980). Characteristics of these slope basins and similarities to the Makara Basin are:

(1) *Size.* The now exposed Makara Basin measures 20 by 30 km, which is similar

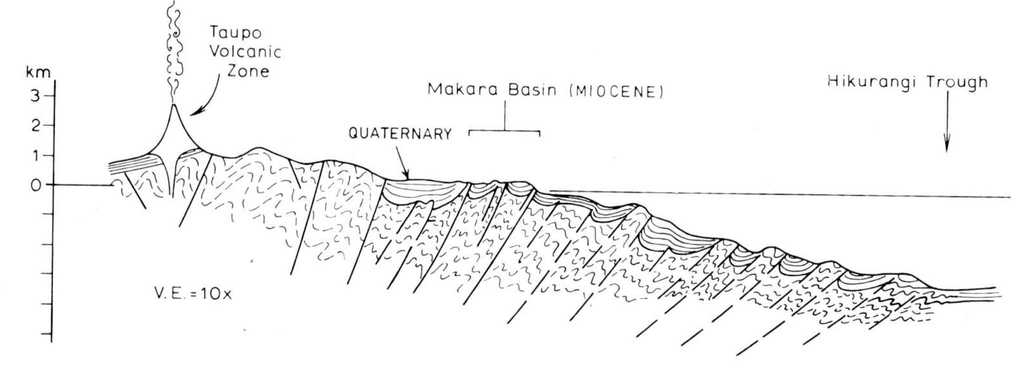

Fig. 12. Schematic cross section through the North Island Subduction System at the present day. Ages of basins shown where known (onshore). Offshore basins schematic.

to present day basins of inner-trench slopes elsewhere (e.g. Aleutian, Sunda, Shikoku & Hikurangi).

(2) *Palaeobathymetry.* Microfaunas from the hemipelagites in the Makara Basin indicate an outer neritic to bathyal depth of deposition.

(3) *Bounding highs.* The Otane Anticlinal Complex and Waimarama–Mangakuri Coastal High which bound the Makara Basin are narrow linear thrust zones, consisting of strongly deformed lithologies older than the basin sediments. A similar situation is described for offshore basins (see Lewis, 1980). These thrust faults dip away from the Hikurangi Trough.

(4) *Sediments.* Derivation of flysch deposits was from up-slope, and of debris flow deposits and slumps from the thrust ridges. Tuff beds in the Makara Basin have analogies in the offshore basins, where dateable ash layers derived from the Taupo Volcanic Zone occur (Lewis & Kohn, 1973).

(5) *Present-day basin geometry.* The structure of the basin is that of an asymmetric syncline, tilting occurring due to progressive growth and rotation of thrust ridges (imbrication). During sedimentation this resulted in a landward shift of the depocentre, and hence asymmetry. The late development of the Elsthorpe Anticline has divided the basin into two synclines. The gradual development of similar basins is recognized elsewhere (Moore & Karig, 1976; Lewis, 1980).

The Australian–Pacific plate boundary, of which the North Island Subduction System forms a part, was propagated southwards from the Tonga–Kermadec Subduction System into the New Zealand continental area in Late Oligocene time (Ballance, 1976; Walcott, 1978a, b; Spörli, 1980).

Although subduction took place, and is still taking place, obliquely to the East Coast Deformed Belt, no field evidence of strike-slip movements exists in the Hawke's Bay area of the belt. Only compressive folding and thrust faulting are evident. Walcott (1978b) suggested that the strike-slip component of the oblique subduction is being taken up further to the west, in a zone now occupied by the Axial Ranges, forming the continuation of the South Island Alpine Fault (see also Spörli, 1980; Lewis, 1980). Dextral offset along this fault zone since the Miocene may amount to approximately 230 km (Ballance, 1976).

In Miocene time the Hikurangi Trough was situated closer to the present-day

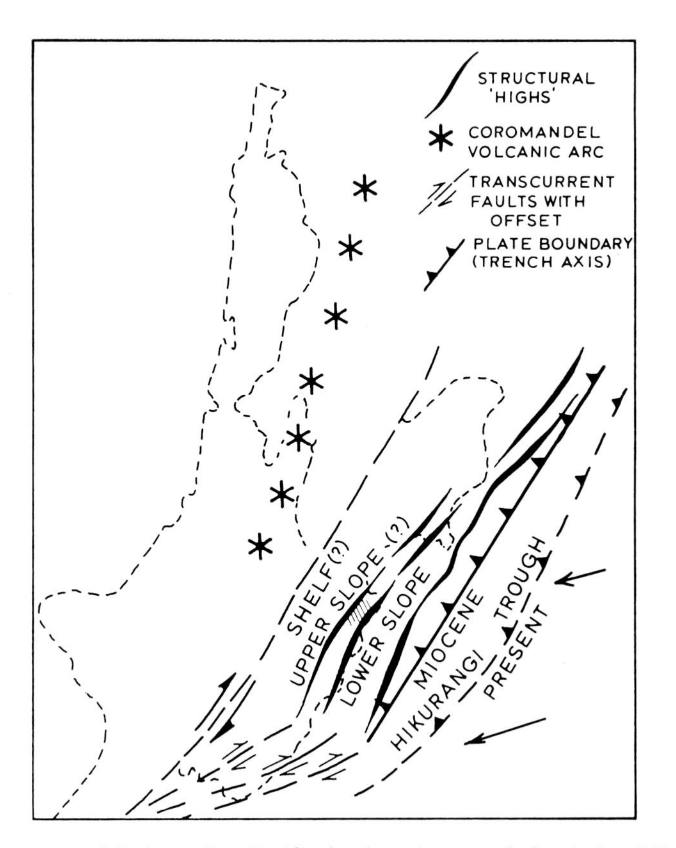

Fig. 13. Reconstruction of the Australian–Pacific plate boundary area during the Late Miocene. Arrows indicate direction of plate convergence (after Walcott, 1978a). Makara Basin area indicated by hatching.

coast line (Fig. 13). The increased separation since that time is due to progressive widening of the East Coast Deformed Belt with continued accretion and imbricate thrust deformation of the inner trench-slope.

During the development of the Makara Basin, the present-day volcanic arc (Taupo Volcanic Zone) did not yet exist. At that time, the Coromandel Arc was active (Ballance, 1976), contributing volcanic detritus to the slope basins. Recent dating of rhyolitic volcanics in the Coromandel area (Rutherford, 1978) supports this suggestion.

Taking the 230 km dextral offset and anticlockwise rotation of the northern part of the North Island (Ballance, 1976) into consideration, a satisfactory palaeogeographic reconstruction for the setting of the Neogene inner trench-slope flysch basins in the East Coast Deformed Belt can be obtained (Fig. 13).

ACKNOWLEDGMENTS

Over the years, many people have assisted us with our Hawke's Bay studies, either in the field, laboratory and darkroom, or by discussing concepts and data. They come from different institutions, the New Zealand Geological Survey (GS), the University of Canterbury, Christchurch (UC), and the University of Auckland (UA). We are very grateful for their help, and would like to use this opportunity to mention them by name: Mr E. T. H. Annear (GS), Dr A. G. Beu (GS), Mr A. Downing (UC), Dr G. W. Gibson (UA), Mr R. M. Harris (UA), Dr N. de B. Hornibrook (GS), Dr R. H. Hoskins (GS), Miss C. A. Hulse (GS), Ms Lee Leonard (UC), Mr W. M. Prebble (UA), Mr G. W. Richards (GS), Dr G. H. Scott (GS), Mr D. Smale (GS), Dr K. B. Spörli (UA), Mr K. M. Swanson (UC), Dr G. P. L. Walker (UA).

The manuscript was read by Dr D. G. Howell (U.S. Geological Survey, Menlo Park, California), Dr P. F. Ballance (UA), and Dr P. B. Andrews (GS), who made many useful suggestions for its improvement.

One of us (J.R.P.) received financial assistance from the National Water and Soil Conservation Organisation, Ministry of Works and Development, and from a Postgraduate Scholarship, awarded by the New Zealand University Grants Committee.

Miss Christina M. Johnstone (GS) typed the various versions of the manuscript.

REFERENCES

BALLANCE, P.F. (1976) Evolution of the Upper Cenozoic Magmatic Arc and Plate Boundary in Northern New Zealand. *Earth Planet. Sci. Lett.* **28,** 356–370.

BOUMA, A.H. (1972) Recent and ancient turbidites and contourites. *Trans. Gulf Cst. Ass. geol. Socs.* **22,** 205–221.

ENOS, P. (1977) Flow regimes in debris flow. *Sedimentology*, **24,** 133–142.

FOLK, R.L., ANDREWS, P.B. & LEWIS, D.W. (1970) Detrital sedimentary rock classification and nomenclature for use in New Zealand. *N.Z. J. Geol. Geophys.* **13,** 937–968.

GROW, J.A. (1973) Crustal and upper mantle structure of the central Aleutian arc. *Bull. geol. Soc. Am.* **84,** 2169–92.

HÄNTZSCHEL, W. (1975) *Trace fossils and problematica.* Part W, supplement 1 (miscellanea) of Treatise on Invertebrate Paleontology (Ed. by C. Teichart). *Geological Society of America*, 269 pp.

KARIG, D.E. & SHARMAN, G.F. (1975) Subduction and accretion in trenches. *Bull. geol. Soc. Am.* **86,** 377–389.

KINGMA, J.T. (1958a) The Tongaporutuan Sedimentation in Central Hawkes Bay. *N.Z. J. Geol. Geophys.* **1,** 1–30.

KINGMA J.T. (1958b) Possible origin of piercement structures, local unconformities, and secondary basins in the Eastern Geosyncline, New Zealand. *N.Z. J. Geol. Geophys.* **1,** 269–274.

KINGMA, J.T. (1960) The tectonic significance of graded bedding in geosynclinal sedimentary systems. *Int. geol. Congr.* 21. Copenhagen, 205–214.

KINGMA, J.T. (1971) *Geology of Te Aute subdivision. Bull. geol. Surv. N.Z.* **70,** 173 pp.

KUENEN, PH.H. (1960) Turbidites in Makara Basin, New Zealand. *Proc. K. Ned. Akad. Wet. Ser. B.* **63,** 127–34.

KUENEN, PH.H. (1970) The turbidite problem: some comments. *N.Z. J. Geol. Geophys.* **13,** 852–57.

LEWIS, K.B. (1980) Quaternary sedimentation on the Hikurangi oblique-subduction and transform margin, New Zealand. In: *Sedimentation in oblique-slip mobile zones* (Ed. by P. F. Ballance and H. G. Reading). *Spec. Publ. int. Ass. Sediment.*, **4,** 171–189.

LEWIS, K.B. & KOHN, B.P. (1973) Ashes, turbidites, and rates of sedimentation on the continental slope off Hawke's Bay. *N.Z. J. Geol. Geophys.* **16,** 439–454.

LILLIE, A.R. (1953) *Geology of the Dannevirke Subdivision. Bull. geol. Surv. N.Z.* **46,** 156 pp.

LINGEN, G.J. VAN DER (1968a) Preliminary sedimentological evaluation of some flysch-like deposits from the Makara Basin, Central Hawkes Bay, New Zealand. *N.Z. J. Geol. Geophys.* **11,** 455–477.

LINGEN, G.J. VAN DER (1968b) Volcanic ash in the Makara Basin (Upper Miocene), Hawke's Bay, New Zealand. *N.Z. J. Geol. Geophys.* **11,** 693–705.

LINGEN, G.J. VAN DER (1969) The Turbidite Problem. *N.Z. J. Geol. Geophys.* **12,** 7–51.

LINGEN, G.J. VAN DER (1970) The turbidite problem: a reply to Kuenen. *N.Z. J. Geol. Geophys.* **13,** 858–72.

LINGEN, G.J. VAN DER (1971) Granulometry of the Upper-Miocene flysch-type sediments of the Makara Basin, New Zealand. (Abstract). *VIII Intern. Sed. Congress,* Heidelberg. Programme with abstracts, p. 59.

MARTINI, I.P. & SAGRI, M. (1977) Sedimentary fillings of ancient deepsea channels: two examples from Northern Apennines (Italy). *J. sedim. Petrol.* **47,** 1542–1553.

MIDDLETON, G.V. & HAMPTON, M.A. (1976) Subaqueous sediment transport and deposition by sediment gravity flows. In: *Marine sediment transport and environmental management* (Ed. by D. J. Stanley and D. J. P. Swift), pp. 197–218. John Wiley, New York.

MOORE, G.F. & KARIG, D.E. (1976) Development of sedimentary basins on the lower trench slope. *Geology,* **4,** 693–697.

PETTINGA, J.R. (1976) Structure and slope failures in Southern Hawkes Bay. *Abstr. geol. Soc. N.Z.* Hamilton Conference, 1976.

PETTINGA, J.R. (1977) Geology and regional significance of the Elsthorpe Anticline, Southern Hawkes Bay. *Abstr. geol. Soc. N.Z.* Queenstown Conference, 1977.

PETTINGA, J.R. (1980) *Geology and landslides of the eastern Te Aute District, southern Hawke's Bay.* Unpublished Ph.D. thesis, University of Auckland. 602 pp.

RICCI-LUCCHI, F. (1975) Depositional cycles in two turbidite formations of Northern Apennines (Italy). *J. sedim. Petrol.* **45,** 3–43.

RUTHERFORD, N.F. (1978) Fission-track age and trace element geochemistry of some Minden Rhyolite obsidians. *N.Z. J. Geol. Geophys.* **21,** 443–448.

SMITH, G.W., HOWELL, D.G. & INGERSOLL, R.V. (1979) Late Cretaceous trench slope basins of central California. *Geology,* **7,** 303–306.

SPÖRLI, K.B. (1980) New Zealand and oblique-slip margins: tectonic development up to and during the Cainozoic. In: *Sedimentation in oblique-slip mobile zones* (Ed. by P. F. Ballance & H. G. Reading). *Spec. Publ. int. Ass. Sediment.,* **4,** 147–170.

WALCOTT, R.I. (1978a) Present tectonics and Late Cenozoic evolution of New Zealand. *Geophys. J.R. astr. Soc.* **52,** 137–164.

WALCOTT, R.I. (1978b) Geodetic strains and large earthquakes in the axial tectonic belt of North Island, New Zealand. *J. geophys. Res.* **83,** 4419–4429.

Spec. Publ. int. Ass. Sediment. (1980) **4**, 217–228

Late Cainozoic sedimentation and tectonics of the East Coast Deformed Belt, in Marlborough, New Zealand

WARWICK M. PREBBLE

Department of Geology, University of Auckland, New Zealand

ABSTRACT

The character of sedimentation in northeast Marlborough changed markedly during the Miocene, when long-established, dominantly calcareous deposition gave way to an influx of clastic detritus. Conglomerate, breccia, olistostromes, sandstones and turbidites then predominated and are described from the coastal section of the shallow fold and fault belt, between the Kekerengu and Waima Rivers.

The change in sedimentation was preceded by submarine volcanism and accompanied by rapid tectonic uplift, a marked increase in sedimentation rate, exposure of older indurated undermass rocks and localized sedimentation of the very coarse clastics. These often chaotic deposits, collectively referred to as the Great Marlborough Conglomerate, formed from partly coalescing submarine slides and debris flows. The sequence is conformable, with substantial variations in thickness of the clastic units and interfingering of the rudites with sandstones, turbidites and mudstones. Localization of the type and size of megaclasts is a conspicuous feature.

The Conglomerate is restricted to the major fault-angle depressions and thickens rapidly towards the faults. Massive slope failure of the rising fault blocks supplied the coarse debris to the fault-controlled basins. A transition from marine to non-marine deposition through the Miocene is seen from the southeast to the northwest.

Sedimentation continued into the Pliocene, after which the covering strata were subjected to imbricate thrust faulting and folding along a northeast trend, parallel to the oblique-slip margin. Subsequently, dextral faulting on the same trend has rotated the covering slab of the shallow thrust and fold belt, which was uncoupled from the undermass as a tectonic décollement in the Ben More block. This décollement is related to a major dextral fracture—the Kekerengu Fault. Another décollement is postulated to occur over the northernmost sector of the Clarence Fault.

A transition occurs in the Marlborough region from late Oligocene subduction and compression to Miocene normal faulting, Pliocene–early Pleistocene normal oblique-slip and late Pleistocene–Recent reverse oblique-slip. The Miocene–Recent movements have been concentrated on the major faults of the region: Wairau (Alpine), Awatere, Clarence, Hope and Kekerengu.

INTRODUCTION

The tectonically active, mountainous region of Marlborough forms the north-eastern portion of the South Island of New Zealand (Fig. 1). The terrain is

0141-3600/80/0904-0216$02.00

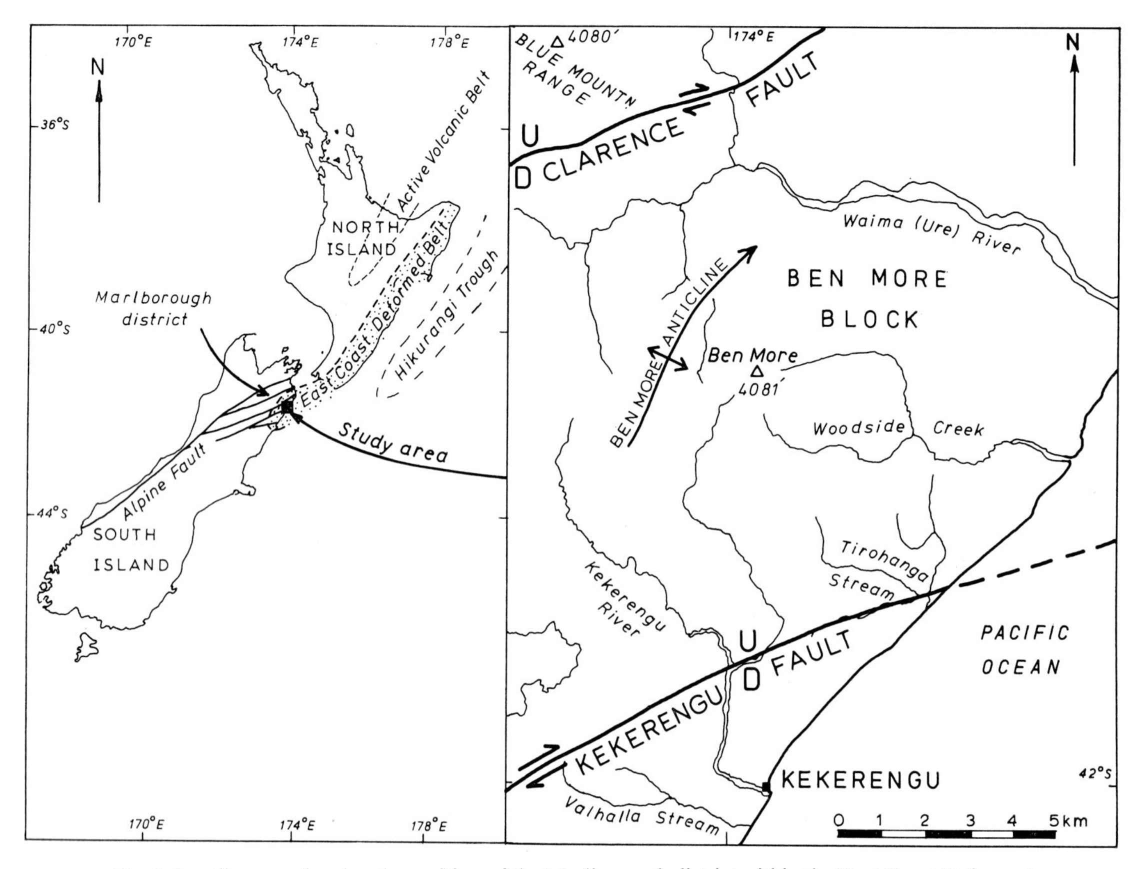

Fig. 1. Locality map showing the position of the Marlborough district within the East Coast Deformed Belt and important features of the study area.

controlled by major northeast trending dextral faults, each of which has a high-standing upthrown block of Mesozoic undermass rocks on its northwestern side. In the fault-angle valleys Upper Cretaceous and Cainozoic covering strata are preserved in elongate strips which are complexly faulted and folded. These cover beds mark the southern end of the East Coast Deformed Belt (Spörli, 1980), a zone of Cainozoic tectonism which parallels the Hikurangi Trough and, by inference, the oblique-slip margin.

Between the Kekerengu and Ure Rivers (Fig. 1), the covering strata are most extensive and probably attain their greatest thickness for the Marlborough region. They form a complex plunging anticline, the Ben More anticline (Prebble, 1976), which is wrapped around a core of indurated Mesozoic undermass strata, forming the Ben More Block (Figs 1 and 2).

This paper describes the Miocene rocks, and their deformation as part of the cover, in the Ben More Block and near Kekerengu (Figs 1 and 2). Similar Miocene strata to the south have been described by Laird & Lewis (1979). An extensive strip of covering strata which includes Miocene rocks occurs to the southwest along the fault-angle Clarence Valley, a small sector of which has been described by Hall (1965).

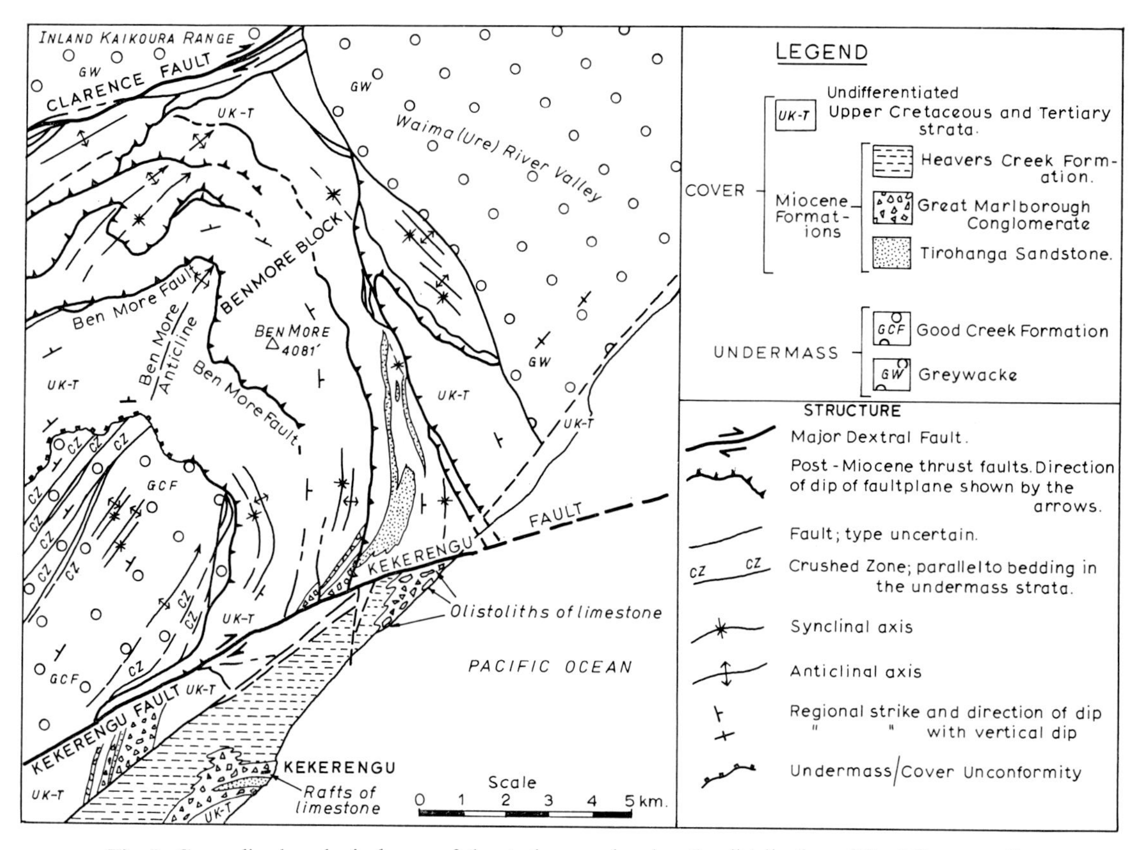

Fig. 2. Generalized geological map of the study area showing the distribution of the Miocene units.

New Zealand stage names are given in parentheses after the international equivalent.

THE UNDERMASS STRATA AND THEIR RELATIONSHIP TO THE COVER BEDS

Structurally, and lithologically, the cover in this block is fundamentally different from the undermass and much less indurated, although only some 6–10 m.y. separate them.

The break between the undermass and the cover is, in part, an unconformity and, in part, a complex fault, the two merging into each other. The fault is interpreted as an undermass/cover décollement in the eastern half of the Ben More Block. Within this block, several more or less parallel faults and folds (Fig. 2) which are confined to the Upper Cretaceous–Tertiary cover swing in an arc around the anticlinal core of undermass strata. This arcuate pattern is exceptional in Marlborough and perhaps throughout the whole of the East Coast Deformed Belt.

THE COVERING STRATA PRIOR TO THE MIOCENE

Throughout the East Coast Deformed Belt, Upper Cretaceous through to Oligocene strata are characterized by widespread fine-grained facies, in which siliceous shale, chert and micritic argillaceous limestone are dominant. In Marlborough, calcareous deposition concluded with an Upper Oligocene (Waitakian) marine calcareous mudstone and was accompanied by andesitic and basaltic sills, breccias, and pillow lavas. Oligocene tectonic activity is indicated by intraformational disharmonic folding which is common in the middle units of the Amuri Limestone (Hall, 1965; Prebble, 1976). Attributed to slump folding, it is considered to have occurred at the same time as a late Oligocene unconformity recorded by Lewis, van der Lingen & Smale (1977) from North Canterbury. These events are minor compared to later tectonism. Similarly, the Upper Cretaceous–Oligocene strata, although widespread, attained only 1500 m of thickness in approximately 60 m.y. By contrast, Miocene sedimentation reached a similar figure in 10 m.y., but is much more localized.

THE MIOCENE STRATA

The uppermost Lower Miocene (Altonian) is marked by a sudden influx of conglomerate, breccia, diamictite and olistostromes. These rudites are known as the Great Marlborough Conglomerate (McKay, 1886). Thick beds, lenses and tongues occur along a 10 km coastal strip centred at Kekerengu, where it attains its greatest known thickness and interfingers with calcareous mudstone (Waima Formation), massive sandstone (Tirohanga Sandstone) and thick mudstone, turbidites, laminated sandstone and marl of the Heavers Creek Formation.

Upper Miocene freshwater conglomerate is widespread in the north of Marlborough, being derived from the rising mountain blocks north of the Awatere Fault (Fig. 3).

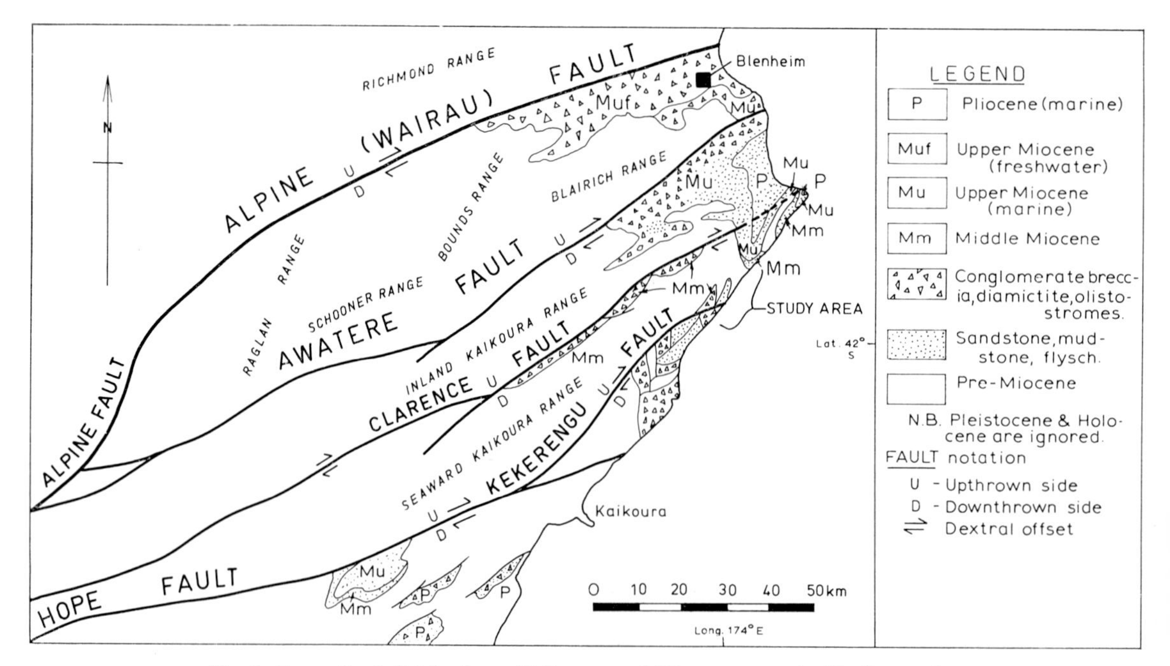

Fig. 3. Generalized distribution of Miocene and Pliocene strata in Marlborough.

Great Marlborough Conglomerate

The Conglomerate is massive, unsorted and well consolidated. Clasts vary in size from pebbles to large angular blocks and rafts longer than 10 m. Most are less than 5 m long. In some exposures pebbles and boulders dominate; in others large angular blocks. Rafts up to hundreds of metres in length have been mapped by Prebble (1976) near Kekerengu, and Hall (1965) several kilometres to the northwest. The largest occur in groups, with their greatest surface area parallel to bedding within the raft and to the crude bedding of the surrounding conglomerate. Smaller clasts generally have a random orientation. (Fig. 4).

Fig. 4. Diamictite of the Great Marlborough Conglomerate, in Heavers Creek near Kekerengu, Marlborough District.

The matrix of the conglomerate is a slightly muddy sandstone. In many exposures, the deposit is essentially a diamictite.

The most common clast lithologies are Amuri Limestone, and indurated Lower Cretaceous sandstone. Minor lithologies include a range of igneous rocks derived from both the Cretaceous undermass plutons of the Inland Kaikoura Block and the Late Oligocene volcanics of the cover, splintery shale, chert, glauconitic sandstone, soft calcareous mudstone and soft calcareous bentonitic shale. One or two lithologies are usually dominant at any particular locality, suggesting a very local derivation. Clasts of undermass origin are absent at some localities, but are common to predominant at others.

The diamictite texture, and the presence of large blocks of low-strength and fractured rock, imply a depositional mechanism with matrix support and without turbulence, i.e. debris flow.

Crude discontinuous stratification is evident in the conglomerate. Truncation by other beds, although common, is indistinctly shown, suggesting deposition in a series of multiple and complex channels and lobes.

Beds of laminated, carbonaceous sandstone and graded sandstone–mudstone beds occur in units up to 45 m thick which interfinger with the conglomerate. The conglomerate fines gradationally upwards into sandstone and then repetitive sequences of graded beds. The basal contact of conglomerate overlying these units is sharp, undulating and channelled, with large ripped-up blocks of the distinctive laminated sandstone immediately above the contact. Some erosive contacts are highly complex, with numerous rip-up clasts and deep channels, the sides of which are often highly irregular, presumably as a consequence of multiple superposition of channels and debris lobes. The interfingering is indicated by detailed mapping (Fig. 5).

The thickness of the Great Marlborough Conglomerate ranges up to 300 or 400 m near Kekerengu and up to 550 m further inland. In general the conglomerate thickens westwards, towards the Kekerengu Fault and thins eastwards especially in the northeast where it dwindles rapidly to a few metres and is replaced by sandstones and finer clastics. A few of the large rafts have been deposited beyond the main body of conglomerate, within the finer grained sediments.

The Great Marlborough Conglomerate is conformably underlain by, and interfingers with, siltstone of the Waima Formation which is late Oligocene–early Miocene in age.

Foraminifera from sandstone and mudstone beds within the conglomerate indicate that its age is latest early Miocene and mid-Miocene (Altonian to Waiauan), and that it is marine.

Heavers Creek Formation

Carbonaceous sandstone, graded sandstone–mudstone beds and massive mudstone overlie and interfinger with the Great Marlborough Conglomerate. They are 650 m thick in the Kekerengu area and are widespread throughout northeast Marlborough.

The sandstone beds are considered to be mass flow deposits and the graded sandstone–mudstone beds indicate deposition by turbidity currents. The massive mudstone, which contains calcareous lenses, has been deposited from suspension.

Foraminifera from near the base of the formation at Kekerengu give an early to mid-Miocene age. Further to the northeast, flysch, conglomeratic sandstone, massive mudstone and shell beds are late Miocene and Pliocene in age (Tongaporutuan–Waitotaran).

Massive sandstone bodies

In the northeast of the region, the Lower Miocene medium quartzose Tirohanga

Sandstone interfingers with calcareous mudstone and pebbly mudstone (Waima Formation) in large elongate units 80 m thick, and also as a large multiple lens-shaped body up to 250 m thick (Fig. 5). Thin beds of conglomerate containing indurated

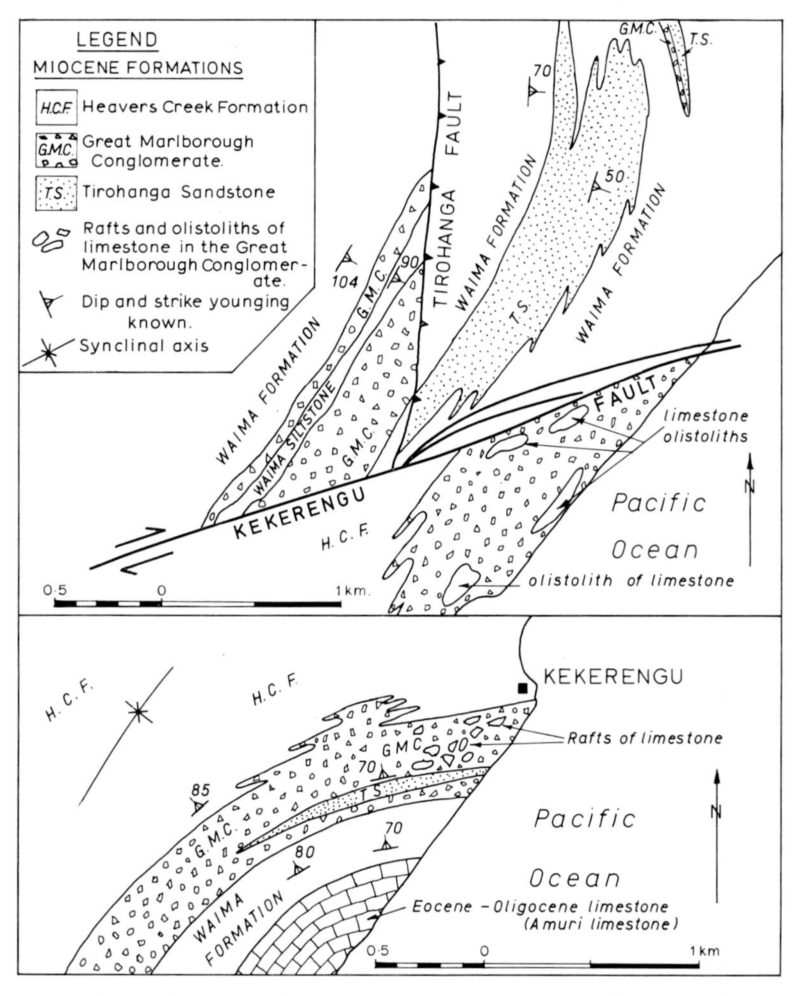

Fig. 5. Relationships of the Miocene Units to each other and to the underlying Waima Formation, near Kekerengu on the northeastern coast of the Marlborough District.

sandstone pebbles occur within the Tirohanga sandstone, which is otherwise massive and lacks persistent or regular bedding. It is thought to have been deposited by non-turbulent sediment gravity flows. Graded beds of sandstone–mudstone showing typical turbidite features conformably overlie the massive sandstone.

Upper Miocene conglomerates

Upper Miocene (Tongaporutuan and Kapitean) freshwater and marine conglomerates are widespread in northern Marlborough where they rest with angular unconformity on indurated Mesozoic undermass strata, adjacent to the Alpine and Awatere Faults (Fig. 3). Large angular blocks are not as common as in the Great Marlborough

Conglomerate, but these Upper Miocene conglomerates are very thick, especially in the Awatere fault-angle depression where approximately 1000 m of conglomerate is indicated (Lensen, 1962).

PLIOCENE AND PLEISTOCENE STRATA

Massive marine mudstone, flysch, conglomeratic sandstone, breccia and shellbeds of Pliocene age are restricted to the northeast and far south of the region where they conformably overlie Upper Miocene marine conglomerates and flysch. The sequence is regressive and is unconformably overlain by Upper Pleistocene gravels.

No Lower Pleistocene deposits are recorded in the region. Upper Pleistocene deposits are almost exclusively terrestrial and everywhere are unconformable on the Mesozoic and Tertiary rocks. Several different formations of glacial and interglacial origin have been distinguished in the region by Lensen (1962) and Suggate (1965). These consist of moraine and widespread glacial outwash gravels with minor sands and silts. Estuarine and beach deposits are recorded only from localities adjacent to the present coast.

DEFORMATION OF THE COVER

A sequence of post-Miocene events has been established. Deformation commenced in the Miocene but was most pronounced in Plio-Pleistocene time, in common with the rest of New Zealand.

Block faulting and differential uplift, originally very rapid in the lower Miocene, persisted into the late Miocene and Pliocene. In Plio-Pleistocene time the character of deformation changed. Thrust faulting, folding, dextral faulting and complete emergence followed in swift succession.

Imbricate thrust faulting of the cover is directed northwestwards towards the Inland Kaikoura Range and was followed by folding along northeast striking axes. The folding is locally tight, with dips exceeding the vertical. All formations of the cover were deformed by the folding which was followed by dextral faulting, on the northeast striking Kekerengu Fault. It is postulated that the undermass was displaced by the fault but that the cover initially remained contiguous across it. Cover on the southern side of the Kekerengu Fault remained coupled to the basement, but the cover on the northern side, in the eastern half of Ben More Block, was dragged along the unconformity surface which separates the undermass and cover. The faults and fold axes were uniformly bent around in a clockwise direction and considerable thrusting, shearing and crushing was created, by the northeast to southwest directed compression in the décollement slab.

Finally, the cover slab was forced up the northeastwards tilted surface of the undermass and was wrenched apart along Kekerengu Fault, incurring a dextral displacement of at least 5 km. Dextral movement on both the Kekerengu and Clarence Faults has continued through to the Recent, with uplift continuing on the northern side of each fault.

Direct evidence for substantial dextral motion of Plio-Pleistocene age in Marlborough is provided by matching belts of Cainozoic facies and structures across the major faults, in particular the Kekerengu and Hope Faults.

The northern end of the Hope Fault displaces a prominent syncline of Upper Cretaceous–Miocene strata, as shown on the geological map of the region (Lensen, 1962). The dextral offset is 12 km. A further 20 km of displacement is indicated on Lensen's map along the southern section of the Kekerengu Fault. Freund (1971) has identified 20 km of dextral displacement along the Hope Fault, in north Canterbury, to the south of the Marlborough region. Lensen (1960) suggested at least 20 km dextral displacement on the Awatere Fault.

Evidence for 'active' Quaternary dextral motion can be found on all the major faults from offset Pleistocene terraces and Recent landforms. Late Quaternary displacements on all the faults, including some of several hundred metres, are listed by Wellman (1953), and a displacement of 300 m along the Kekerengu Fault on Upper Pleistocene Winterholme Formation gravels is suggested by Prebble (1976). Lensen (1968) has identified 70 m of lateral offset on the Alpine (Wairau) Fault which he attributes to several movements in the last 20 000 years.

UPPER CAINOZOIC TECTONICS

Upper Oligocene andesitic and basaltic volcanics are restricted to four main localities of marine pillow lavas, sills and breccias. The enclosing clastic rocks are far more extensive and persist into the lower Miocene, while the volcanics do not. Thus limited subduction may have been initiated in Marlborough during late Oligocene time, but it ceased abruptly and was superseded by Lower Miocene deposition of coarse clastics.

The upper part of the Waima Formation, although still calcareous, demonstrates an increase in tectonic activity in Lower Miocene time by the incoming of sandstone beds and small pebbles of Amuri limestone and undermass sandstone. The overlying Great Marlborough Conglomerate was deposited during rapid uplift in mainly marine conditions.

In considering the probable origin of the Great Marlborough Conglomerate, an instructive parallel is seen today in the East Coast Deformed Belt in the North Island. Tectonic activity is high and on-land slope failures involving huge block glides, earth flows, debris flows and semi-rotational slides are common. Large regional slumps have been identified in southern Hawke's Bay by Pettinga (1979), probably extending offshore (Lewis, 1971), and smaller failures involving several square kilometres abound (Prebble, 1979). Similar failures leading to debris slides and flows beneath the sea in the Miocene, in Marlborough, could account for the presence of the large blocks, rafts and olistoliths of limestone amongst the conglomerate.

Regional semi-rotational and translational slides are most likely to occur in an extensive pile of gently dipping covering strata, where slope angles exceed those of the beds, as in present-day southern Hawke's Bay. This situation does not exist in Marlborough today, where the cover is very steeply dipping and restricted in extent, and the few slope failures are small and rarely reach the coast.

The Great Marlborough Conglomerate is essentially conformable within the cover strata, and these form a continuous sequence. Hence it is contended that only minor tilting, if any, occurred at this time. However, it has been shown that initial shear failure, in a terrestrial environment, can occur in very gently dipping strata (Lensen & Suggate, 1968; Prebble, 1979; Thompson, 1979). Thus massive terrain failures and the associated myriad of debris slides and debris flows, are postulated to have characterized Marlborough in the early Miocene.

A series of multiple and complexly arranged distributary channels, offshore and down-slope from major areas of terrestrial landsliding, is envisaged, as well as sediment input from major river systems. Extensive submarine slope failure, both of the terrestrially derived debris and also submarine bedrock (the Upper Cretaceous–Cainozoic cover strata) is similarly considered to have contributed debris, including the large rafts and olistoliths.

The conglomerate thickens rapidly westwards towards both the Kekerengu and Clarence Faults (Fig. 3) and was deposited only adjacent to the highstanding fault blocks. Hence the Miocene heralds the initiation of the Marlborough fault system, with evidence for rapid dip-slip movement. All the faults are currently dextral. Although the Alpine Fault may be as old as Mesozoic (Spörli, 1980), there is no evidence that the other faults existed before the Miocene. Later vertical movement on the Marlborough faults has been concentrated in the northeast, where it has created the Kaikoura Ranges. Substantial dextral movement in Marlborough has certainly occurred in post-Miocene time, but a question of particular interest in this study is whether it may have been initiated during the Miocene.

Late Miocene and Plio-Pleistocene tectonics

An initial compressive phase accompanying limited subduction in the late Oligocene was replaced by rapid dip-slip movement on the new-formed Marlborough Fault system in the early and mid-Miocene. These strong dip-slip movements persisted into the Late Miocene with substantial block faulting and tilting continuing, especially along the Wairau (Alpine) and Awatere Faults. The Clarence and Kekerengu fault-angle depression in the East received mainly finer grained clastics and even calcareous mudstones in the Late Miocene, indicating a localized quieter phase, synchronous with the rapid uplift and accumulation of conglomerate in the west.

During the Pliocene, emergence of the fault-angle depressions commenced and was complete during the early Pleistocene. Pliocene and early Pleistocene faulting was dominantly normal, with substantial tilting and a dextral component (Lensen, 1962).

Late Pleistocene–Holocene faulting has been dominantly dextral, with a compressional component appearing on all the major faults (Lensen, 1962; Prebble, 1976).

Hence a transition can be recognized in Marlborough from compression and subduction (late Oligocene) to dip-slip block faulting (Miocene) to oblique slip, normal faulting (Pliocene–Early Pleistocene) and oblique slip, reverse faulting (late Pleistocene–Holocene). Hence compression is a comparatively new feature, a conclusion reached also from sea floor spreading data by Walcott (1978).

DISCUSSION

It is interesting to consider the Miocene of the Marlborough sector of the East Coast Deformed Belt in the light of criteria discussed by Reading (1980) for the recognition of old strike-slip orogenic belts. Those which are clearly recognizable in Marlborough (*in addition to the direct evidence for later oblique dextral motion*) include the following:

(1) Thick, but not laterally extensive, sedimentary piles deposited rapidly adjacent

to faults. In particular, the Great Marlborough Conglomerate but also the other Miocene clastic units conform to this criterion.

(2) Localized uplift and erosion giving rise to unconformities of the same age as the thick sedimentary fills nearby. The unconformity described by Lewis, van der Lingen & Small (1977) in north Canterbury was created at a time when the Great Marlborough Conglomerate was being deposited to the northeast. Similarly, Upper Miocene conglomerates unconformably overlie Mesozoic undermass strata in the Wairau and Awatere fault angle depressions but at the northern end of the Clarence and Kekerengu Faults there is a continuous thick conformable sequence of Miocene and Pliocene marine strata.

(3) Extreme lateral facies variations.

(4) No metamorphism has occurred within the covering strata and little within the undermass.

(5) Igneous activity is absent, except in the pre-Miocene units, where it is restricted in time and volume.

Evidence for Miocene dextral faulting in the study area is lacking, but may well have been masked by the undermass/cover décollement. Lensen (1962) suggests that a dextral component was initiated in Marlborough in the late Miocene–Pliocene and this can be inferred by reference to the geological map, sheet 16 (Lensen, 1962) in which larger displacements are generally seen in mid-Miocene and older rocks than in late Miocene–Recent strata.

ACKNOWLEDGMENTS

The author is indebted to Dr P. F. Ballance for his review of the manuscript and to Mr R. M. Harris for drafting of the figures.

Most of the data in the Kekerunga–Waima area was drawn from the author's thesis (1976), which was reviewed by Professor H. W. Wellman of Victoria University of Wellington, New Zealand. In particular the author wishes to thank Professor Wellman for his constructive discussion on structure and tectonics in the study area.

REFERENCES

Freund, R. (1971) The Hope fault, a strike slip fault in New Zealand. *Bull. geol. Surv. N.Z.* **86,** 49 pp.

Hall, W.D.M. (1965) *The Geology of Coverham and the Upper Waima Valley, Marlborough.* Unpublished M.Sc. Thesis, Victoria University of Wellington, New Zealand.

Laird, M.G. & Lewis, D.W. (1979) Deposition of Great Marlborough Conglomerate and associated sediments, Deadman Stream Marlborough. In: *Geology in New Zealand* 1978. (*Abstr. geol. Soc. N.Z.*). 16.

Lensen, G.J. (1960) A 12-mile lateral drag along the Awatere Fault. *Rep. 9th Sci. Congr. R. Soc. N.Z.*, **47** (Abstr.).

Lensen, G.J. (1962) Sheet 16 Kaikoura (1st edn) *Geological Map of New Zealand* 1 : 250 000. *N.Z. D.S.I.R.*

Lensen, G.J. (1968) Analysis of progressive fault displacement during downcutting at the Branch River terraces: South Island, New Zealand. *Bull. geol. Soc. Am.* **79,** 545–556.

Lensen, G.J. & Suggate, R.P. (1968) Inangahua Earthquake—Preliminary account of the geology. In: Preliminary Reports on the Inangahua Earthquake, New Zealand, May 1968 (Ed. by R. D. Adams *et al.*). *Bull. N.Z. D.S.I.R.* 193.

Lewis, D.W., van der Lingen, G.J. & Smale, D. (1977) Sand intrusions into the Amuri Limestone. *Abstr. geol. Soc. N.Z. Queenstown Conference.*

Lewis, K.B. (1971) Slumping on a continental slope inclined at 1°–4°. *Sedimentology*, **16,** 97–110.

McKay, A. (1886) Reports on Geological Explorations. *Rep. N.Z. colon. Mus. and geol. Surv.* 17.

Pettinga, J.R. (1979) The effects of geologic structure and lithology on slope failure in southern Hawke's Bay. *Abstr. 49th ANZAAS Congr.* **1,** 160.

Prebble, W.M. (1976) *The Geology of the Kekerengu-Waima River district, north-east Marlborough.* Unpublished MSc thesis, Victoria University of Wellington, New Zealand.

Prebble, W.M. (1979) Aspects of Engineering Geology for Hydro-Electric Development in steep, mountainous soft rock terrain, Hawke's Bay, New Zealand. *Abstr. 49th ANZAAS Congr.* **1,** 163.

Reading, H.G. (1980) Characteristics and recognition of strike-slip systems. In: *Sedimentation in oblique-slip mobile zones* (Ed. by P. F. Ballance & H. G. Reading). *Spec. Publ. int. Ass. Sediment,* **4,** pp. 7–26.

Spörli, K.B. (1980) New Zealand and oblique slip margins. Tectonic development up to and during the Cainozoic. In: *Sedimentation in oblique-slip mobile zones* (Ed. by P. F. Ballance & H. G. Reading). *Spec. Publ. int. Ass. Sediment.* **4,** pp. 147–170.

Suggate, R.P. (1965) Late Pleistocene Geology of the Northern part of the South Island, New Zealand. *N.Z. geol. Surv.* **77,** 91 pp.

Thompson, R.C. (1979) Landsliding in Soft Rocks of the Utiku-Mangaweka Area. *Abstr. 49th ANZAAS Congr.* **1,** 161.

Walcott, R.I. (1978) Present tectonics and Late Cenozoic evolution of New Zealand. *Geophys. J.R. astr. Soc.* **52,** 137–164.

Wellman, H.W. (1953) Data for the study of Recent and Late Pleistocene Faulting in the South Island of New Zealand. *N.Z. J. Sci. Technol.* **34B,** 270–288.

Spec. Publ. inst. Ass. Sediment. (1980) **4,** 229–236

Models of sediment distribution in non-marine and shallow marine environments in oblique-slip fault zones

P. F. BALLANCE

Department of Geology, University of Auckland, New Zealand

ABSTRACT

In oblique-slip fault-zones strike-slip movement on a rising block carries a sequence of source rocks past a locus of accumulating sediment on the adjacent, stationary, sinking block. A diachronous stratigraphy of sediment from distinctive source rocks may result, getting younger in the direction of strike-slip motion on the rising block. Four models of sedimentation in non-marine and shallow-marine environments are presented, and illustrated by New Zealand examples. Interpretation of oblique-slip faulting in ancient non-marine sediments requires recognition of (a) clasts from distinctive sources, (b) diachroneity of earliest and/or latest appearance of clasts, and (c) palaeo-flow direction. In zones of known oblique-slip, knowledge of one or two of those three factors would allow prediction of the others.

INTRODUCTION

Zones of oblique-slip motion are widespread both at plate margins and within plates (Reading, 1980). Where seen on land, the strike-slip motion is frequently dominant and accompanied by lesser dip-slip motion, for example in the San Andreas Fault Zone (Crowell, 1974) and in the Alpine Fault Zone of New Zealand (Walcott, 1978; Rynn & Scholz, 1978; Spörli, 1980).

This paper considers various models of non-marine and adjacent shallow-marine sedimentation in an active oblique-slip zone, and illustrates them with possible New Zealand examples. Increased tectonic activity in the New Zealand oblique-slip zone since the late Pliocene has resulted in high-standing, fault-bounded mountain ranges, and in voluminous non-marine and shallow-marine deposits in adjacent tectonic depressions.

Since little detailed work has been done on these rocks, the models are hypothetical and largely untested. However, general considerations suggest that these sediments have the potential to record oblique-slip movements clearly in their stratigraphy.

THE GENERAL SITUATION

Strike-slip movement on a rising block carries a sequence of source rocks past a

0141-3600/80/0904-0229$02.00

locus of accumulating sediment on an adjacent sinking block. A diachronous stratigraphy will normally result, in which beds derived from a particular source rock get younger in the direction of strike-slip motion of the rising block. The simple situation can be complicated by many factors, but this paper concentrates on sediment distribution factors, especially variations in the direction of drainage on the sinking block.

A source of sediment on the sinking block will be stationary with respect to the accumulating sediment, and will not give rise to diachronous beds.

Four models of non-marine/shallow-marine sediment distribution in the oblique-slip zone are presented, and illustrated by examples from New Zealand (Fig. 1).

Model 1. Rivers on a sinking block flow parallel to the fault and in the same direction as strike-slip motion on the rising block (Fig. 2a).

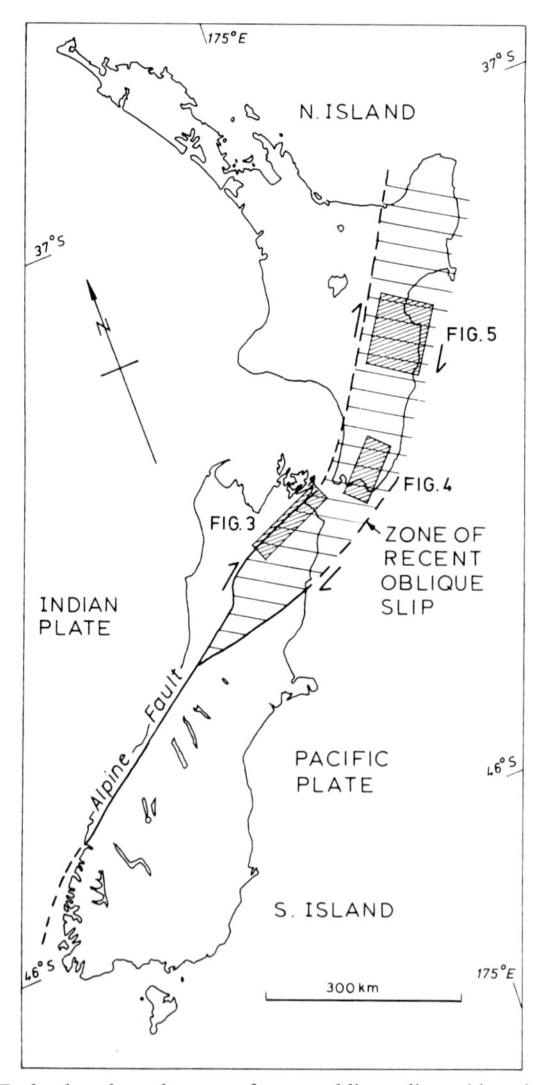

Fig. 1. Map of New Zealand to show the zone of recent oblique-slip and locations of Figs 3, 4 and 5.

Sediment sources are moved progressively downstream and the earliest appearance of sediment from a given source is isochronous throughout the basin. The latest appearance, however, is diachronous, getting younger downstream and in the direction of a strike-slip motion, from different sources on both the rising and sinking blocks. The stratigraphy of mixtures of sediment from different sources is shown diagrammatically in Fig. 2a, where sediment supplied from source rocks upstream on the rising block is mixed with sediment supplied locally from the sinking block. As source rocks on the rising block are carried downstream, their contributions successively disappear from the mixture.

A New Zealand example of Model 1 is shown in Fig. 3.

Model 2. Rivers on a sinking block flow parallel to the fault and in the opposite direction to strike-slip motion on the rising block (Fig. 2b).

Sediment is carried in the opposite direction to the strike-slip motion on the rising block. Sediment sources are moved progressively upstream. In contrast to Model 1, the earliest appearance of sediment from a given source is diachronous, getting younger upstream and in the direction of strike-slip motion of the rising block, while the latest appearance is isochronous.

In the same way, the stratigraphy of mixtures of sediments is the reverse of that in Model 1 (cf. Fig. 2a and 2b). As source rocks on the rising block are carried upstream, their contributions are *added* to the mixture.

A New Zealand example of Model 2 is shown in Fig. 4.

Model 3. Sediment dispersal is now transverse to the fault line and funnelled through a gorge in the sinking block (Fig. 2c).

Sediment from a particular source on the rising block has a restricted distribution in the proximal part of the fault-angle depression. In contrast to Models 1 and 2, both the earliest and latest appearances of such sediment are diachronous, getting younger in the direction of strike-slip motion on the rising block. Sediments from different sources occur in a stratigraphic sequence which is the same as the order of arrival and departure of those sources.

Mixing of sediment increases distally from the rising block, and the contribution from the downsinking block is minimal.

The Ruataniwha Depression is an example of Model 3 (Fig. 5).

Model 4. Rivers flow away from the rising block directly to the sea (Fig. 2d).

This model is a modification of Model 3, and all aspects of stratigraphy, diachronism, and mixing of sediment are the same. There is no contribution of sediment from a downsinking block. However, such tectonic depressions tend to be larger than the simple fault-angle depressions considered in Models 1 to 3, and may receive drainage along the strike from a wide variety of terrains.

The most striking New Zealand example of Model 4 is the Heretaunga Plains tectonic depression, Hawke's Bay, eastern North Island (Fig. 5). This large tectonic depression truncates a number of fault blocks and folds on its southern margin, and gathers a number of major rivers. Other examples occur on the West Coast of the South Island, forming extensive alluvial plains between the Alpine Fault and the Tasman Sea.

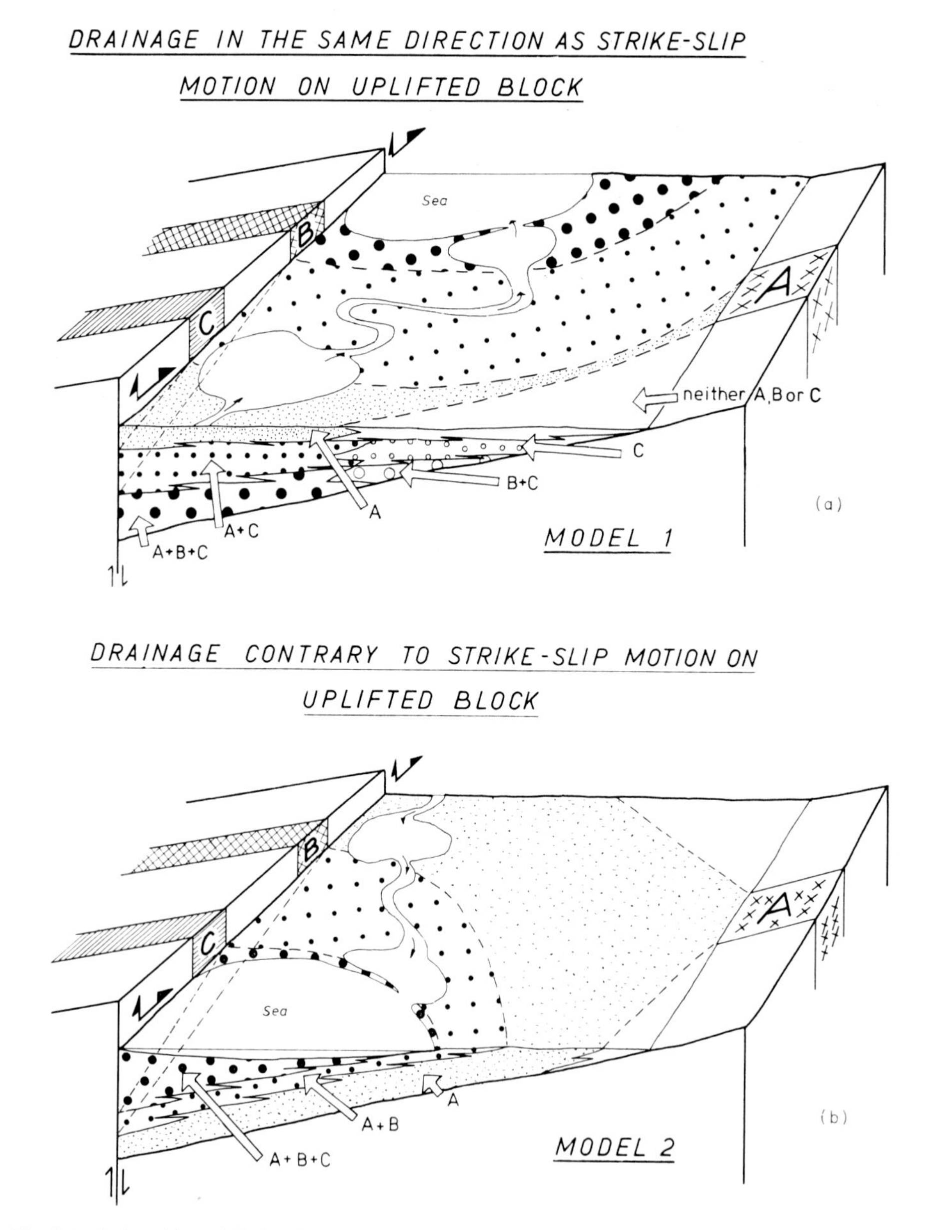

Fig. 2. Relationships of distinctive petrographic units in alluvial sediments in fault-angle depressions in oblique-slip zones. (a) Drainage in the fault-angle depression in the same direction as strike-slip motion on the rising block. (b) Drainage contrary to strike-slip motion on the rising block. (c) and (d). Drainage transverse to strike-slip motion on the rising block.

Influence of the shallow marine environment

In contrast to the non-marine environment, where transport is normally unidirectional, transport in the shallow-marine environment is often variable in direction. Thus the mixing of sediment from different sources will normally be greater in the marine environment. This is indicated in Models 1, 2 and 4 (Fig. 2).

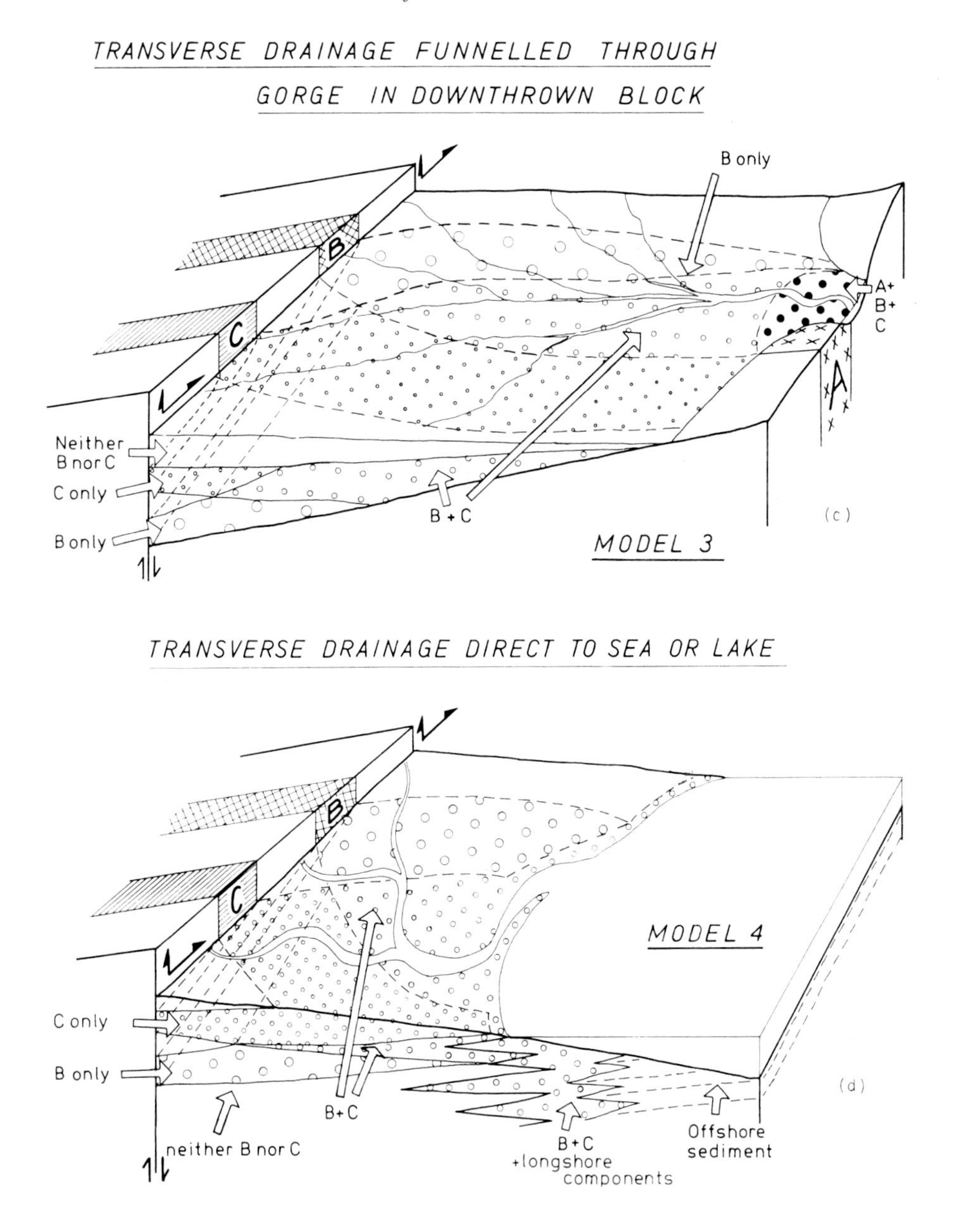

This homogenization of sediment will tend to destroy the diachronous relationships postulated for non-marine sediments, but should not destroy the vertical stratigraphic sequences that result from the movement of a sequence of source rocks past the marine basin.

Influence of complex tectonic histories on the models

The idealized sedimentational models described above are subject to many complicating influences. The chief of these is the complex tectonic history of active oblique-slip zones which can cause erosion to alternate with deposition. The shoreline

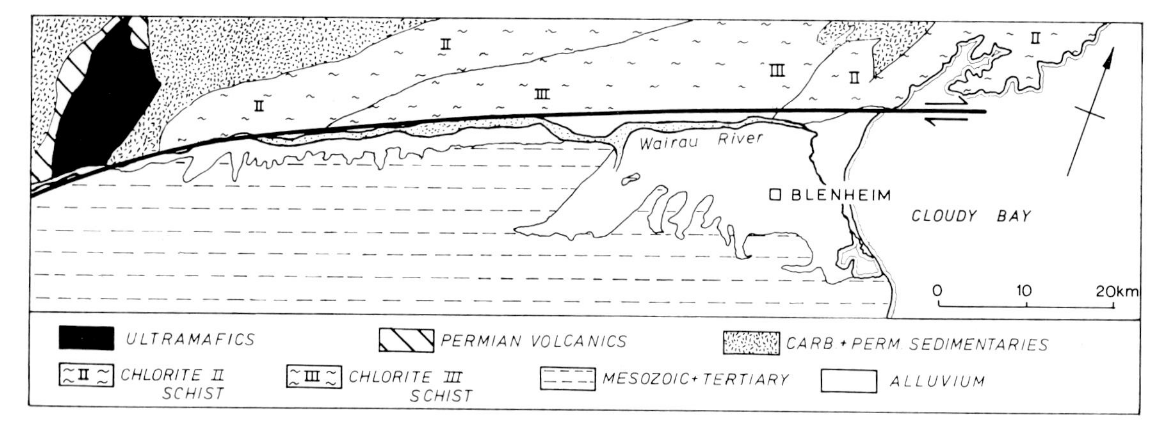

Fig. 3. An example of Model 1; location shown on Fig. 1. The north side of the Alpine/Wairau Fault is rising, and a wide variety of distinctive source rocks is being carried past the fault-angle depression.

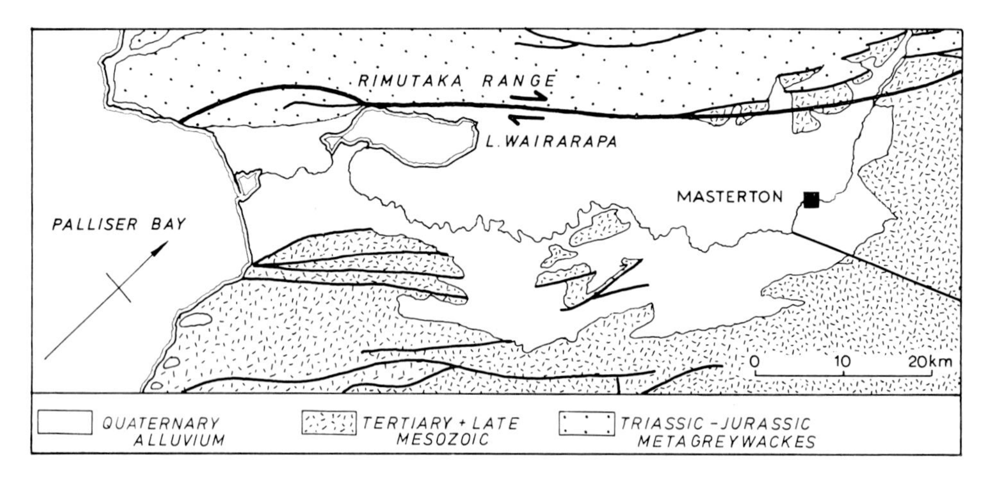

Fig. 4. An example of Model 2; location shown on Fig. 1. The rising Rimutaka Range on the north side of the Wairarapa Fault is composed of metagreywackes with little variation.

in Models 1, 2 and 4 is very sensitive to vertical movements. For example, during the Napier earthquake of 1931 an uplift of 2 m caused some 13 km² of the shallow tidal Ahuriri Lagoon to be drained (Marshall, 1933).

A mid-Pleistocene sequence which was deposited on an early version of the Heretaunga Plains (Fig. 5) is now exposed dipping at 10° in the coastal cliffs at Clifton, east of Hastings (Kingma, 1971). It contains many alternations of fluvial and shallow-marine sediment. In such a sequence, incursions of fluvial sediment can result from pulses of sediment supply following distant tectonic episodes or growth of new delta lobes. In addition, a Pleistocene sequence such as this one bears the effects of glacio-eustatic fluctuations (Kamp, 1977). Although, as indicated in the captions to Figs 4 and 5, the basement metagreywackes of the rising fault blocks are of generally uniform rock type, variations in the accessory lithologies such as chert do occur, and, on that basis, preliminary field inspection suggests that different conglomerate beds can be

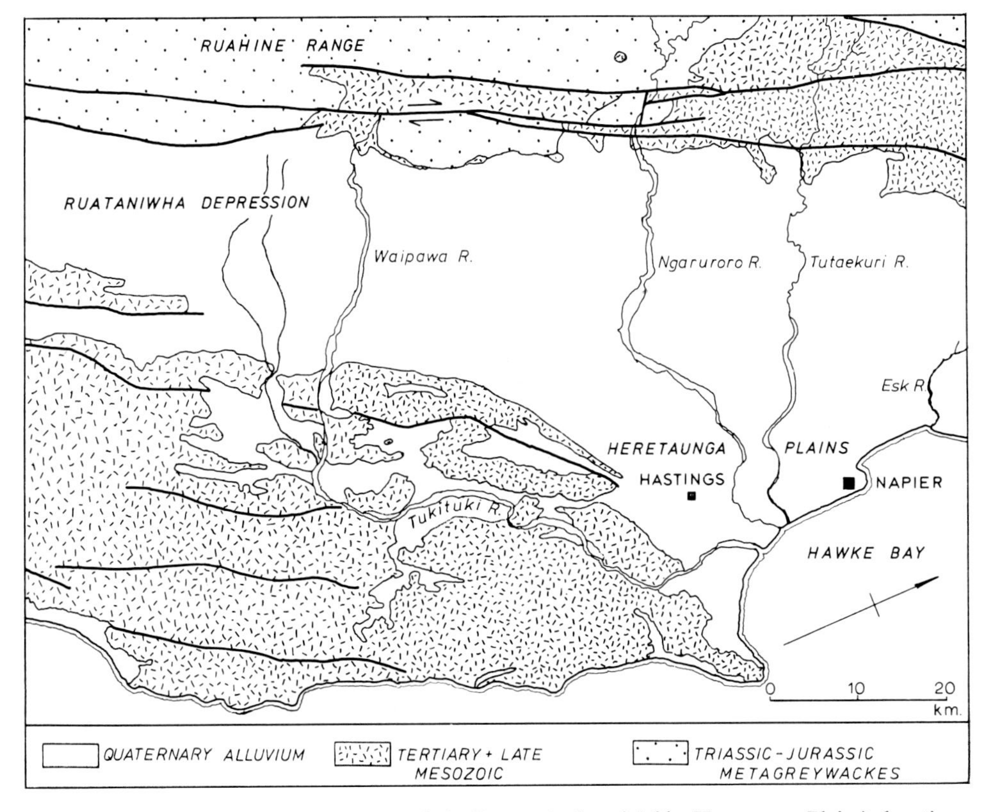

Fig. 5. Examples of Models 3 (the Ruataniwha Depression) and 4 (the Heretaunga Plains); location shown on Fig. 1. The rising Ruahine Range is composed of metagreywackes with little variation.

traced back to different basement fault blocks (K. B. Spörli, personal communication, 1979).

DISCUSSION

The models erected in this paper ideally would allow both the recognition of oblique-slip motion in an active fault zone, and a determination of its direction. It is known from the New Zealand stratigraphic record that oblique-slip has persisted for at least 3 m.y., and may have persisted for more than 20 m.y. (Ballance, 1976; Carter & Norris, 1976). Similar time spans have been suggested for other strike-slip zones, e.g. the San Andreas Fault (Blake *et al.*, 1978) and the Dead Sea Rift (Schulman & Bartov, 1978).

The models suggest that knowledge of three factors is necessary to determine the direction of motion of a specific oblique-slip zone:

1. distinctive petrographic units derived from different sources
2. demonstrably diachronous upper and/or lower limits to such units
3. direction of palaeo-flow.

Work at the necessary level of detail has not yet been done on any New Zealand examples to test the models. To date, stratigraphic sequences recording the unroofing of the Alpine Schists are known to be widespread in the South Island (e.g. Nathan, 1978), and lateral displacement of sediment sources by the Alpine Fault has been recorded (e.g. Cutten, 1975). However, with respect to (1) there may be difficulties in New Zealand because of the uniformity of the Permian to Jurassic basement rocks; in Figs 4 and 5 the basement rocks on the rising blocks show little variation. With respect to (2), non-marine rocks are notoriously difficult to date on a fine scale. In eastern North Island of New Zealand, however, all Neogene sequences contain rhyolitic tephra layers derived from the Taupo Volcanic Zone or its precursors; they make ideal, dateable isochronous markers.

In situations where the direction of oblique slip is already known, determination of one or two of the three factors listed above would give the others. This could then be an important predictor for a wide variety of stratigraphic, palaeogeographic and mineralogic relationships.

ACKNOWLEDGMENTS

I am grateful to P. B. Andrews, R. M. Carter, M. R. Gregory, R. J. Norris and H. G. Reading for their reviews of the paper.

REFERENCES

Ballance, P.F. (1976) Evolution of the Upper Cenozoic magmatic arc and plate boundary in northern New Zealand. *Earth Planet. Sci. Lett.* **28,** 356–370.

Blake, M.C., Campbell, R.H., Dibblee, T.W., Howell, D.G., Nilsen, T.H., Normark, W.R., Vedder, J.C. & Silver, E.A. (1978) Neogene basin formation in relation to plate-tectonic evolution of San Andreas fault system, California. *Bull. Am. Ass. Petrol. Geol.* **62,** 344–372.

Carter, R.M. & Norris, R.J. (1976) Cainozoic history of southern New Zealand: an accord between geological observations and plate-tectonic predictions. *Earth Planet. Sci. Lett.* **31,** 85–94.

Crowell, J.C. (1974) Origin of Late Cenozoic basins in southern California. In: *Tectonics and Sedimentation* (Ed. by W. R. Dickinson). *Spec. Publ. Soc. econ. Paleont. Miner. Tulsa.* **22,** 190–204.

Cutten, H.N.C. (1975) Provenance of the conglomerates of the upper Maruia Syncline pair and their bearing on the movement on the Alpine Fault. *Abstr., Geol. Soc. N.Z. Conf.*, Kaikoura.

Kamp, P.J.J. (1977) *Stratigraphy and sedimentology of conglomerates in the Pleistocene Kidnappers Group, Hawkes Bay.* Unpublished M.Sc. Thesis, University of Waikato, New Zealand.

Kingma, J.T. (1971) Geology of Te Aute Subdivision. *Bull. geol. Surv., N.Z.*, **70**.

Marshall, P. (1932) Effects of earthquakes on coast-line near Napier. *N.Z. J. Sci. Techn.,* **B15,** 79–92.

Nathan, S. (1978) Upper Cenozoic stratigraphy of South Westland, New Zealand. *N.Z. J. Geol. Geophys.* **21,** 329–361.

Reading, H.G. (1980) Characteristics and recognition of strike-slip fault systems. In: *Sedimentation in oblique-slip mobile zones* (Ed. by P. F. Ballance & H. G. Reading). *Spec. Publ. int. Ass. Sedim.* **4,** 7–26.

Rynn, J.M.W. & Scholz, C.H. (1978) Seismotectonics of the Arthur's Pass region, South Island, New Zealand. *Bull. geol. Soc. Am.* **89,** 1373–1388.

Schulman, N. & Bartov, Y. (1978) Tectonics and sedimentation along the Rift Valley. *10th. Int. Congr. Sediment. Guidebook Pt. II: Postcongress, Israel,* pp. 35–94.

Spörli, K.B. (1980) New Zealand and oblique-slip margins: tectonic development up to and during the Cainozoic. In: *Sedimentation in oblique-slip mobile zones* (Ed. by P. F. Ballance & H. G. Reading). *Spec. Publ. int. Ass. Sedim.* **4,** 147–170.

Walcott, R.I. (1978) Present tectonics and late Cenozoic evolution of New Zealand. *Geophys. J. R. astr. Soc.* **52,** 137–164.

Spec. Publ. int. Ass. Sediment. (1980) **4**, 237–265

Offshore sedimentary basins at the southern end of the Alpine Fault, New Zealand

R. J. NORRIS *and* R. M. CARTER

University of Otago, Dunedin, New Zealand

ABSTRACT

Cainozoic sedimentation in southwest New Zealand was strongly influenced by concurrent tectonic activity from the late Eocene onwards. Fault-bounded flysch basins were filled with thick wedges of Oligo-Miocene redeposited and associated deep marine sediments. Changes in tectonic regime at about 10 m.y. BP (late mid-Miocene) led to folding and uplift of the sediments. Due to this mid-Miocene and later tectonism it is difficult to reconstruct the original size, shape or disposition of the onland Oligo-Miocene basins. Seismic profiles of the offshore region, in the head of the Solander Trough, show the presence of two major basins, the Balleny and Solander Basins, of similar size and style to the onshore basins, and separated from each other across the offshore extension of the Moonlight fault system. Four seismic units are recognized in each basin, with stratigraphic control provided by HIPCO's 3628 m deep PARARA-1 exploration well. Regional time-isopach maps for the lower seismic units (Eocene to early Miocene) show both basins were filled largely from their western margins; since no source regions exist to the west of either basin today, it is inferred that they have been shifted northeast by lateral movement on the Alpine and Moonlight fault systems. Time-isopach maps for the upper seismic units (late Miocene to Recent) show the Balleny Basin to have received a thin, mainly pelagic drape, whereas the Solander Basin has been filled by a thick shelf sequence, prograding from off the southern end of the New Zealand landmass.

The development of the basins of southwest New Zealand is related to the behaviour of the Fiordland microplate, a rigid block of continental crust located between the Moonlight and Alpine fault systems. During the Eo-Oligocene, extensional oblique-slip (transtension) on these fault systems led to the subsidence and final submergence of Fiordland. Compressive oblique-slip (transpression) since the middle Miocene has caused a northwards movement of the microplate, with associated deformation and uplift reaching a maximum in the north and decreasing in intensity southwards. The main modern drainage systems have thereby developed from north to south, more or less along the line of the Moonlight tectonic zone.

INTRODUCTION

The Indo-Australian/Pacific plate boundary crosses the New Zealand continental plateau as the Alpine oblique transform fault system (e.g. Walcott, 1978). Plate tectonic arguments, based on marine magnetic anomalies in the southwest Pacific Ocean,

0141-3600/80/0904-0237$02.00

date the inception of this plate boundary during the Eocene (Molnar *et al.*, 1975; Weissel, Hayes & Herron, 1977).

Between the Late Cretaceous and Eocene the New Zealand continent was emergent but of generally low relief; shallow seas transgressed onto the landmass, restricting non-marine, coal measure sedimentation to a progressively narrowing central belt (Crooks & Carter, 1976; Norris, Carter & Turnbull, 1978). In the latest Eocene and Oligocene this regionally uniform marine transgression was disrupted by tectonic activity in western New Zealand, followed in the early Miocene by arc-volcanism in northern North Island (Pilaar & Wakefield, 1978; Nelson & Hume, 1977). Clearly, movement had commenced on the Alpine plate boundary. The subsequent tectonic history of the plate boundary inferred from geology is in close agreement with that inferred from geophysics (Ballance, 1976; Carter & Norris, 1976).

In southwestern South Island, southeast of the Alpine Fault, Cretaceous-Eocene coal measure facies are sharply followed by thick flysch sequences of latest Eocene and younger ages. Post-Eocene sedimentation occurred in active, fault-bounded marine basins, often at bathyal depths. Individual basins were often only a few to a few tens of kilometres across, but regionally they formed a linked sedimentation system controlled by the Moonlight Tectonic Zone. The most recent manifestation of the Moonlight Tectonic Zone is a system of Late Miocene to Holocene high-angle reverse faults which extend from north of Lake Wakatipu southwestwards to the coast at Te Waewae Bay (Turnbull *et al.*, 1975). The modern intermontane Te Anau and Waiau basins lie respectively to the northwest and southeast of this Moonlight Fault System (Fig. 1).

Due to mid-Miocene and later tectonism, resulting from major transpression along the main Alpine transform (Molnar *et al.*, 1975), it is difficult to reconstruct the original size, shape or disposition of the Oligo-Miocene flysch basins. The Moonlight Fault System joins the Fiordland Boundary Fault System near Lake Monowai (Figs 1, 2 and 4), and continues offshore as the boundary zone between the Balleny and Solander Basins (Figs 4, 5). Seismic profiles of the offshore region show a pattern of fault-bounded basins similar to that envisaged for the onland examples prior to their eversion, with the intensity of deformation of the sedimentary sequence decreasing southwestwards. Basinal sedimentation continues offshore today in the head of the Solander Trough.

Though few detailed studies have been published, the onland sedimentary history is now moderately well understood. In contrast, there are virtually no published data available for the offshore region. We present in this paper the results of a preliminary analysis of oil exploration sesimic data from some 18 000 square miles of ocean offshore from southwestern South Island, interpreting the data in the light of our knowledge of nearby onland geology.

Sources of data

During 1970–1, seismic data collected by Hunt International Petroleum Company (HIPCO) from the offshore Te Waewae Bay and Solander Trough regions (Fig. 2) were used to locate wildcat well PARARA-1 in 484 feet of water in the head of the Solander Trough. In late 1977 more than 3000 km of seismic data (Fig. 2) and the log of PARARA-1 became available on open-file (HIPCO, 1970, 1971, 1976).

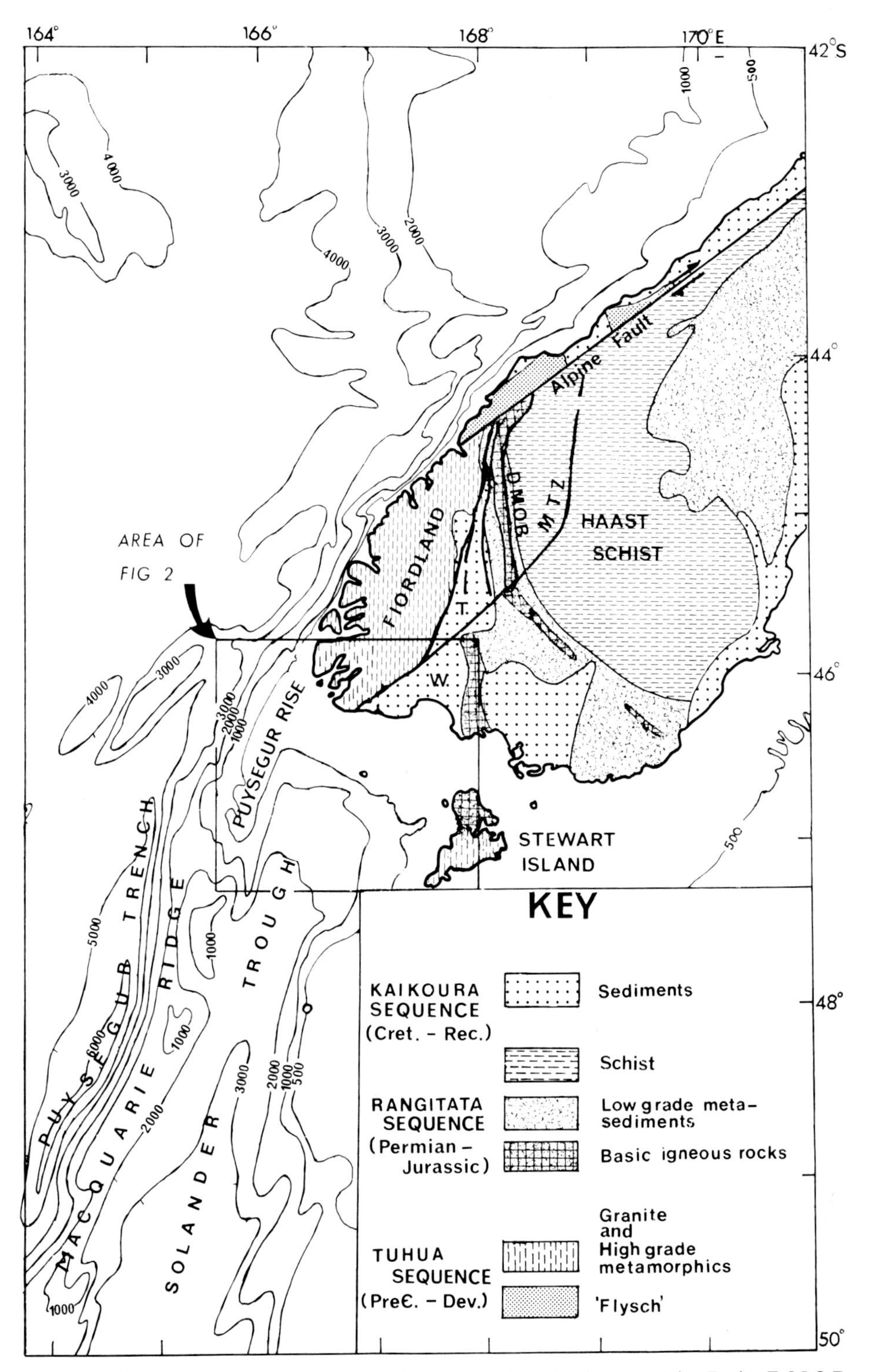

Fig. 1. Locality map of southwestern South Island. T: Te Anau Basin; W: Waiau Basin; D.M.O.B.: Dun Mountain Ophiolite Belt; M.T.Z.: Moonlight Tectonic Zone. Sequence nomenclature after Carter *et al.* (1974).

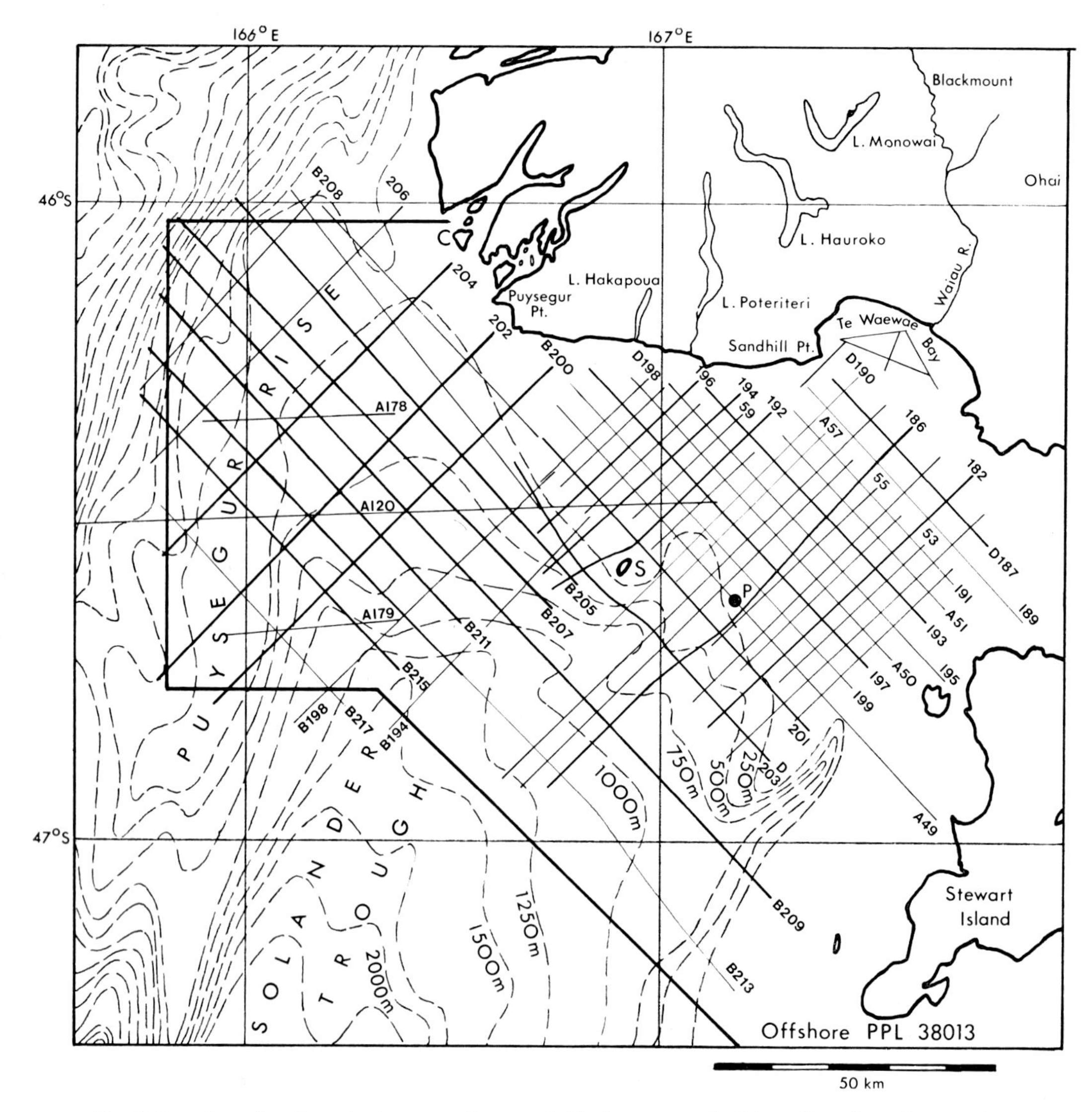

Fig. 2. Location of major seismic survey lines available from area of offshore Petroleum Prospecting Licence 38013, southwestern South Island. Lines reproduced in this paper (Figs 6–10) shown in heavier ink. C: Chalky Island; S: Solander Island; P: PARARA-1 well.

Most of the seismic cover is 12-fold CDP with an Aquapulse source, a lesser amount being in single or several channel form. Though the seismic data is available in field processed form only (i.e. after preliminary stacking and deconvolution), in most cases interpretation is clear, with penetration of stratified sequences to depths in excess of 4 s two-way travel time. Some areas of intense faulting, particularly along basin margins, are so overlain with diffraction hyperbolae that unambiguous interpretation is not possible in the absence of migrated sections.

Two main scales of seismic line are available. We have prepared typical cross-sections (Figs 6–10) and time–isopach maps (Fig. 12a–d) largely from an analysis of the smaller scale B and D lines of Fig. 2. The larger scale A lines were used for checking important stratigraphic or structural detail. In addition, we show all regional structural features on a composite 'basement' map (Fig. 4).

REGIONAL STRATIGRAPHIC HISTORY

The tectonic and sedimentary complexity of the Neogene succession in southwestern South Island makes it difficult to devise a simple and economical nomenclature. In the summary of regional stratigraphy presented below (cf. Table 1) we have therefore adopted a primarily chronostratigraphic subdivision. Our summary is based on more detailed work described in McKellar (1956), Bowen (1964), Wood (1969), Sutherland (1969), Mutch (1972), Landis (1974), Turnbull *et al.* (1975), Carter & Lindqvist (1977), and Carter & Norris (1977a, b); less detailed, regional discussions are presented by Katz (1968), Carter & Norris (1976) and Norris *et al.* (1978).

Cretaceous and Early Cainozoic non-marine strata

Locally, as at Ohai (Fig. 2), the basal sediments in southwest South Island are thick sequences of immature, mid-late Cretaceous coal measures, located in post-Rangitata fault-angle depressions (Bowen, 1964). At Sand Hill Point on the south coast of Fiordland, a non-marine, red bed fanglomerate sequence over 300 m thick may be as old as early Cretaceous. These older non-marine beds are followed, unconformably at Ohai, by younger Eocene–Oligocene coal measures of the Nightcaps Group, which lap out of the fault-bounded troughs to form a regional blanket of non-marine deposits unconformably overlying weathered basement (Wood, 1966).

During the Late Cretaceous and early Cainozoic, therefore, initial relief on a tectonically stable post-Rangitata landscape was reduced by erosion and infilling of the block-faulted topography. By late Eocene time peneplanation was well advanced over wide areas, particularly in the northeast.

Eo–Oligocene tectonism and basin formation

The mature Eocene land-surface was abruptly affected by a phase of tectonism in the latest Eocene and early Oligocene. Rapid marine incursion was accompanied by deposition of thick breccia, arkosic sandstone and flysch in fault-bounded, deep marine basins. In the west, at Chalky Island, the lower mass-emplaced sediments pass upwards into nanno-chalk marls; the Balleny Group, though almost 1 km thick, entirely falls within a single faunal zone of early Oligocene age. Further east, in the fault-controlled Waiau and Te Anau Basins, Oligocene redeposited marine sequences over 2 km thick occur (Fig. 3), but lack the chalk marl facies of Chalky Island.

Basin margin sequences of Oligocene age are known, particularly along the western side of the Te Anau and Waiau Basins (Fig. 3). Shallow water, calcareous arkosic sands, and molluscan, algal and bryozoan limestones are typical lithologies (Point Burn and Tunnel Burn Formations; McKellar, 1956). Sequences are thin, usually several tens of metres, and contain Oligocene paraconformities (cf. Hyden, 1975; Carter & Norris, 1977b).

Early Miocene depression

Basin-margin sequences on the west side of the Waiau and Te Anau Basins are followed abruptly in the early Miocene by deep marine mudstones (Garden Point Formation) which grade upwards into early to middle Miocene flysch of northern

Table 1. Summary lithostratigraphic table for Kaikoura Sequence deposits of southwestern South Island. Unconformities are represented by oblique hachuring; asterisks in left-hand column refer to phases of tectonism

	AGE	S. FIORDLAND	TE ANAU	WAIAU	S. PLAINS	SOLANDER	BALLENY
	PLEIST.	Terrace deposits	Glacial deposits	Terrace deposits		D	4
	PLEIST.	Shelly conglomerate (high terrace)	Prospect Fm.	Te Waewae Fm.	Terrace deposits	D	4
	PLIO.			Waikoau Fm. Port Craig beds		C	3
★ 10 my	MIOCENE		? ? ?	Prospect Fm.		B	2
★ 15 my	MIOCENE	Deep marine mdst, marl, cglt & minor flysch (Wairaurahiri Syncline)	Deep marine mdst & flysch (similar to the Waiau Basin) BASIN MARGIN Borland Fm. Garden Pt. Fm.	Duncraigen Fm. Monowai Fm. Borland Fm. McIvor Fm.	Forest Hill Lst.	B	2
★ 22 my	OLIG.	? ? ? chalk Balleny Gp.: breccia	Tunnel Burn Fm Point Burn Sst breccia	Blackmount Fm. breccia	Chatton Fm. Winton Hill Fm.	A	1
★ 38 my	EOCENE	Puysegur Fm.		Nightcaps Gp.	(=Mako c.m.)	A	1
	CRET.	(?parts of Puysegur Fm.)		Ohai Gp. Sand Hill breccia			

provenance (Table 1); thus further regional depression, including former basin margins, was accompanied by uplift of an area northeast of Fiordland. Meanwhile, in the central Waiau Basin at Blackmount, McIvor calcflysch accumulated as carbonate fans derived from early Miocene bryozoan shelf limestones to the east, followed by further pulses of northerly derived terrigenous flysch of the Borland Formation (Table 1; Fig. 3).

Middle Miocene shoaling

During the late early and middle Miocene the Waiau Basin sequence at Blackmount exhibits a shoaling trend, interrupted by brief episodes of renewed subsidence. A northerly derived deltaic wedge (Monowai Formation; Carter & Norris, 1977a) was followed by further flysch (Duncraigen Formation) which again shoals to deltaic deposits (Prospect Formation; Table 1).

Mid-to-late Miocene unconformity

The youngest formations of the Blackmount column are of late Middle Miocene age; they are extensively folded and faulted and overlain with right-angle unconformity by Pleistocene deposits. Further south, in the southern Waiau Valley, late Middle Miocene strata are overlain by late Miocene sediments with angular unconformity (Wood, 1966). Earliest late Miocene (Tongoporutu Zone) sediment is

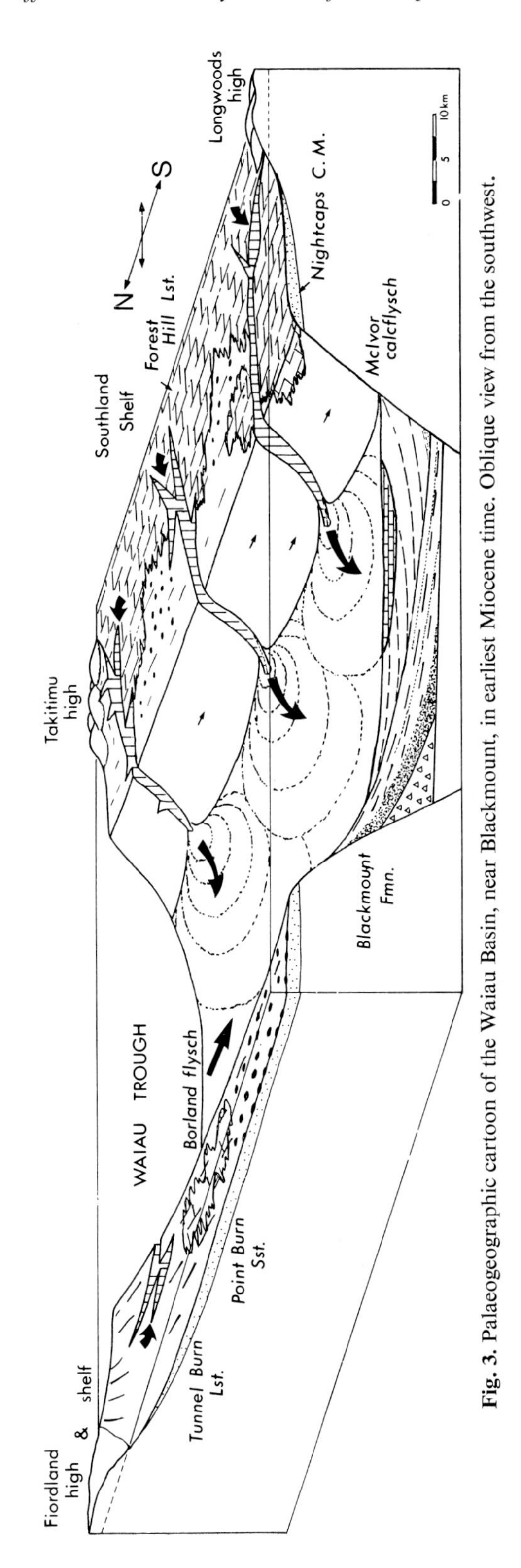

Fig. 3. Palaeogeographic cartoon of the Waiau Basin, near Blackmount, in earliest Miocene time. Oblique view from the southwest.

reduced in thickness or absent across this unconformity, indicating another regionally significant tectonic pulse.

Late Miocene to Recent sedimentation

Following the late Middle Miocene tectonism, marine sedimentation was restricted to areas within and southwards of the southern Waiau Basin. In the Te Anau and northern Waiau Basins, fluvial molasse, and later glacial, deposits constitute the sedimentary record up to the Recent. In the southern Waiau Basin, coarse-grained shallow marine sediments of latest Miocene age (Port Craig beds, Table 1), pass up into deep marine mudstones with bathyal faunas (Waikoau Formation), which are in turn overlain by shallow marine sediments of Pliocene age (Te Waewae Formation). Further west, in the Wairaurahiri Syncline (Fig. 4), thick late Miocene mudstones and siltstones occur (Wood, 1960).

Clearly, in the south, further active basin depression took place during the latest Miocene, but except near major faults this has been followed by relative stability during the final regressional sedimentary phase.

STRUCTURE AND SEISMIC STRATIGRAPHY

The offshore tectonic and summary time-isopach maps (Figs 4, 5), and the seismic profiles (Figs 6–10), reveal that two major sedimentary basins occur in the offshore area. The two basins, here named the Solander and Balleny Basins, are separated by a major zone of faults and basement uplifts which, when extrapolated to the coast, corresponds to the Hauroko fault and associated faults of Wood (1966). This group of faults, which represents the southwest continuation of the Moonlight Fault System (Norris *et al.*, 1978; cf. Fig. 13), is referred to hereafter as the Fiordland Boundary Fault Zone.

Both offshore sedimentary basins are complex in detail, with intrabasinal highs and sub-basins. Though the dominant faulting is oriented NE–SW, both basins are bounded to the north by zones of NW-oriented cross-faults; a significant consequence is that the onland Waiau Basin is seen to be clearly distinct from the offshore Solander Basin (cf. Katz, 1974).

In addition to the structural interpretation of the seismic profiles, presented in Fig. 4, the sedimentary fill of the basins has been divided into a number of unconformity-bounded seismic units, following principles discussed by Mitchum, Vail & Thompson (1977). Because of the difficulty of tracing these units across the Fiordland Boundary Fault Zone, a different designation is used for the seismic units recognized in the Balleny (Seismic Units 1–4) and Solander (Seismic Units A–D) Basins, even though the same number of units is recognized in each.

Solander Basin

Structure. The Solander Basin is bounded to the west by the Big River Highs and the Fiordland Boundary Fault Zone, to the north by the Hump Ridge–Midbay High and to the east by the Stewart Island shelf (Fig. 4). Southwards the basin opens out into the head of the Solander Trough.

The Hump Ridge–Midbay High is a complexly faulted basement high which

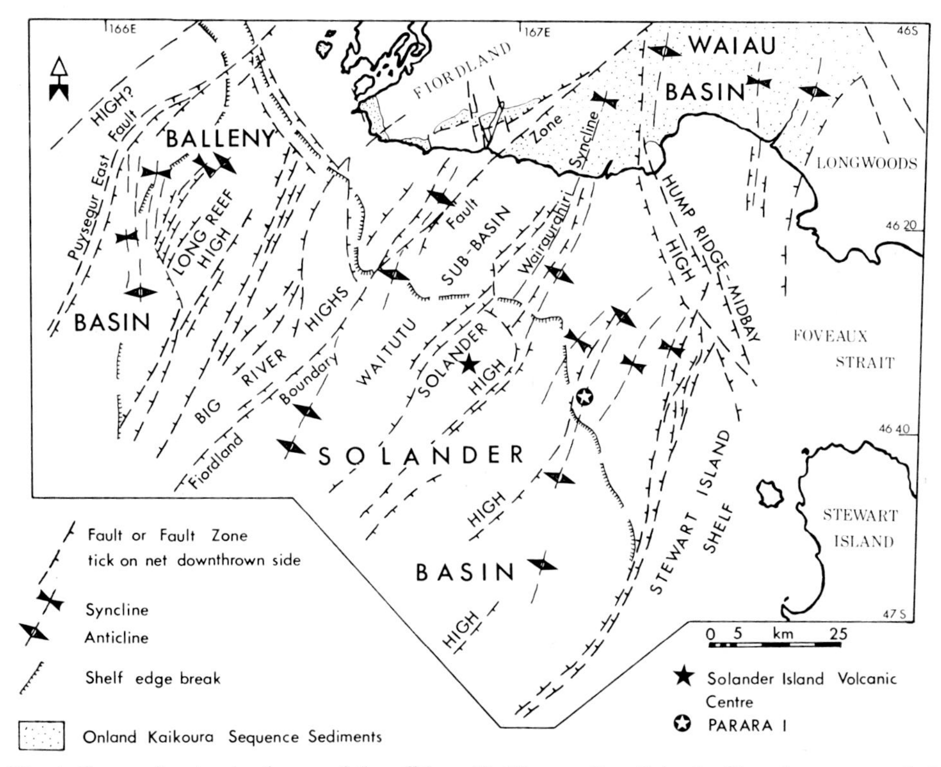

Fig. 4. Composite structural map of the offshore Te Waewae Bay–Solander Trough area, compiled from the seismic lines shown in Fig. 2. Structures shown include those affecting only the lower units.

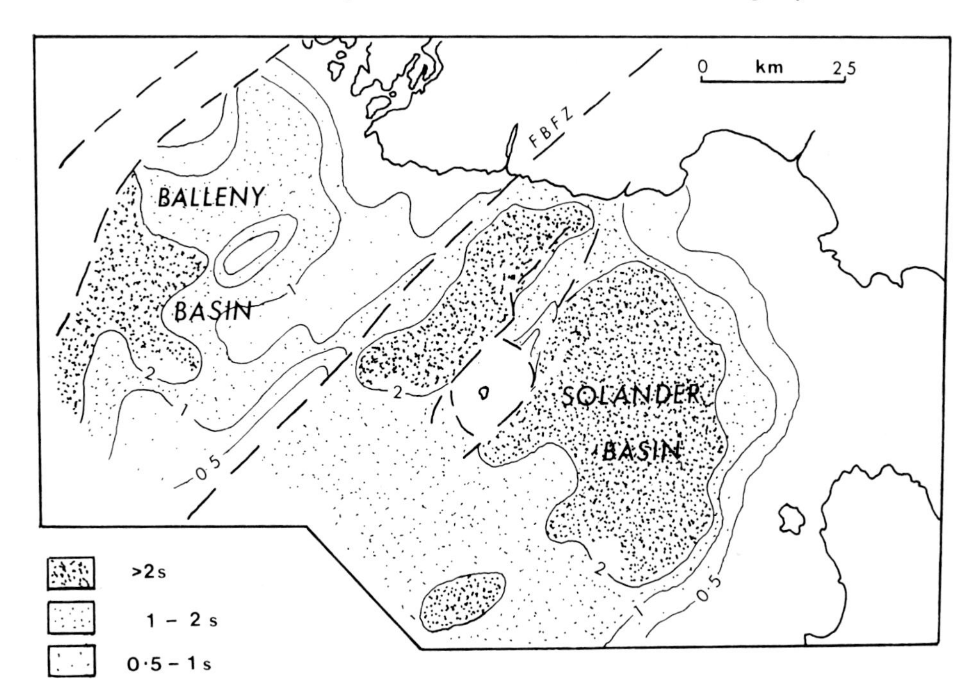

Fig. 5. Time–isopach map for total sediment thicknesses in the Te Waewae Bay–Solander Trough area.

separates the Solander Basin from the Waiau Basin, and against which the sediments of the Solander Basin lap out northeastwards. To the northwest, the high extends onland as Hump Ridge, formed of folded and uplifted early Cainozoic sediments underlain by crystalline basement. Similarly, the Stewart Island shelf comes to the surface in the crystalline rocks of Stewart Island.

The dominant structure within the Solander Basin is the Solander High, an elongate zone of uplifted basement bounded by NE–SW trending faults, and separating the Waitutu Sub-basin from the rest of the Solander Basin. To the north and south the two basins merge, and part of the sedimentary fill is exposed on land in southern Fiordland (cf. Wood, 1960). The Solander High is buried by sediments for most of its length except where a steep-sided bathymetric feature rises above sea level as the andesitic volcanics of Solander Island. Profiles close to the island suggest that intrusive igneous rocks form a 15 km diameter subcircular edifice above and to the east of an older basement high, the intrusives cutting steeply through almost all of the sedimentary sequence (Fig. 6). No radiometric ages dating the volcanic activity are available, but an age of Pleistocene has generally been inferred because of the freshness of the rock (Harrington & Wood, 1958).

A number of open folds trending 10–30° east of the dominant faults are developed within the Solander Basin, and one of these folds was the target for the PARARA-1 exploration well. Many of the anticlines are associated with faults on their western side which show an opposite sense of uplift in the sedimentary rocks compared with their offset of the basement surface (Fig. 7, profile D197), suggesting that these faults have been re-activated in a reversed sense.

Seismic Stratigraphy. The sedimentary fill of the Solander Basin locally exceeds 4 s two-way travel time, and is in excess of 2 s over a large area of the basin (Fig. 5). Insufficient velocity data exist to translate travel times to accurate thicknesses, but using the velocities measured in PARARA-1, a thickness range of 2–6 km is indicated, similar to thicknesses measured in the onland Waiau Basin (Norris *et al.*, 1978). The four major seismic units recognized within the sediment fill are designated A, B, C and D in order of superposition (Figs 7, 8).

Seismic Unit A is markedly variable in thickness, reflecting infilling of an irregular and faulted basement surface (e.g. Fig. 7). Except in the deepest parts of the section, where reflections become weak and may be swamped in the background, the Unit A-basement contact is conspicuous. Locally, the unit may be subdivided into A1 and A2 on the basis of internal seismic signature, with A1 showing very poorly developed, discontinuous and irregular reflecting horizons (e.g. Fig. 7, profile D197).

The boundary between Units A and B is commonly marked by a change from short discontinuous reflections in A to longer, better developed reflectors in B. In places, however, discordances occur between A and B, suggesting a prior tilting of A, and faults cutting Unit A do not always extend into B (e.g. Fig. 7, profile D201).

The B–C boundary corresponds over large areas of the basin with a marked unconformity (e.g. Fig. 7, profile D197; Fig. 8, profile D194), Unit B having been tilted, folded and eroded prior to the deposition of C. This boundary can be traced onland from the head of the Waitutu Sub-basin, and corresponds to the widespread late Mid-Miocene unconformity mapped by Wood (1960, 1968).

Seismic Unit D is a series of seaward dipping foresets which culminate in the present day shelf edge (Fig. 8, profile D182) and pass further seawards into long,

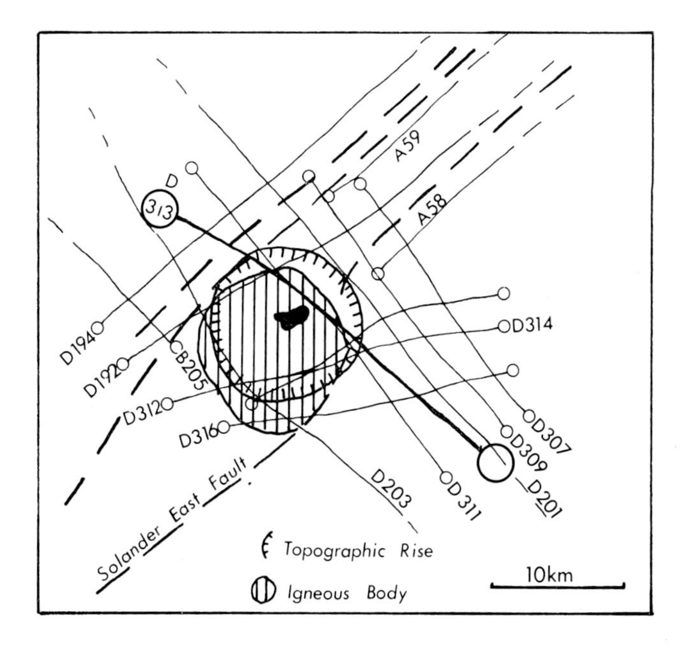

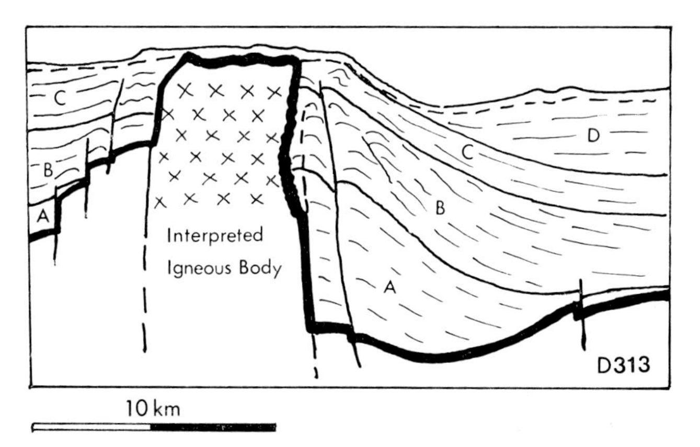

Fig. 6. Map of area around Solander Island and Seismic reflection profile along line D 313.

continuous reflecting units of the upper slope; the underlying Unit C has a similar seismic signature to the latter. Locally within D, the reflectors lose their regular seawards dip, dissolving into irregular and upward concave reflections which probably represent the site of filled shelf-incising channels and canyons (cf. Mitchum, Vail & Sangree, 1977). In the Waitutu Sub-basin, tilting against the Fiordland Boundary Fault Zone appears to affect even the youngest deposits, so the shelf here is largely erosional, with only the thinnest cap of shelf-wedge Unit D (Fig. 8, profile D194).

Balleny Basin

Structure. The Balleny Basin is structurally more complex, but of broadly similar

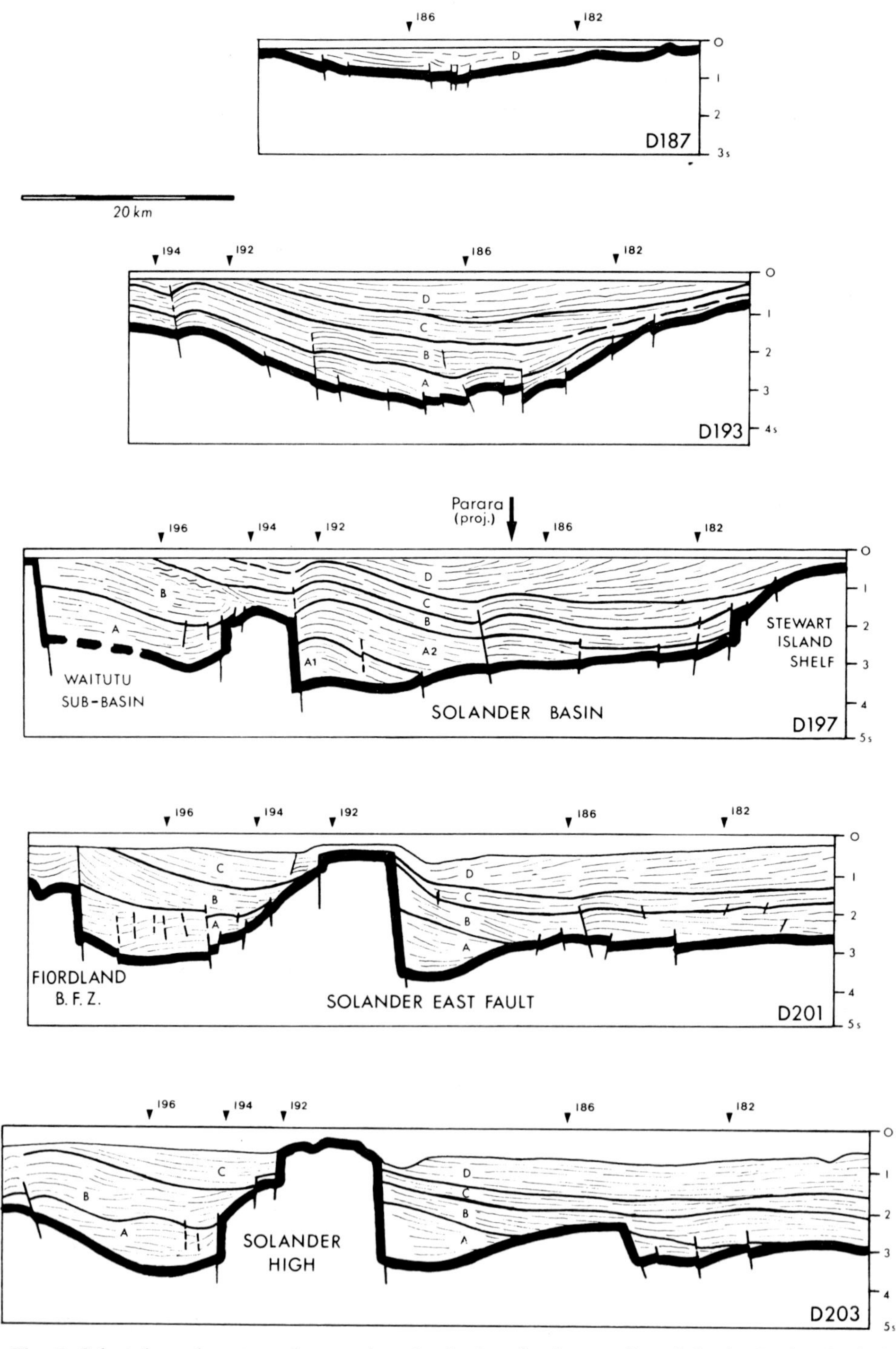

Fig. 7. Selected northwest–southeast oriented seismic reflection profiles, Solander Basin. Vertical exaggeration approximately $\times 2\frac{1}{2}$.

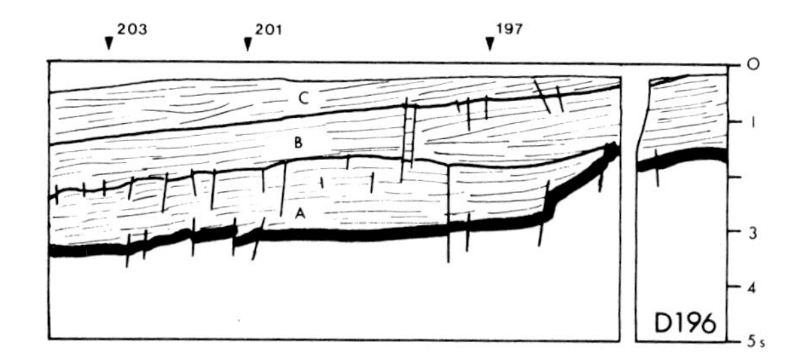

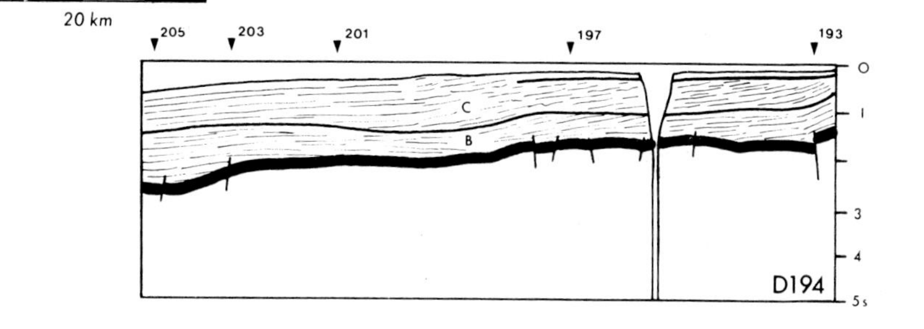

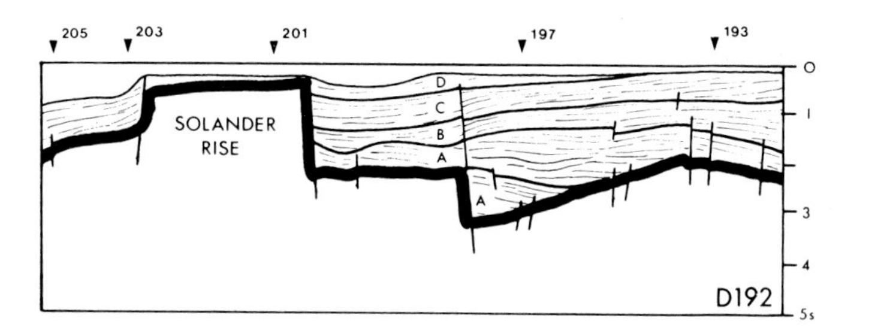

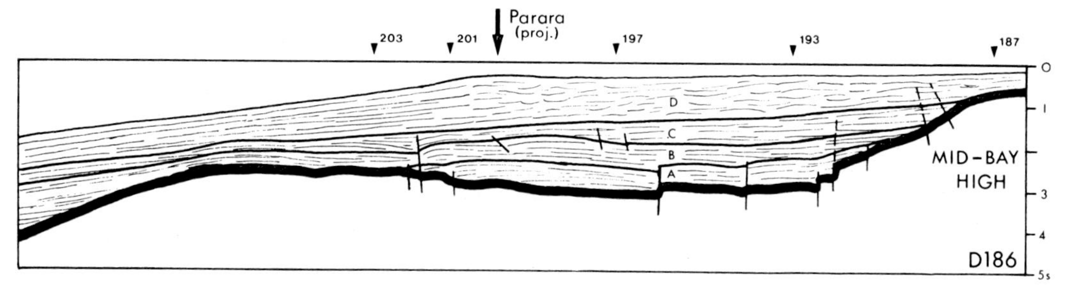

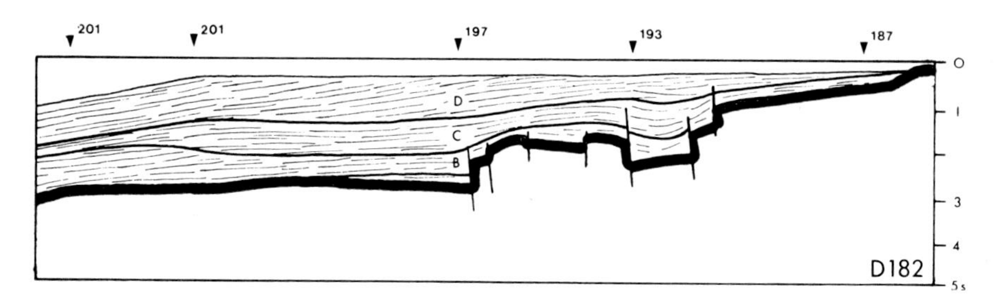

Fig. 8. Selected northeast–southwest oriented seismic reflection profiles, Solander Basin. Vertical exaggeration approximately $\times 2\frac{1}{2}$.

tectonic style to the Solander Basin (Fig. 4). On the western side of the basin the Puysegur Rise is a flat-topped submarine bank whose upper surface is largely cut across the upturned edges of eastward dipping strata of the Balleny Basin; the Rise carries only a thin veneer of modern sediment (Fig. 9). Under the western edge and slopes of the Puysegur Rise the Balleny Basin sediments abut seismically opaque material which is probably basement and may mark the site of an original basement 'High' (Fig. 9, profile B213); seismic signatures are poorly developed under the

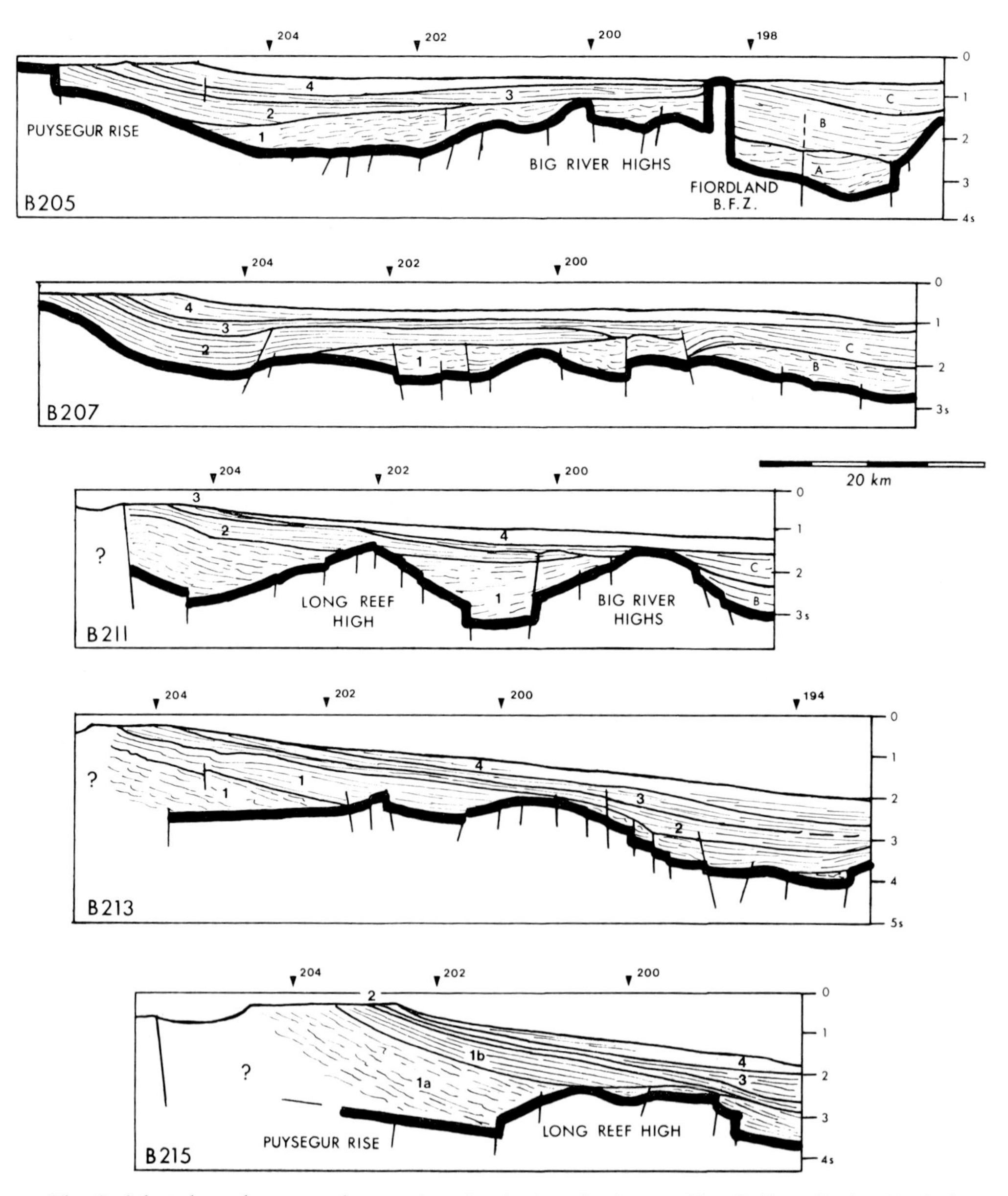

Fig. 9. Selected northwest–southeast oriented seismic reflection profiles, Balleny Basin. Vertical exaggeration approximately $\times 2\frac{1}{2}$.

western side of the Puysegur Rise, but a fault has been mapped between sediments and basement (Fig. 4). A number of NW trending faults form the northern margin of the basin and separate it from Fiordland, while to the south the basin opens out and merges with the Solander Basin.

A number of faults also occur within the basin, largely following a trend of NNE, about 10–20° west of the Fiordland Boundary Fault Zone. These faults define a series of highs and sub-basins (Figs 4, 9). The important highs are the conspicuous Long Reef High in the centre of the basin, and the Big River Highs along the Fiordland Boundary Fault Zone. In places the eastern boundary of the basin is marked by a clearly recognizable major fault (e.g. Fig. 9, profile B205), in others by a zone of faults and basement uplifts (Fig. 9, profile B213) which give rise to a confusing zone of diffractions on the seismic lines. The vertical separation along this fault zone varies from bringing basement right to the surface (Fig. 8, profile B211) to hardly any change in basement level at all (Fig. 9, profile B207). The dip of the sediments generally increases east of the fault zone.

Seismic Stratigraphy. The thickness of sediments is reflected by two-way travel times exceeding 3 s in the western part of the basin. However, the eastward tilting but westward thickening of the sediments (Fig. 9) makes it difficult to measure a composite stratigraphic thickness. As in the Solander Basin, four seismic units are recognized,

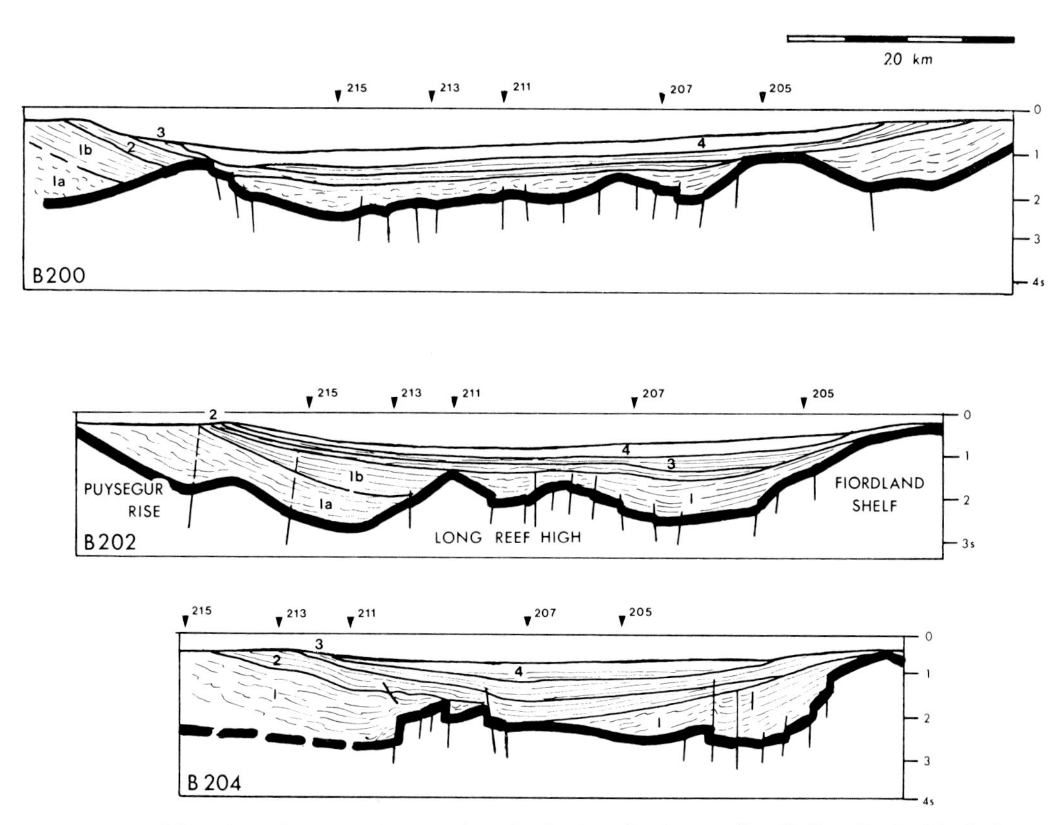

Fig. 10. Selected northeast–southwest oriented seismic reflection profiles, Balleny Basin. Vertical exaggeration approximately $\times 2\frac{1}{2}$.

designated 1, 2, 3 and 4 in ascending order; direct correlation cannot be demonstrated between the two basins.

Criteria for the recognition of Seismic Units 1–4 are similar to those used to characterize Units A–D of the Solander Basin. Unit 1 may in places be subdivided into a lower subunit 1a, with discontinuous and irregular reflections, and subunit 1b, with well developed reflectors. Unit 2 commonly overlaps Unit 1 across uplifts (e.g. Fig. 9, profile B205).

The boundary between Units 2 and 3 is a marked unconformity, particularly conspicuous near the Fiordland Boundary Fault Zone where uplift and tilting has resulted in Units 1–2 being truncated and overlapped by Unit 3 (Fig. 9, profiles B205, B211). Unlike Unit D of the Solander Basin, Unit 4 does not form a prograding shelf wedge, but appears as a thin sedimentary blanket; despite its probable youth, it appears to be tilted eastwards on the Puysegur Rise (Fig. 9).

Equivalence of Units between Balleny and Solander Basins

Four seismic lines pass across the full width of both the Balleny and Solander Basins (Fig. 2), but it is difficult to trace any seismic units between the two basins even though individual reflectors are apparently locally continuous across the Fiordland Boundary Fault Zone. The thickness and facies changes in all units across the boundary fault zone indicate that the units developed in the two basins are not directly equivalent, and suggests that lateral movements on the intervening fault zone may have juxtaposed contrasting sequences. Major vertical movements on the Fiordland Boundary Fault Zone clearly took place prior to the deposition of Unit 3, and continued afterwards, and uplift of the western margin of the Waitutu Sub-basin still continues. Late Cainozoic activity in fact appears to be a feature of the western margins of both the Balleny and Solander Basins, as the greatest angular discordance in both basins is between the tilted strata there and the modern erosion surface.

Stratigraphy of PARARA-1

PARARA-1 was drilled to a total depth of 11,883 ft (3628 m). The well stratigraphy conforms to the known regional pattern (Fig. 11). All depths are quoted sub-seafloor.

Seismic Unit A is dominated by poorly sorted, immature, biotitic, feldspar-quartz sands of granitic and high grade metamorphic provenance. The middle and upper parts of the unit contain seams of coal. The first marine fossils appear in the upper parts of Unit A, at around 9100 ft (2780 m) and are shallow marine foraminifera of early Oligocene age (*Amphistegina* sp., *Rotaliatina* sp.). A little below this, the presence of glauconite and dinoflagellates suggests marginal marine conditions. Unit A has yielded diverse pollen assemblages of late Eocene age, including the index forms *Concolpites leptos, Cupanieidites reticularis, Triporopollenites ambiguus,* and *Tricolpites striatus.*

Seismic Unit B is dominated by mudstones and marls, with occasional cryptocrystalline carbonate concretions. Sandstone interbeds become increasingly common above 7500 ft (2290 m), the flysch nature of the upper part of the unit showing clearly on the sonic and caliper logs. Microfaunas contain abundant planktic forms, particularly higher in the unit, and index forms of early Miocene (*Globoquadrina dehiscens, Globigerina woodi, Globigerinoides trilobus*) and middle Miocene (*Globorotalia miozea, G. mayeri, Orbulina universa*) age are present. Both planktic and

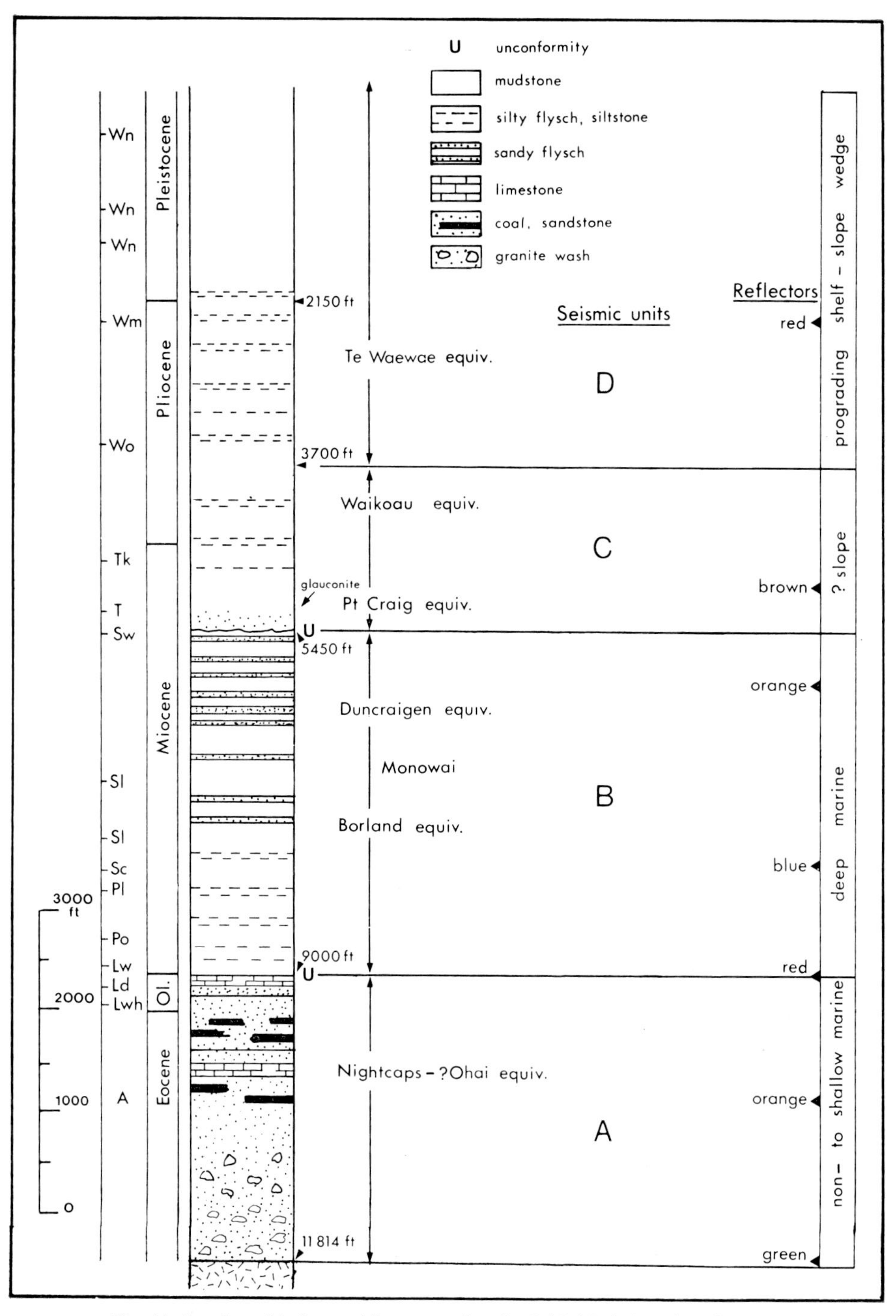

Fig. 11. Stratigraphic log and interpretation for PARARA-1 exploration well.

benthic faunas indicate deep water, probably bathyal conditions; minor admixtures of shallow water forms such as *Amphistegina* are interpreted as redeposited. The sandstones are of wider provenance than those of Unit A; though high grade metamorphic and plutonic detritus still occurs, appreciable low grade metamorphic detritus is present also, and probably derives from a source far to the north.

Seismic Unit C is also mainly fine grained, comprising mudstones and micaceous siltstones with rare fine sandstones. For parts of Unit C the logs suggest flysch-type sand-silt/mud alternations. Foraminiferal faunas are dominated by benthic forms, but also contain common planktic index species, in particular the late Miocene *Globorotalia miotumida, G. conomiozea* and *G. sphericomiozea,* and the early Pliocene *G. puncticulata*. The faunas are consistent with upper slope to outer shelf conditions of deposition.

The base of Unit C, appreciably sandier than the bulk of the unit, is glauconitic, and coincides with a marked unconformity on the seismic lines. Faunal evidence suggests that the late Miocene Tongoporutu zone is missing across this unconformity, which is therefore of similar age to the regionally significant late Miocene unconformity already recognized on land (Wood, 1966, 1969).

Seismic Unit D comprises largely siltstone with lesser mudstone and sandstone. The presence of dextrally coiled *Globorotalia crassaformis* indicates a Late Pliocene age for the lower parts of the unit, the base of the Pleistocene being taken at the appearance of *Globorotalia tosaensis* at around 2150 ft (656 m); *Globorotalia truncatulinoides* occurs above about 1500 ft (460 m). Seismic evidence shows that Unit D represents a prograding shelf wedge (cf. Fig. 8, line D186). Faunas from Unit D on the other hand show alternations of abundant mid-shelf species of *Notorotalia* and *Astrononion*, and rich faunas of outer shelf–upper slope aspect, suggesting that late Cainozoic sea-level changes have had a complicating effect on the sequence of microfaunas within the prograding wedge.

SEDIMENT THICKNESSES AND DISTRIBUTION

Solander Basin

As discussed earlier, the seismic units described from the PARARA-1 well can be recognized throughout the Solander Basin. Two seismic lines that pass close to the site of PARARA-1 have been used as reference lines; they are line D197 (Fig. 7) across the Solander Basin, and line D186 (Fig. 8) along the axis of the Solander Basin.

Seismic Unit A, of which only the upper part is represented in PARARA-1, is thickest adjacent to the Solander East Fault, downlapping and thinning southeastwards away from the fault (Figs 7, 12a); line 186 (Fig. 8) shows Unit A to pinch out against the Hump Ridge–Midbay high at the head of the Solander Basin, and also against a broad basement high in the centre of the basin. Further west, thick Unit A occurs in the Waitutu Sub-basin; downlap and thickness relationships suggest derivation largely from the Fiordland Boundary Fault (Figs 7, 12a).

Downlapping and thinning relations of Unit A away from fault-bounded highs suggest fault controlled sedimentation. For the Unit A2 sequence penetrated by PARARA-1, the faulting was of Eocene–Oligocene age; however, the lower parts of Unit A, particularly A1 as seen along the Solander East Fault, may have accumulated during the earlier mid-late Cretaceous phase of faulting that controlled deposition of the onland Sand Hill and Ohai Group non-marine strata. Though not conclusive, the seismic evidence also suggests that the strata in the Waitutu and main Solander Basin may initially have formed a single eastward thinning wedge, entirely derived from west of the Fiordland Boundary Fault but later disrupted by uplift of the now-intervening Solander High.

Seismic Unit B has a similar distribution to Unit A, though it is more uniform in thickness and overlaps the lower basement highs (Fig. 8, D186). The unit abuts the major basement highs such as the Hump Ridge-Midbay High and the Solander High (Fig. 7, D201), and also thickens towards the west, especially in the Waitutu Sub-basin. Downlap and thickness variations (Fig. 12b). together with the petrographic data from PARARA-1, suggest derivation of Unit B sediments from both sides of the basin (but particularly from the west), and also from the north.

Unit B represents a continuation of sediment accumulation in the fault-bounded basins formed prior to and during the accumulation of Unit A. However, the sudden deepening from marginal marine to bathyal conditions seen across the A–B boundary in PARARA-1 requires a further major phase of downfaulting at that time, as already inferred for the onland Waiau Basin sequence (Norris *et al.*, 1978).

Unit B can be traced onland from the head of the Waitutu Sub-basin (Fig. 8, D196), and is equivalent to the deep marine mudstones of Oligocene through middle Miocene age that outcrop on the west side of the Wairaurahiri Syncline (Wood, 1960). By comparison, the Oligocene is thin in the PARARA-1 well, suggesting that the well location is marginal to the major early Oligocene depocentres. Away from the PARARA-1 site, particularly where Units A or B are thick, it is likely that sediments equivalent to the onland Blackmount Formation will occur either in upper Unit A or lower Unit B. Regionally, however, Unit B is predominantly equivalent to the Borland and Duncraigen formations of the Waiau column.

Seismic Units C and D occur above an unconformity of regional extent which occurs between sediments of middle and late Miocene age. Unit C is thickest in the Waitutu Sub-basin (Fig. 12c), longitudinal profiles of which show distinct foresets downlapping to the south (Fig. 8, D194, 196). Unit D is thickest beneath the constructional shelf of the Solander Basin, indicating a north to south progradation. Two modern channel-systems cut into Unit D and lead from the shelf edge into the head of the Solander Trough (Fig. 12d).

The presence of a demarcated Waitutu Sub-basin during the accumulation of C, together with the apparent local onlap and overlap of Unit C over the Solander High (Fig. 7, D193, 201), suggests that initial uplift of the High occurred between deposition of Units B and C, i.e. concomitant with the phase of tectonism that produced the Mid-Late Miocene unconformity. The foresetting shown by C in the Waitutu Basin is consistent with sedimentary wedges filling the basin from the north, but these wedges are not necessarily related to the modern shelf-slope profile. Units C and D in the main Solander Basin, as well as the thin cover of D in the Waitutu Sub-basin, were also derived from the north (cf Fig. 12c, d) as a shelf-slope wedge that successively advanced to its present position.

Unit C is equivalent to the Port Craig and Waikoau beds (Table 1), which occur at the coast around the western side of Te Waewae Bay, and to Late Miocene silts in the axis of the Wairaurahiri Syncline. Though the upper parts of Unit D were not sampled during the drilling of PARARA-1 (cf. Fig. 11), D can be viewed as equivalent

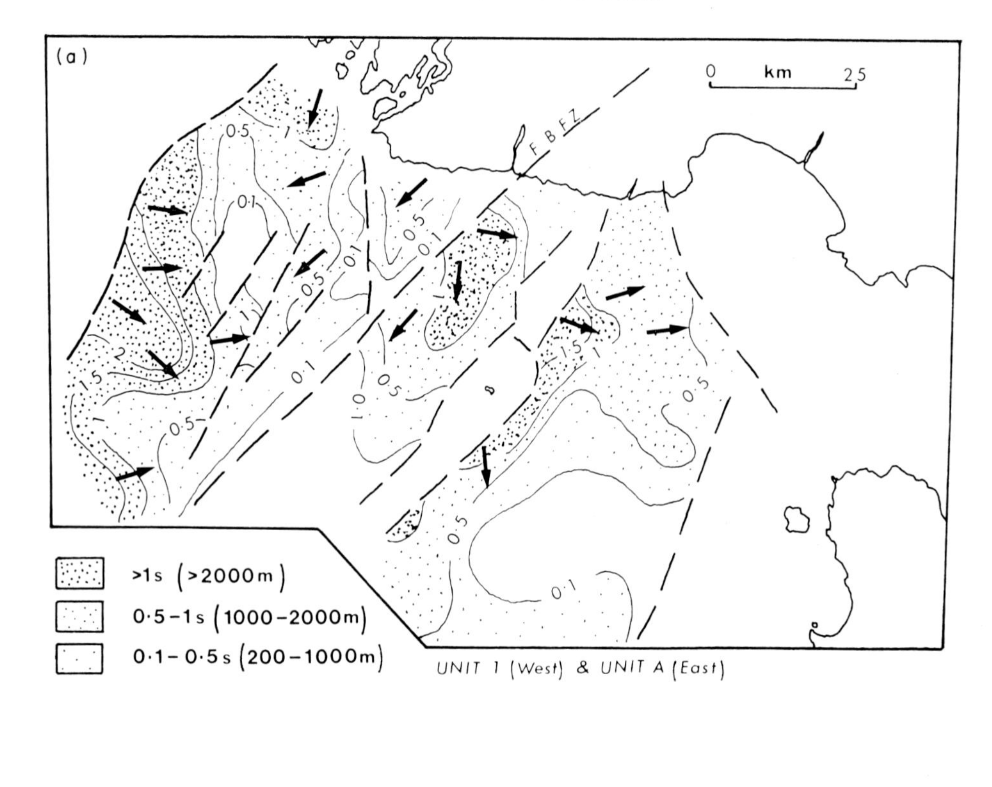

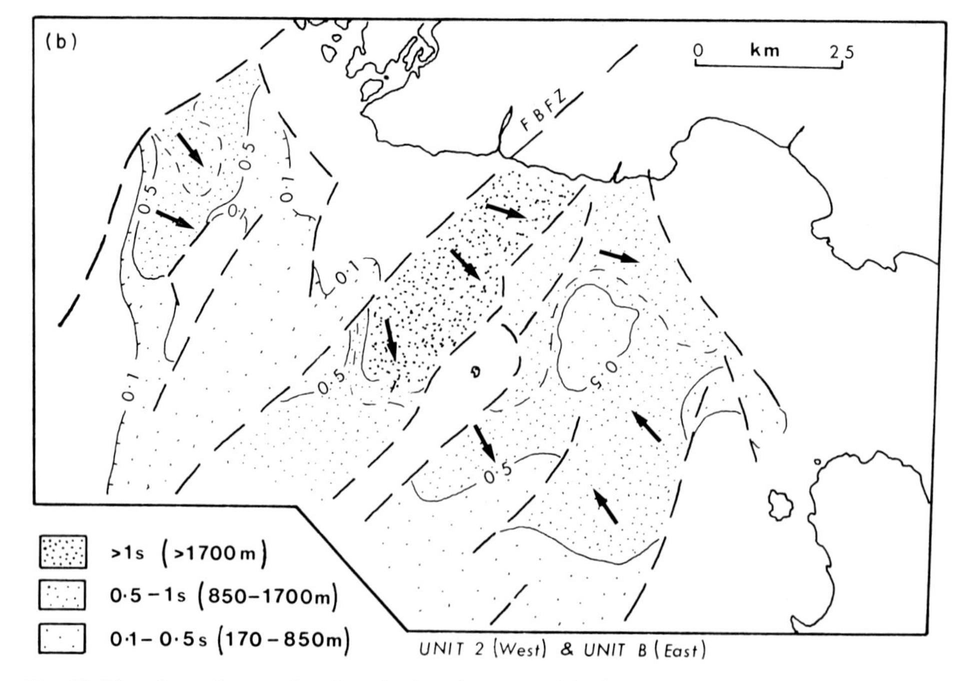

Fig. 12. Time–isopach maps for the seismic units mapped in the Te Waewae Bay–Solander Trough area. Arrows show the direction of sedimentary downlap.

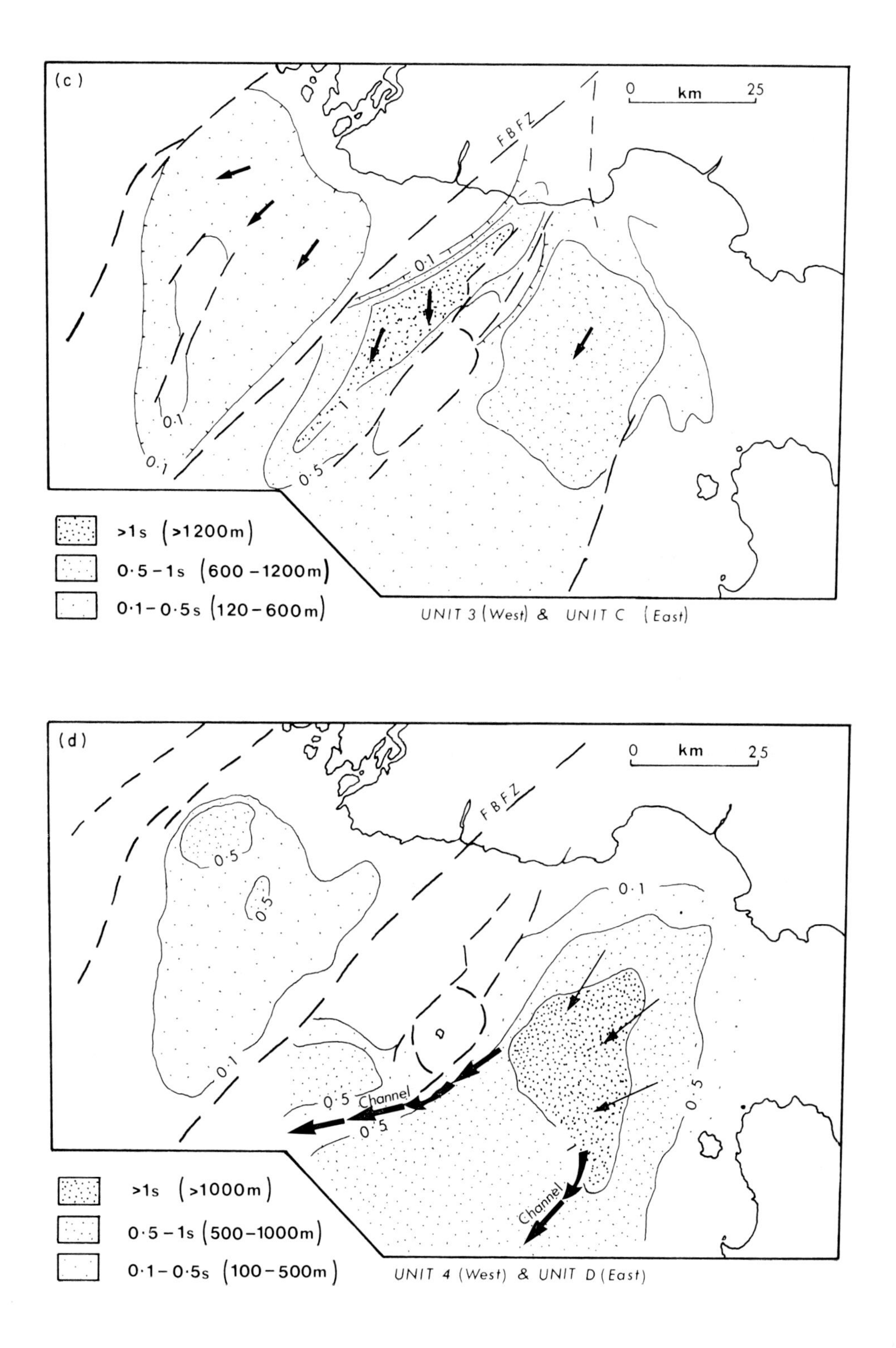
(c)
0 km 25
FBFZ
0·1
1
0·1
0·1
0·5
>1s (>1200m)
0·5 – 1s (600 – 1200m)
0·1 – 0·5s (120 – 600m)
UNIT 3 (West) & UNIT C (East)
(d)
0 km 25
FBFZ
0·5
0·5
0·1
0·1
0·5
Channel
0·5
0·5
Channel
>1s (>1000m)
0·5 – 1s (500–1000m)
0·1 – 0·5s (100 – 500m)
UNIT 4 (West) & UNIT D (East)

to the onland shelf deposits laid down during the maximum sea-level advances of the late Pleistocene. Such deposits include the Te Waewae Formation, and also the sediments beneath coastal terraces in Te Waewae Bay. Much of Unit D, being laid down on an outer shelf or slope, has no strict equivalent on land, but the Unit D shelf silts must have accumulated seawards from contemporaneous molasse phases, such as the Prospect Formation of the Waiau and Te Anau Basins.

Balleny Basin

The absence of any stratigraphic well control makes interpretation of the seismic stratigraphy in the Balleny Basin difficult, particularly as seismic units cannot be directly matched to PARARA-1 across the Fiordland Boundary Fault Zone.

Seismic Unit 1 reaches its maximum thickness along the Puysegur East Fault (Fig. 9, B215; Fig. 10, B204), though significant accumulations also occur in the northeastern head of the Balleny Basin (Fig. 10). Time-isopach and downlap data (Fig. 12a) show Unit 1 to be derived from varying parts of the basin margin, probably as a series of discrete fans.

Unit 1 can be traced onland from the head of the Balleny Basin, where it passes into the thick early Oligocene, redeposited breccia and flysch sequence (Balleny Group) of Chalky Island. Thus Unit 1 in the northern parts of the basin was certainly deposited in response to Eo-Oligocene tectonism (cf. Carter & Lindqvist, 1977). The presence of the intrabasinal Long Reef High makes it difficult to demonstrate that Unit 1 in the north is an exact equivalent of Unit 1 to the west under the Puysegur Rise. The clearest profiles (e.g. B202, Fig. 10) suggest that Unit 1 in the north may be represented mainly by Unit 1b under the Puysegur Rise. Unit 1b is therefore probably equivalent to the Balleny Group and of Eo-Oligocene age. Unit 1a of the Puysegur Rise may be appreciably older, and possibly includes Cretaceous non-marine sediments.

Seismic Unit 2 has a similar distribution to the western wedges of Unit 1, being thickest under the Puysegur Rise and downlapping and thinning southeastwards away from the Puysegur East Fault (Fig. 12b). However, the locus of maximum thickness for Unit 2 is northeast of that of Unit 1 (cf. Fig. 12a, b), with the result that Unit 2 comes to rest directly on basement along the northern parts of the Puysegur East Fault (Fig. 9, B213, 211, 207).

The main depositional body of Unit 2 is again fan-shaped, and requires the presence of a major source area to the west of the Puysegur East Fault (Fig. 12b). Since Unit 2 overlaps onto a wedge of Unit 1 derived from the northeast (Fig. 9, B205), it must be slightly younger than the Balleny Group and probably comprises mid-late Oligocene deep marine marl and flysch deposits.

Seismic Units 3 and 4 are separated from each other mainly by the acoustically transparent nature of Unit 4. Although much thinner than their counterparts, Units C and D, in the Solander Basin, they also overlie the earlier sediments unconformably (Figs 9, 10).

Slight foresetting (Fig. 10), minor thickness variations and downlap directions (Figs 12c, d) suggest some coarser detritus was supplied to Units 3–4 from Fiordland. The major source area to the west of Balleny Basin was no longer present. The acoustic nature of these units, together with their distribution, suggests they represent a hemipelagic drape which has accumulated in a largely inactive late Neogene Balleny

Basin. Assuming the Unit 2–3 unconformity to be equivalent to the B–C unconformity in the Solander Basin, then Units 3–4 are probably Late Miocene through Recent in age.

REGIONAL SYNTHESIS

Four major sedimentary basins developed in southwestern New Zealand during the early Cainozoic, their boundaries controlled by substantial fault systems (Fig. 13). The present Alpine Fault plate boundary is located to the west of these basins, passing offshore near Milford Sound and continuing southwards along the Fiordland margin and into the Puysegur Trench, which lies immediately west of the Balleny Basin.

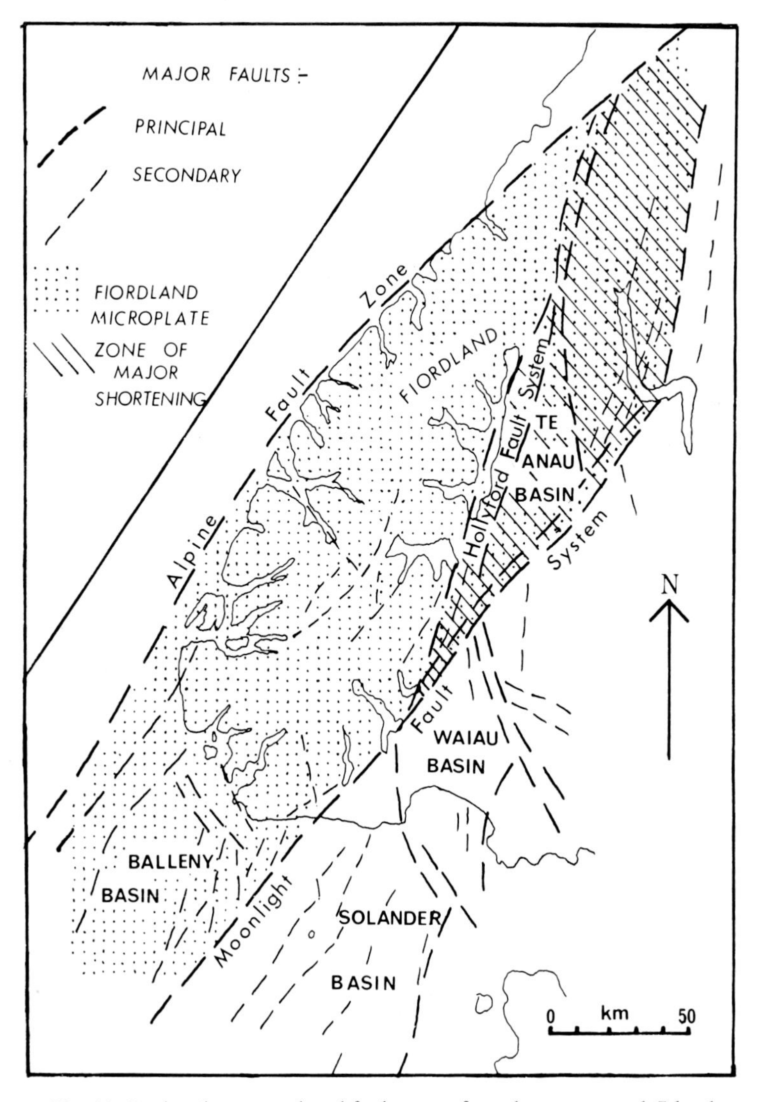

Fig. 13. Regional structural and fault map of southwestern south Island.

Structural tends

The dominant regional fault trend is northeasterly, subparallel to the main Alpine plate boundary. More northerly or northwesterly trending faults are also important, however, and the combination of the two sets of faults, both active simultaneously during basin development, has led to the basins being broadly rhomb-shaped, a feature also characteristic of fault-controlled basins within the San Andreas Fault System of southern California (Moore, 1972; Crowell, 1974; Blake *et al.*, 1978).

The orientation of the faults and folds in the Balleny and Solander Basins (Fig. 4) is consistent with their development within a dextral strike-slip regime (Wilcox, Harding & Seely, 1973; Junger, 1976). Many of the faults were reactivated and virtually all the folds were initiated after deposition of Unit 2 in the Balleny Basin and Unit B in the Solander Basin, that is, post-early Miocene. Prior to this, onlapping sedimentary sequences and the nature of faulting on the seismic profiles generally suggest extension and subsidence, with little evidence for contemporaneous compression and uplift. This is in broad agreement with the record in the onland basins (Norris *et al.*, 1978) where block faulting during the late Cretaceous, probably related to the rifting of the New Zealand plateau away from Gondwanaland (Carter *et al.*, 1974), was followed in the late Eocene and Oligocene by sudden and rapid subsidence of marine basins as the Alpine fault was initiated as an extensional transform plate boundary. During the development of the basins, the crystalline basement of Fiordland appears to have acted as a relatively rigid and uplifted block, thus having an important influence on sedimentation in the surrounding basins (Norris *et al.*, 1978). By early Miocene time Fiordland had submerged sufficiently to cease being a major source of clastic detritus, and uplift of the low-grade schists to the northeast provided a new sediment source for basins along the Moonlight Tectonic Zone. Change to a compressive transform regime during the mid-Miocene gave rise to folding and uplift (Molnar *et al.*, 1975; Carter & Norris, 1976).

Disappearing source areas

In the Balleny Basin, Unit 1 shows evidence of derivation as sediment fans from two source areas. One was southern Fiordland; the other, the major source, was west of the Puysegur Rise, a locality now occupied by the Puysegur Trench and Tasman Basin. Unit 2, though thinner than Unit 1, also appears to be derived from this western source area. Units 3 and 4 on the other hand show no evidence for the existence of a source area to the west. This western source area, which was a prodigious supplier of detritus during Unit 1 deposition, weakened during Unit 2 times and finally disappeared before Unit 3 was deposited, that is, probably some time during the middle Miocene.

The disappearance of a clastic source area as recorded in the sedimentary succession may take place purely by erosion or subsidence. In the present case, however, no suitable submerged or buried source area exists today, so that it must have been either destroyed, say by subduction below Fiordland, or rafted laterally along the Alpine transform. The latter appears the most likely explanation, and is consistent with other arguments for major lateral movements on the Alpine transform since early Miocene (Wellman, 1971; Carter & Norris, 1976). Unfortunately, no samples of the westerly-derived sediments of the Balleny Basin are available on which to carry out provenance studies to identify the source, but it was presumably part of the West Coast block of the South Island.

The pattern of sedimentation in the Solander Basin is similar. Major sediment fans of Units A and B were derived largely from an area to the west of the Fiordland Boundary Fault. Since only small areas of the present highs in this region (the Big River Highs; cf. Fig. 4) have no cover of Units 1 or 2, they are unlikely to have acted as a major source of clastic detritus for Units A and B in the Solander Basin. Hence here, as in the Balleny Basin, a major source seems to have disappeared. Cuttings from PARARA-1 show the source of Unit A and the lower part of Unit B to have been crystalline basement as found in Fiordland. Evidence from further north also indicates that Fiordland was a major source of detritus until the early Miocene (Norris *et al.*, 1978). Sediments in the Waitutu Sub-basin suggest uplift, tilting and erosion adjacent to the Fiordland Boundary Fault following deposition of Units 2 and B; and though Unit C shows a more northerly source direction, it is again from west of the fault zone. These facts together would be consistent with a small amount, say 20–30 km, of post Middle Miocene dextral strike-slip along the Fiordland Boundary Fault Zone, sufficient to move the Fiordland block away from an original position adjacent to the sediment fans in the Solander Basin. Since the Pleistocene, the Waiau River has been the main source of clastic detritus for the offshore area, providing sediment for the construction of a prograding shelf sequence, Unit D, and supplying sediment to the upper Solander Trough via submarine channels.

Combinations of displaced source areas, localized sedimentation, relationship between 'proximal' sediments and major faults and repeated movements through time, such as has been described above, are typical of sedimentation in zones of active strike-slip faulting (e.g. Reading, 1975; Heward & Reading, 1980).

The Fiordland microplate

The Fiordland Boundary Fault forms part of the Moonlight Fault System, which to the north crosses the steeply dipping Dun Mountain Ophiolite Belt (Fig. 1) with only a kilometre or two of dextral separation (Norris *et al.*, 1978). However, the strike of the ophiolite belt swings northwards across the fault system. Cainozoic sediments of the Te Anau and northern Waiau Basins are highly deformed, the deformation decreasing in intensity in the southern Waiau Basin and becoming much less in the offshore basins (Carter & Norris, 1977b). Thus there is a crude inverse relationship between the amount of deformation of the Cainozoic sediments and the amount of inferred strike-slip movement of the Moonlight Fault system. We suggest, therefore, that the 20 or so km of dextral strike-slip on the Fiordland Boundary Fault segment of the Moonlight Fault System is translated northwards into more distributed deformation and shortening of sediments in the northern Waiau and Te Anau Basins, clockwise rotation of the ophiolite belt and uplift of northern Fiordland.

In this manner, the area between the Moonlight–Hollyford fault systems and the Alpine transform has acted as a large, fault-bounded wedge, or microplate, that was driven northwards during the late Cainozoic dextral fault movements. The tectonic and sedimentary history of the basins of southwest New Zealand may be explained in terms of the behaviour of this microplate within the overall context of movements on the plate boundary (Molnar *et al.*, 1975; Weissel *et al.*, 1977). During the Eo-Oligocene phase of extensional transform movement, rifting around the margins of the microplate (cf. Moore, 1976) led to rapid subsidence and the formation of a number of small, interlinked marine basins. Erosion reduced most of the Fiordland block to sea level by the end of the Oligocene. The cessation of extension during the early Miocene, accompanied by dextral slip on the plate boundary, led to the start of northward movement of the microplate between its converging boundary faults; the resulting uplift of the northern plate edge provided a source of low-grade metamorphic detritus, and associated southwards tilting gave rise to further depression of the areas to the south. During this time, the strike-slip movement on the Alpine transform removed the source area west of the Balleny Basin. The change to a compressive transform regime during the middle Miocene was accompanied by increased northward movement of the microplate, with rapid uplift of its northern parts, deformation of the sedimentary rocks in the basins along its northeastern margin and related strike-slip movement along its southeastern edge. The associated tilting may also have caused the subsidence inferred for middle to late Miocene time in the southern Waiau Basin. Only along the northeastern margin of the microplate has true eversion of the basins so far occurred (Turnbull *et al.*, 1975); elsewhere, intermontane depressions still ring the microplate along the Moonlight Tectonic Zone, and continue offshore as the present day Solander Trough.

CONCLUSIONS

(1) Two major sedimentary basins occur offshore from southwestern South Island, immediately southeast of the Alpine Fault sector of the Indo-Australian/Pacific plate boundary. The basins are separated by the offshore extension of the Moonlight Fault System, a major tectonic lineament subparallel to the Alpine Fault.

(2) Sedimentation in the Balleny and Solander Basins was controlled by movements on the Moonlight, Alpine and other ancillary fault systems; fault-controlled sedimentation commenced in the late Eocene–Oligocene, and the offshore pattern of faults and folds is consistent with the development of a dextral strike-slip regime. Up until the middle Miocene, extension and subsidence were dominant; after the middle Miocene, folding and uplift were characteristic, with reversal of movement on some faults.

(3) A substantial source terrain was present immediately west of the Balleny Basin during the early Cainozoic. This source region had largely disappeared by the late middle Miocene, due to its northwards displacement along the Alpine Fault system.

The similar disappearance of an early Cainozoic source terrain from immediately west of the Solander Basin is consistent with 20–30 km northwards displacement of the Fiordland block along the Moonlight Fault System since the middle Miocene. This displacement was partially compensated in the north by rotation and shortening of the onland Te Anau basin.

(4) The late Miocene to Recent fill of the Balleny and Solander Basins has been derived mainly from the north, by progradation of shelf-slope sequences off the southern end of the New Zealand landmass.

(5) The tectonic and sedimentary pattern in the offshore basins throughout the Cainozoic corresponds closely to that already established for onland sequences in southwestern Fiordland and in the Te Anau and Waiau Basins.

(6) The sedimentary and tectonic history of the southwest South Island is best explained in the context of geophysical reconstructions of oblique-slip movements

along the Alpine Fault sector of the Indo-Australian/Pacific plate boundary, and in particular in terms of the relatively rigid behaviour of the Fiordland microplate.

ACKNOWLEDGMENTS

The basis of this paper is the large amount of unpublished data previously collected by Hunt International Petroleum Company. We appreciate very much the permission for publication granted by R. M. Sandford, and the encouragement and assistance we received at all times from Jack Lichtenwalter, both of HIPCO. We acknowledge also the use of unpublished faunal data from the PARARA-1 Well Completion Report, palynology by D. M. Mildenhall and foraminiferal determinations by N. de B. Hornibrook. D. G. Bishop and G. M. Gibson kindly provided information regarding fault orientations in Fiordland. P. Anderton, formerly of the N.Z. Geological Survey, provided much excellent advice regarding the interpretation of seismic sections, and we thank also R. Cook for his assistance with the provision of the seismic sections. The draft manuscript was critically read by P. F. Ballance, D. Howell and H. G. Reading; we thank these colleagues for their advice, and particularly thank Peter Ballance for his kind invitation to participate in this symposium volume. This research has been supported by funds granted by the University of Otago.

REFERENCES

BALLANCE, P.F. (1976) Evolution of the Upper Cenozoic magmatic arc and plate boundary in northern New Zealand. *Earth Planet. Sci. Lett.* **28,** 356–370.

BLAKE, M.C. Jr, CAMPBELL, R.H., DIBBLEE, T.W. Jr, HOWELL, D.G., NILSEN, T.H., NORMARK, W.R., VEDDER, J.C. & SILVER, E.A. (1978) Neogene Basin formation in relation to Plate-Tectonic Evolution of San Andreas Fault System, California. *Bull. Am. Ass. Petrol. Geol.* **62,** 344–372.

BOWEN, F.E. (1964) Geology of Ohai Coalfield. *Bull. geol, Surv. N.Z.* **51.**

CARTER, R.M., LANDIS, C.A., NORRIS, R.J. & BISHOP, D.G. (1974) Suggestions towards a high-level nomenclature for New Zealand rocks. *J. R. Soc. N.Z.* **4,** 5–18.

CARTER, R.M. & LINDQVIST, J.K. (1977) Balleny Group, Chalky Island, Southern New Zealand: an inferred Oligocene submarine canyon and fan sequence. *Pacific Geol.* **12,** 1–58.

CARTER, R.M. & NORRIS, R.J. (1976) Cainozoic history of southern New Zealand: an accord between geological observations and plate-tectonic predictions. *Earth planet. Sci. Lett.* **31,** 85–94.

CARTER, R.M. & NORRIS, R.J. (1977a) Redeposited conglomerates in a Miocene flysch sequence at Blackmount, western Southland, New Zealand. *Sedim. Geol.* **18,** 289–319.

CARTER, R.M. & NORRIS, R.J. (1977b) Blackmount, Waiau Basin. *Post-conference Field Guide, Geol. Soc. N.Z. Conf.* (Queenstown), 31 pp.

CROOKS, I. & CARTER, R.M. (1976) Stratigraphy of Maruia and Matiri Groups at their type section, Trent Stream, Matiri River, Murchison. *J. R. Soc. N.Z.* **6,** 459–478.

CROWELL, J.C. (1974) Origin of Late Cenozoic basins in Southern California. In: *Tectonics and Sedimentation* (Ed. by W. R. Dickinson). *Spec. Publ. Soc. econ. Paleont. Miner. Tulsa,* **22,** 190–204.

HARRINGTON, H.J. & WOOD, B.L. (1958) Quaternary andesitic volcanism at the Solander Islands. *N.Z. J. Geol. Geophys.* **1,** 419–431.

HEWARD, A.P. & READING, H.G. (1980) Deposits associated with a Hercynian to late-Hercynian continental strike-slip system, Cantabrian Mountains, northern Spain. In: *Sedimentation in oblique-slip mobile zones* (Ed. by P. F. Ballance & H. G. Reading). *Spec. Publ. int. Ass. Sedim.* **4,** 105–125.

HIPCO (1970) Marine Seismic Survey PPL's 695–707 and 740–743 (Stewart Island area). *Unpublished report,* available on open file at New Zealand Geological Survey, Lower Hutt.

HIPCO (1971) Preliminary results of geophysical work in the offshore waters of the South Island—October 1970 to June 1971. *Unpublished report,* available at New Zealand Geological Survey, Lower Hutt.

HIPCO (1976) Well completion report—PARARA-1. *Unpublished report,* available on open file at New Zealand Geological Survey, Lower Hutt.

HYDEN, F.M. (1975) Mid-Tertiary sedimentation patterns east of the Longwood-Takitimu Ranges, Southland. *Abstr. geol. Soc. N.Z. conf.* (*Kaikoura*), 31.

JUNGER, A. (1976) Tectonics of the Southern California Borderland. In: *Aspects of the Geologic History of the California Continental Borderland* (Ed. by D. G. Howell). *Misc. Publ. Am. Ass. Petrol. Geol.* (*Pacific Section*), **24,** 486–498.

KATZ, H.R. (1968) Potential oil formations in New Zealand, and their stratigraphic position as related to basin evolution. *N.Z. J. Geol. Geophys.* **11,** 1077–1133.

KATZ, H.R. (1974) Margins of the Southwest Pacific. In: *The Geology of Continental Margins* (Ed. by C. A. Burk & C. L. Drake), pp. 549–565. Springer Verlag, New York.

LANDIS, C.A. (1974) Stratigraphy, lithology, structure and metamorphism of Permian, Triassic and Tertiary rocks between the Mararoa River and Mt. Snowdon, western Southland. *J. R. Soc. N.Z.* **4,** 229–251.

MCKELLAR, I.C. (1956) Geology of the Takahe Valley district, eastern Murchison Mountains. *N.Z. J. Sci. Technol.* **B38,** 120–128.

MITCHUM, R.M. Jr, VAIL, P.R. & SANGREE, J.B. (1977) Seismic stratigraphy and global changes of sea level, Part 6: Stratigraphic interpretation of seismic reflection patterns in depositional sequences. In: *Seismic Stratigraphy—application to hydrocarbon exploration* (Ed. by C. E. Payton). *Mem. Am. Ass. Petrol. Geol.* **26,** 117–133.

MITCHUM, R.M. Jr, VAIL, P.R. & THOMPSON, S. (1977) Seismic stratigraphy and global changes of sea level, Part 2: Depositional sequence as a basic unit for stratigraphic analysis. In: *Seismic Stratigraphy—application to hydrocarbon exploration* (Ed. by C. E. Payton). *Mem. Am. Ass. Petrol. Geol.* **26,** 53–62.

MOLNAR, P., ATWATER, T., MAMMERICX, J. & SMITH, S.M. (1975) Magnetic anomalies, bathymetry and the tectonic evolution of the South Pacific since the late Cretaceous. *Geophys. J.R. astr. Soc.* **40,** 383–420.

MOORE, D.G. (1972) Reflection profiling studies of the California Continental Borderland: Structure and Quaternary turbidite basins. *Spec. Publ. geol. Soc. Am.* **107,** 142 pp.

MOORE, G.W. (1976) Basin development in the California Borderland and the Basin and Range Province. *In: Aspects of the Geologic History of the California Continental Borderland* (Ed. by D. G. Howell). *Misc. Publ. Am. Ass. Petrol. Geol.* (*Pacific Section*), **24,** 383–391.

MUTCH, A.R. (1972) Geology of the Morley Subdivision. *Bull. N.Z. geol. Surv.* **78.**

NELSON, C.S. & HUME, T.M. (1977) Relative intensity of tectonic events revealed by the Tertiary sedimentary record in the North Wanganui Basin and adjacent areas, New Zealand. *N.Z. J. Geol. Geophys.* **20,** 369–392.

NORRIS, R.J., CARTER, R.M. & TURNBULL, I.M. (1978) Cainozoic sedimentation in basins adjacent to a major continental transform boundary in southern New Zealand. *J. geol. Soc. Lond.* **135,** 191–205.

PILAAR, W.F.H. & WAKEFIELD, L.L. (1978) Structure and stratigraphic evolution of the Taranaki Basin, offshore North Island, New Zealand. *J. Ass. Petrol. Engin. Aust.* 1978, 93–101.

READING, H.G. (1975) Strike-slip fault systems; an ancient example from the Cantabrians, *IXth Int. Congr. of Sedim. Nice, 1975,* Thème **4,** 287–91.

SUTHERLAND, J.I. (1969) *Palaeontology and ecology of some Tertiary sediments in Te Waewae Bay.* Unpublished Dip. Sci. Thesis, Otago University, Dunedin.

TURNBULL, I.M., BARRY, J., CARTER, R.M. & NORRIS, R.J. (1975) The Bobs Cove beds and their relationship to the Moonlight Fault Zone. *J.R. Soc. N.Z.* **5,** 355–394.

WALCOTT, R.I. (1978) Present tectonics and Late Cenozoic evolution of New Zealand. *Geophys. J. R. astr. Soc.* **52,** 137–164.

WEISSEL, J.K., HAYES, D.E. & HERRON, E.M. (1977) Plate tectonic synthesis: the displacements between Australia, New Zealand and Antarctica since the Late Cretaceous. *Mar. Geol.* **25,** 231–277.

WELLMAN, H.W. (1971) Age of the Alpine Fault, New Zealand. *Proc. int. geol. Congr.* (*India*), *Sect.* **4,** 148–162.

WILCOX, R.E., HARDING, T.P. & SEELY, D.R. (1973) Basic wrench tectonics. *Bull. Am. Ass. Petrol. Geol.* **57,** 74–96.

WOOD, B.L. (1960) Sheet 27—Fiord: Geological Map of New Zealand, 1 : 250,000. *N.Z. Dept. Sci. Ind. Res., Wellington.*

WOOD, B.L. (1966) Sheet 24—Invercargill: Geological Map of New Zealand, 1 : 250,000. *N.Z. Dept. Sci. Ind. Res., Wellington.*

WOOD, B.L. (1969) Geology of Tuatapere Subdivision, Western Southland. *Bull. geol. Surv. N.Z.* **79**.